PLANT INVASIONS:
ECOLOGICAL MECHANISMS AN
HUMAN RESPONSES

PLANT INVASIONS: ECOLOGICAL MECHANISMS AND HUMAN RESPONSES

Edited by U. Starfinger, K. Edwards, I. Kowarik and M. Williamson

Backhuys Publishers, Leiden, 1998

ISBN 90-5782-005-6

Printed in the Netherlands

TABLE OF CONTENTS

Case studies II: Biotopes/regions

Plant-insect interactions

PREFACE

Plant invasions are still a major issue, in ecological science, in public awareness, and in applied fields such as the conservation of bioversity. Consequently, the series of conferences, held at Loughborough, UK (de Waal *et al.* 1994), Kostelec nad Cernými lesy, Czech Republic (Pyšek *et al.* 1995), and Tempe, AZ, USA (Brock *et al.* 1997), was continued with the 4th International Conference on the Ecology of Invasive Alien Plants, 1-4 October 1997, in Berlin. This book builds on material presented at the Berlin conference.

The flora of central Europe is rich in non-native plant species, and the introduction and naturalization of plants has long been studied here. Berlin has a particularly rich tradition in plant invasion studies (Sukopp 1998). Beginning with "adventive floristics" in the 19th century, research in this field was very active after WW II, when Berlin West was isolated from its surroundings by the famous wall. Differently from many other parts of the world, non-native species were not primarily perceived as undesired aliens, but rather their role in the urban vegetation and their naturalization process were the main focus.

This book, like the ones before it, contains papers on case studies of single invaders, on general questions in invasion biology, and on control. A new aspect is that several papers focus on questions regarding the perception of invasion events. Several authors stress the importance of assessment and evaluation of invasion effects as a prerequisite for management actions.

We gratefully acknowledge the help of institutions and individuals in the organization of the conference which has made this book possible: The conference was organized by the "Botanischer Verein von Berlin und Brandenburg, gegründet 1859" (Botanical Society of Berlin and Brandenburg, founded in 1859) under its chairman Prof. Dr. Herbert Sukopp, and the Institut für Ökologie und Biologie, Technische Universität Berlin. Many members of the institute have given practical help. The DFG, the German National Science Fund, has funded the conference and travel expenses for East-Europeans. We also wish to thank Wil Peters of Backhuys Publishers for his advice and his patience during the editing process.

Uwe Starfinger, Keith Edwards, Ingo Kowarik, Mark Williamson

References

Brock, J.H., Wade, M., Pyšek, P. and Green, D. 1997. Plant Invasions: Studies from North America and Europe. Backhuys, Leiden.

de Waal, L.C., Child, L.E., Wade, P.M. and Brock, J.H. 1994. Ecology and management of invasive riverside plants. John Wiley & Sons, Chichester.

Pyšek, P., Prach, K., Rejmánek, M. and Wade, M. 1995. Plant Invasions - general aspects and special problems. SPB Academic Publ., Amsterdam

Sukopp, H. 1998. On the study of anthropogenic plant migrations in central Europe. In: Starfinger, U., Edwards, K., Kowarik, I. and Williamson, M. (eds.) Plant invasions: Ecological Mechanisms and Human Responses pp. 43-56 Backhuys Publishers, Leiden.

GENERAL ASPECTS

INVASIVE ALIEN PLANTS AND VEGETATION DYNAMICS

Janusz Bogdan Faliński
Geobotanical Station of Warsaw University, PL - 17-230 Białowieża, Poland

Abstract

Plant communities which are subject to different ecological processes such as primary and secondary succession, regression, degeneration and regeneration, as well as fluctuation, are vulnerable to varying degrees to the penetration of alien species and their entry into the neophyte phase. Degeneration followed by regeneration of the natural community occurs, creates particularly favourable conditions for the establishment and permanent maintenance of alien species in plant communities, which stabilise over time. Secondary succession may be similar significant in plant communities on the ruins of settlements, unused transportation routes and, rarely on abandoned fields. As a process determining the permanence of a given community and its links with the biotope and biochore, the fluctuation ongoing in natural, undamaged plant communities is a process which basically prevents the encroachment and establishment of alien species, in spite of the fact that there is – at the same time and in the same conditions – a constant renewal of populations of the primary component species proper to the community.

Introduction

The success of alien species in migrating beyond their primary range may be considered from the biogeographical, genetic and biocoenotic points of view:

1) biogeographical (phytogeographical) success entails migration which extends the primary range or establishes a secondary one.

2) genetic success ensues when a species in a new homeland adapts to new conditions by way of a change in life strategy, mutation or hybridisation with a closely-related species that is present, and/or when it obtains a permanent ability to compete, etc.

3) biocoenotic (phytocoenotic) success has occurred where a species in a new homeland finds a permanent place within a previously-existing (natural) plant community; playing a part in its structure, internal dynamics and functioning; being able to pass through the whole cycle of reproduction and being maintained thanks to the development of a secondary, autochthonous population (i.e. neophytism). A further degree of biocoenotic success may be said to have been achieved when an alien species is able to alter the species composition, structure and dynamics of a previously-existing community, changing it into one of entirely new properties (post-neophytism; Tab. 1; Fig. 10).

In this paper I deal with alien invasive plants biocoenotic success only.

The aim of this article is to present the factors and processes leading to neophytism, on the basis of examples of the behaviour of selected alien species in the processes of the regeneration and succession of communities. The results and conclusions from this research provide a basis for a discussion of the nature of neophytism and its role in the synanthropization of plant cover.

Plant Invasions: Ecological Mechanisms and Human Responses, pp. 3–21
edited by U. Starfinger, K. Edwards, I. Kowarik and M. Williamson
© *1998 Backhuys Publishers, Leiden, The Netherlands*

Table 1. Neophyte (sensu Rikli 1903/1904); Neophytism

Most important studies; Theory and terminology:
Sukopp (1995); 1962, 1976; Lohmeyer and Sukopp (1992); Faliński 1968, 1969, 1986, 1991a; Kornaś 1968a, 1968b; Kornaś and Medwecka-Kornaś 1968; Castri di 1990, Kowarik 1992, etc; Trepl 1990, Trepl and Sukopp 1993; Rejmánek 1989, 1996 and others.

Definition:

Neophytism is a component of the extensive transformation of plant cover under the impact of man (synanthropisation). It is the process of the establishment of foreign species in natural plant communities.

Neophytes as species of foreign origin belong to the anthropophyte group (Thellung 1915, 1918/1919) and form what is called the Flora adventiva (Rikli 1903/1904).

Source: J.B. Faliński (1968, 1969, 1986)

Neophytism is one of the consequences of the active behaviour of a species beyond its primary geographical range. It entails the penetration and establishment (naturalization) of individuals in natural or near-natural plant communities. Neophytism involves an alien species integrating itself into the structure of an existing plant community and impacting upon its functions and dynamics thanks to the creation of a secondary, autochthonous population capable of effective fertility and self-maintenance in these conditions.

Source: J.B. Faliński (npbl.)

Explanation:

1. Single or multiple penetration of a plant community by an alien species (seeding, development not ending in the emergence of an autochthonous population) is the initial (*pro-neophyte*) phase in the development of neophytism, but does not always lead to the establishment (naturalization) of the arrival in that community (the *eu-neophyte* or *para-neophyte* phases).
2. Neophytism is thus the new, but permanent, internal relationship between a population of an alien species and a natural plant community and the components present in it hitherto.

Study object. Materials and methods

The results and concepts presented here derive from research done mainly in the persistent natural forest communities of "Puszcza Białowieska" (the Białowieża Primeval Forest) of NE Poland, as well in non-forest communities undergoing secondary succession in the foreland of this forest complex. The work was thus done in a relatively large (c. 1300 km^2) forest complex dominated by broad-leaved forest (communities of the *Carpinion* and *Alno-Padion* alliances) and meso-oligotrophic forests (*Dicrano-Pinion*). The entry and naturalization of alien species into the flora and forest communities occurred later here than elsewhere in the European Lowland, as a result of various natural barriers (a peripheral location in relation to the main colonisation routes; the intactness of the forest complex; the vertical structure of forest communities; the

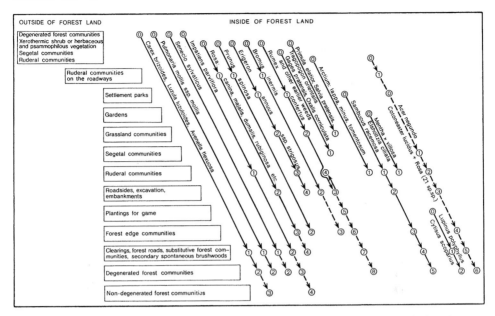

Fig. 1. Consecutive phases of colonization by alien species penetrating from the periphery into forest complex of Białowieża. (After Faliński 1986, supplemented).

advanced age, multi-species and multi-aged character of stands; the absence of larger rivers; the later onset of orchard fruitgrowing and the exploitation of forest resources and the earlier application of different kinds of protective regime (Faliński 1966, 1968, 1986). It is possible to trace the success of many alien species around the complex, and the attempts of these species in penetrating (Fig. 1).

The basic method was long-term observation on permanent plots with repeated photographing, large-scale mapping and experimentation. The work was done in forest and non-forest communities penetrated by alien species of plant (Faliński 1977, 1986). Here, I just show the behaviour in the forest of one Northern-American species (*Lupinus polyphyllus*), two south-east European species (*Gypsophila paniculata* and *Rumex confertus*) whose primary ranges lie far from Białowieża, and two Central European montane species (*Carex brizoides* and *Sambucus racemosa*) whose primary ranges do not include the complex.

Research on *Carex brizoides* and *Sambucus racemosa* in the oak-lime-hornbeam forest community (*Tilio-Carpinetum*), and on *Lupinus polyphyllus* in thermophilous oak forest (*Potentillo albae-Quercetum*), lasted more than 20 years, as did that on *Rumex confertus* in the meadow community *Arrhenatheretum elatioris*. Work on the invasion of *Gypsophila paniculata* was of 5 years' duration. With *G. paniculata* and *S. racemosa*, the development of local populations and the course of the invasion were also reconstructed dendrochronologically.

Results

Gypsophila paniculata in abandoned fields in communities undergoing **secondary succession** *towards continental oligotrophic pine forest (Peucedano-Pinetum).*

Gypsophila paniculata is a native of the vegetation of south-east Europe's sandy steppes. It is often also cultivated as an ornamental. It appeared abruptly en masse in Belarus, Ukraine, Slovakia and Poland (Faliński 1986; Nikolaeva and Zefirov 1971; Kozlovskaya and Parfenov 1972). Sokołowski (1981, 1995) regards this species as native to Polish flora.

The species was observed several times in the south-western foreland of the Białowieża Forest between 1980 and 1997. It appeared en masse, took over newly-abandoned fields and verges (Fig. 2), as well as gardens, and now extends into cereal fields and young forests. It penetrates psammophilous grassland, more rarely also xerothermic grassland and juniper-aspen scrub, where it develops as secondary succession proceeds (Table 2, Fig. 2).

The species is most abundant in abandoned fields and gardens after 5-7 years of succession. Here the density may reach 156 individuals per 100 m², with the cover being up to 50%. The maximum height of specimens is often close to 1 m, with the tap roots of the largest and oldest extending down into the soil as far as 2 m. New shoots are formed each year on each individual (between 9 and 62) and it is usually possible to count the bases of shoots from the previous year (Table 2). Lignification of the upper part of the tap root allows the determination of age, and shows that the oldest specimens are between 12 and 15 years old. Age structure was analysed in 5 samples (Fig. 2), and confirmed the timing of colonisation and the maximum density which the species may attain in abandoned fields and gardens. In communities in the early phases of succession (fallows and psammophilous grasslands in the south-western foreland of the Białowieża Forest), *G. paniculata* has become the dominant species in recent years. Its presence, albeit in small numbers, on permanent plots, in communities in later phases of succession (juniper scrub, juniper-aspen brush) shows it can survive for a while, as revealed in observations on permanent plots.

The structure of the final forest community (*Peucedano-Pinetum*) limits the occurrence of a light-demanding and short-lived species. Juniper – the pioneer and promoter

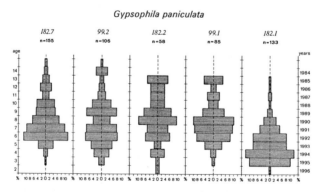

Fig. 2. Age structure pyramids for *Gypsophila paniculata* populations in early-successional communities developing on abandoned fields and gardens (see Table 2). (orig.)

Table 2. Gypsophila paniculata L. in the psammophilous grassland developed on the abandoned fields in Kleszczele (SW of Białowieża Forest; NE-Poland, 1996)

Number of relevé		182.7	182.1	99.2	99.1	182.3	182.5	182.2	182.4
Cover of shrub layer	b	10	.	.	+	.	.	.	+
herb layer	c	90	80	90	90	80	70	90	90
moss & lichen layer	d	50	60	60	50	10	40	30	20
Surface of relevé	m^2	100	100	100	100	100	100	100	100
Total number of vascular plant species		35	29	27	32	33	21	30	30
Psammophilous grassland species:									
Artemisia campestris		2.2	1.2	2.3	+.2	1.2	2.3	1.2	2.2
Berteroa incana		1.3	1.2	1.1	1.3	1.3	1.2	2.2	1.2
Oenothera sp.		1.2	2.2	+	1.2	1.2	2.2	2.2	1.2
Potentilla argentea		.	+.2	1.2	.	+.2	.	+.2	.
Jasione montana		.	+.2	+.2	+	.	+.2	+	+.2
Echium vulgare		1.2	1.1	1.2	2.2	.	.	1.2	.
Centaurea stoebe (*C. rhenana*)		1.2	+	.	1.2	1.2	.	+	+.2
Hieracium pilosella		2.3	1.2	3.3	3.4	2.3	.	1.2	+.3
Poa compressa		1.2	1.2	1.3	1.2	.	.	.	+.2
Sedum acre		+.2	2.3	+.2	.	1.3	2.3	1.3	+.2
Astragalus arenarius		.	1.2	+.2	.	+.2	1.2	+.2	1.2
and others species									
Fallow community relicts:									
Agropyron repens		3.3	2.3	2.3	1.2	3.4	3.4	3.4	3.4
Festuca rubra		2.2	1.2	3.3	+.2	3.4	2.2	2.3	2.2
Achillea millefolium		+	+	+.2	+.2	1.2	1.2	+.2	1.2
Conyza canadensis		.	+.2	.	+	+	+.2	1.3	+
Convolvulus arvensis		.	.	.	+.2	+.2	+.2	+	1.2
and others species									
Forest pioneer species:									
Pinus sylvestris b/c		1.1	+	.	1.1	.	.	.	1.1
***Gypsophila paniculata*:**									
Abundance-dominance, sociability		3.2	3.3	3.2	3.4	2.2	3.3	3.3	2.2
Density: ind./100 m^2		156	133	106	85	77	74	58	34
Size h cm	max	94	111	98	92	102	86	101	83
	mean	52	49	56	60	50	52	67	57
Number of shoots/ind.	max.	9	62	11	19	11	12	18	36
	mean	3	4	2	4	4	3	5	5
Lenght of tap root cm	max.	*	*	*	164	*	*	200	*
	mean	*	*	*	55	*	*	80	*
Individual age yrs.	max.	12	12	13	12	*	*	12	*
	mean	7	4	7	7	*	*	7	*

* not estimated

species in this successional series – lives much longer, often persisting several decades after the closure of the forest canopy (Faliński 1980, 1986, 1995).

The communities of the earlier developmental phases are also penetrated by two North American escapes from cultivation (*Acer negundo* from roadside planting and *Prunus serotina* from forest planting) and one West-European (*Cytisus scoparius*, also during post-fire regeneration).

Table 3. Community with *Rumex confertus* in Białowieża Clearing (Białowieża Forest, NE-Poland, 1997).

Community	Arrnenath. Meadow		Eu.-Arct. Ruderal
Number of relevé	9702	9703	9701
Cover %	100	100	100
Surface of relevé	50	50	50
Number of species	28	24	22
Rumex confertus	2.3	3.3	4.4
Dactylis glomerata	2.2	3.4	+.2
Poa pratensis	+	+.2	+.2
Trifolium pratense	+.2	1.2	+
Ranunculus acris	1.2	1.2	+
Taraxacum officinale	+	+	+
Arrhenatherum elatius	2.2	1.2	.
Festuca pratensis	+.2	1.2	.
Festuca rubra	1.2	+.2	.
Trifolium repens	+.2	1.2	.
Lathyrus pratensis	+.2	+	.
Lotus corniculatus	+	+	.
Leontofon autumnalis	1.2	2.3	.
Tragopogon pratensis	+.2	+	.
Tragopogon orientalis	+	+	.
Plantago lanceolata	+	+	.
Chrysanthemum leucanthemum	+	+.2	.
Rumex acetosa	+	+	.
Avenochloa pubescens	1.3	.	.
Trisetum flavescens	+.2	.	.
Phleum pratense	1.2	.	.
Agrostis stolonifera	+.2	.	.
Centaurea jacea	1.2	.	.
Knautia arvensis	+.2	.	.
Alchemilla sp.	+.2	.	.
Galium mollugo	+.2	.	.
Vicia cracca	+.2	.	.
Vicia sepium	+.2	.	.
Medicago falcata	.	+	.
Medicago lupulina	.	+.2	.
Trifolium medium	.	+.2	.
Carum carvi	.	+	.
Convolvulus pratensis	.	+.2	.
Achillea millefolium	.	+	+
Heracleum sibiricum	.	.	2.2
Artemisia vulgaris	.	.	2.2
Anthriscus sylvestris	.	.	2.2
Urtica dioica	.	.	1.2
Agropyron repens	.	.	1.2
Silene alba	.	.	+.2
Oenothera sp.	.	.	+
Origanum vulgare	.	.	+
Equisetum arvense	.	.	+
Cirsium arvense	.	.	+
Erigeron acris	.	.	+
Sonchus arvensis	.	.	+
Lapsana communis	.	.	+
Galeopsis tetrahit	.	.	+

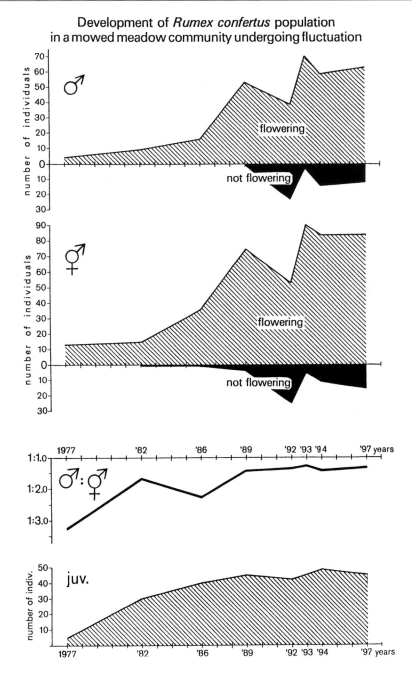

Fig. 3. Development of a *Rumex confertus* population in a mowed meadow community undergoing fluctuation. (orig.).

Rumex confertus in mesophilous hay meadows (Arrhenatheretum elatioris) undergoing fluctuation

This south-eastern European *Rumex* species is one of the largest in of the genus and has both male and bisexual individuals. It was noted in Poland for the first time in 1873 (Trzcińska-Tacik 1963), and on a thrice-mown meadow in Białowieża in 1922 (Wiśniewski 1923).

The initial spread of the species beyond its original range was towards the north-west via river valleys. Later spread was facilitated by railway lines (Trzcińska-Tacik 1963; Ćwikliński 1990). In mesophilous meadow communities, the species forms large aggregations by vegetative growth and prolific seeding. This is maintained by regular mowing. The retention at mowing of fruiting stems (in an experimental garden) gives rise to a steady increase in the population. There is a decline in the preponderance of bisexual individuals found the outset of colonisation, in time (Table 3). After 20 years, a balance between the sexes had still not been achieved (Fig. 3). Bisexual individuals usually function as females, with the full development on them of male inflorescences and flowers being exceptional. Locally, ruderal communities of the *Eu-Arction* and *Onopordion* alliances have *Rumex confertus* permanently, albeit with the appearance of the species in the former being relatively recent (Table 3).

*Lupinus polyphyllus in thermophilous oak forest (Potentillo albae-Quercetum); subject to **regression** a result of strong herbivore pressure*

This species was introduced as a fodder crop during the First World War (Paczoski 1930; Faliński 1966, 1968, 1986). It spread out along woodland edges and the verges of railways and lines transporting timber, and entered light oak forest.

The naturalization of the species in thermophilous oak forest led, over 30-40 years, to a clear increase in the total biomass and primary production of the herb layer (Fig. 4), as well as to the complete disappearance of *Oxalis acetosella*.

Almost 100 years of the regression of thermophilous oak forest (*Potentillo albae-Quercetum* come from changed pressure by large herbivores (European bison *Bison bonasus*, red deer *Cervus elaphus* and domestic cattle) and allowed hornbeam *Carpinus betulus* to become dominant, while the composition and properties of the litter layer changed leaving this community transformed into a shady oak-hornbeam forest (a community of the *Carpinion* type). This led first to an increase, and then to a curtailment, in the abundance of *Lupinus polyphyllus* in the herb layer. The gradual decline of *L. polyphyllus* with an increase in the abundance of other species (*Calamagrostis arundinacea, Viola mirabilis*) indicates change is ongoing.

*Carex brizoides in communities of oak-lime-hornbeam forest (Tilio-Carpinetum) subject to **degeneration** as a result of strong herbivore pressure*

Carex brizoides is widely-known for its mass appearance in European broadleaved deciduous and mixed forests in meso-oligotrophic pine oak forests (*Fagion, Carpinion, Alno-Padion*) and in oak-spruce forests (*Dicrano-Pinion, Vaccinio-Piceion*). The Białowieża Forest is beyond the original range of the species, and the local forest communities adjacent to the oldest roads have only been penetrated by the species since the end of the 19th century (Paczoski 1930). In spite of the constant enlargement of

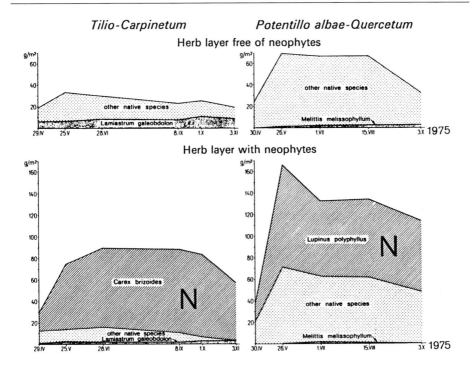

Fig. 4. Increase of biomass production in the herb layer of two forest communities during the vegetation period – as an effect of the establishment of alien species (neophyte = N). (After Faliński 1986, modified).

Potentillo albae-Quercetum

Herb layer without *Lupinus polyphyllus* (a); with *Lupinus polyphyllus* (b)

30 April   26 May   1 July   15 August   3 October 1975 r.

Galium boreale
Hepatica nobilis
Dactylis glomerata
Hieracium umbellatum
Potentilla alba
Peucedanum oreoselinum
Milium effusum
Rubus saxatilis
Lathyrus vernus
Melica nutans
Viola mirabilis
Viola reichenbachiana
Melittis melissophyllum
Carpinus betulus

Tilio-Carpinetum

Herb layer without *Carex brizoides* (a); with *Carex brizoides* (b)

1975 r.

29 April   24 May   28 June   8 September   1 October   3 November

Aegopodium podagraria
Isopyrum thalictroides
Ficaria verna
Acer platanoides
Dentaria bulbifera
Hepatica nobilis
Sanicula europaea
Majanthemum bifolium
Galium odoratum
Viola reichenbachiana
Stellaria holostea
Lamiastrum galeobdolon

Fig. 5. Changes during the vegetation period in the frequency of native species in the herb layer of two forest communities dominated by alien species (*Lupinus polyphyllus* or *Carex brizoides*). (After Faliński 1986, modified).

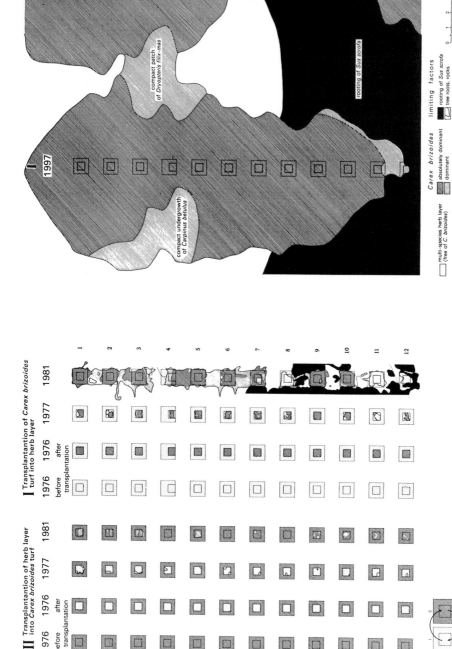

Fig. 6. Transplantation experiment to ascertain the effect of the neophyte *Carex brizoides* on the herb layer of *Tilio-Carpinetum*. The experiment consisted in exchange of soil monoliths with vegetation of 0.5 x 0.5 x 0.2 m in a herb layer invaded by *C. brizoides* (II) and a herb layer free of it (I). Twenty one years after the transplantation (1997) in the experiment I *C. brizoides* overwhelmed the herb layer on the transect and its surrounding. In experiment II the herb layer monolith transplanted into *C. brizoides* turf withered away witkin five years. (After Faliński 1986, supplemented by results of observations in 1997)

sites, and the predominance of *C. brizoides* over ever larger areas of the forest in Białowieża Forest, the link with the proximity of roads remains clear. The species may be presumed to have been brought to the area with hay at a time when the augmentative rearing of game animals was practised (1892-1915). The local excess of herbivores obstructed the process of stand renewal, simplified forest structure and transformed the habitat, herb layer and undergrowth.

In areas dominated by *C. brizoides*, the rooting activities of wild boar *Sus scrofa* limit the spread of what are often extensive patches of the species (covering between several areas and 4 ha), but also facilitate spread via the transfer of fragments of plant material to neighbouring areas. *C. brizoides* has become established most abundantly and most frequently in forests of the *Tilio-Carpinetum* type, though it occurs in all types of forest community other than bog forest.

It is clear from studies of relations between *C. brizoides* and other stable components of the herb layer, carried out several times a year during the growing season, that the relation is a "reducing" one entailing the limitation of the cover, growth and vitality of many species: *Hepatica nobilis, Sanicula europaea, Viola reichenbachiana, Lamiastrum galeobdolon, Galium odoratum* and *Stellaria holostea* (Fig. 5). Seedlings and juvenile specimens of trees – especially hornbeam – are eliminated altogether. *Ranunculus ficaria* and the *Aegopodium podagraria* increase in the presence of *C. brizoides*.

These two phenomena were confirmed in an experiment entailing the transplantation of 0.5 x 0.5 x 0.2 m slabs of *C. brizoides* together with soil to an area of herb-layer vegetation free of the species, as well as the transfer to existing patches of *C. brizoides* of slabs of vegetation free of the species. There was 1.5 m between transplanted slabs (Fig. 6). In the first variant there was, over 4 years, an expansion of *C. brizoides* and a merging of turves, and in the adjacent areas overgrown a decrease of species from the first group and a rise in the second. After 21 years, and in spite of rooting by wild boar, the total area of *C. brizoides* deriving from the twelve 0.25 m² slabs had increased 75-fold (from 3 m² to 223 m²; Fig. 6). Likewise, slabs of herb-layer vegetation without *C. brizoides* were entirely taken over by this alien species after introduction into an existing patch, and the numerical representation, participation, habit and vitality of the initial components of the vegetation were very much reduced (Faliński 1986).

Thus the local predominance of *C. brizoides* in forest previously subject to herbivore-induced prevents regeneration even after its influence has been moderated.

*Sambucus racemosa in a community of oak-lime-hornbeam forest (Tilio-Carpinetum) undergoing **regeneration** after the easing of strong herbivore pressure*

It was Paczoski (1930) who first observed the penetration of *Sambucus racemosa* into communities in Białowieża Forest that had been disrupted by grazing by large wild herbivores or domestic cattle. The phenomenon became more pronounced, but was always seen in the forest adjacent to open land and settlements. It was also easy to point to the source of propagules of this bird dispersed species, namely two parks in the large Białowieża Clearing (Polana Białowieska) and from other settlements also within forest clearings (Fig. 7; Faliński 1966, 1968, 1986; Sokołowski 1981, 1995).

The behaviour of *S. racemosa* was studied in the southern part of Białowieża Na-

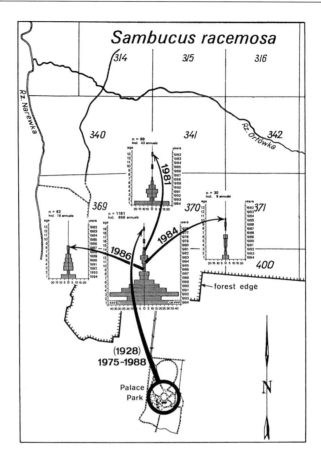

Fig. 7. Analysis of age structure of *Sambucus racemosa* in the population in a forest near Polana Białowieska allowed for reconstruction of the sources of propagules and the times of establishment of successive populations of the species developing in the forest interior. (orig.)

tional Park near Białowieża Clearing, where the species occurs at four larger sites and several smaller ones. The oldest population, and the largest in area and numbers, is where the forest meets Białowieża Clearing. It has at least 1181 individuals (in a 1.3 ha area) where specimens are concentrated and where the shrubs have taken advantage of optimal locations for establishment and later propagation of fruits (Fig. 8).

The establishment of permanent plots, measurement and observations concerning reproduction were followed by the cutting of all individuals with a view to the obtainment of cross-sections from the main trunk by which the age of the bush could be determined. Age pyramids drawn up from these data confirm:

1) the start date and continuity of penetration; shown by there being individuals in all age classes (the oldest – of 19 years; Fig. 7; in populations at the forest edge);

2) the establishment of the species in forest communities once subject to marked degeneration by herbivores; this by the development of a further autochthonous generation (derived from plants of autochthonous origin);

3) the secondary origin of the next 3 populations lying somewhat further into the

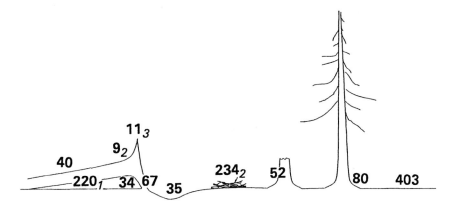

Fig. 8. Forest floor differentiation in oak-line-hornbeam forest (*Tilio-Carpinetum*) used by the alien species *Sambucus racemosa*. (After Faliński and Łuczaj mscr.).

forest than the oldest, peripheral population. This should be taken to indicate that the older populations arose as a consequence of the (also bird-assisted?) transfer of propagules from the first one which developed at the edge of the forest following its attainment of an adequate density, a greater age and the stage at which fruiting became abundant (Fig. 7).

   After the 1994 cutting of all individuals in the four populations there was no renewed colonisation of the same places by the species in the years 1995-1997.

Discussion

These examples of the behaviour of alien species in previously-existing plant communities of the Białowieża Forest provide valuable information on the nature, course and conditioning of neophytism. More examples could have been given, including *Impatiens parviflora* in the forest communities of Puszcza Białowieska and other forest complexes in Central Europe (Ćwikliński 1978; Faliński 1966; Kujawa-Pawlaczyk 1991).

   First consider the circumstances leading to the penetration of plant communities by newcomer species, skipping the obvious question of the presence nearby of individuals of the aliens. The most important factor favouring penetration of a community is disturbance of its structure, with disturbance of soil structure and the introduction to soil of alien substrata. It is not sufficient merely for a gap to appear in the vegetation, or even for the soil to be disturbed, since such conditions are always available in natural forests under the impact of windfalls, rooting by wild boar and the feeding- and reproduction migrations of animals (trails, breeding sites and wallows). Yet except in *Carex brizoides*, these factors are not important for the penetration of communities by aliens. A factor is the previous isolation of the forests and their distance from sources of the propagules of alien species. It would seem enough that small changes in habitat properties occur (as a result of the bringing to the surface and exposure to the sun of soil from deeper layers, supplementary nutrient enrichment, trampling or influxes of chemical substances from dumped refuse), together with pressure imposed by animals

Relations of neophytes to native components of the community

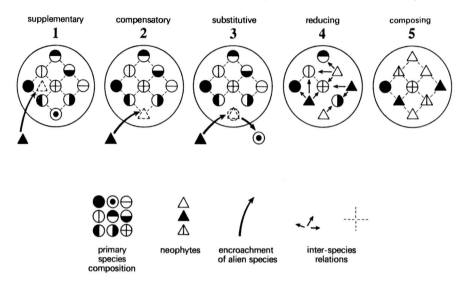

Fig. 9. Relations between neophytes and native components in a plant community. (After Faliński 1968, 1969, 1986, modified).

or man. Also important is the spread of propagules on the limbs and skins of animals or the shoes and clothes of people, or else in the faeces, regurgitated matter or food residues. For these factors to come into play, the free area on the forest floor (gap) need not be at all large. In the colonisation phase, alien species may even utilise unusual substrata like logs or the root systems of windthrown trees, as has been shown for *Sambucus racemosa* (Fig. 7).

Forest management and associated measures are certainly factors favouring the penetration of forest by alien species. Changes in habitat derive first and foremost from species ecologically- (though not necessarily geographically-) alien to the habitat. For example, in north-east Poland, pines planted in areas of fertile broad-leaved forest habitat – especially near settlements – are rapidly and spontaneously dominated by alien species of tree, shrub and herb. Łuczaj and Adamowski (1991) found saplings and seedlings of as many as 21 alien species on a 0.2 ha plot. The species included *Cotoneaster lucidus, Acer negundo, A. pseudoplatanus, A. tataricum, A. ginnala, Quercus rubra* and *Fraxinus pennsylvanica*. Also taking advantage of such circumstances are other aliens widespread in Central Europe *Prunus serotina*, and *Ailanthus altissima* (Starfinger 1990; Auge 1997).

Litter distributed unevenly over the soil surface and decomposing slowly combines with the long-term presence of organic matter (necromass) from plants and animals appearing in the course of succession to create similar conditions in communities subject to these processes to the sudden disturbances induced in permanent natural communities.

Alien species appearing when communities are created or transformed by secondary or primary succession may also establish final species composition. This is illustrated by the example of *Gypsophila paniculata* colonizing abandoned fields where

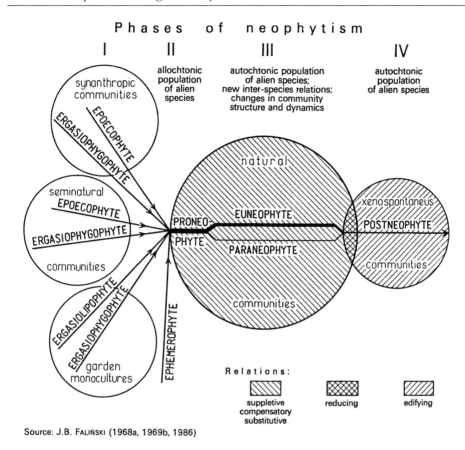

Source: J.B. FALIŃSKI (1968a, 1969b, 1986)

Fig. 10. Neophytism as the phenomenon of the establishment of an alien species in a previously-existing plant community. Of decisive significance for the course of this phenomenon is the development of an autochthonous population of an alien species in a previously-existing community and the relations with its native components. (After Faliński 1968, 1986).

secondary succession leads to the development of continental pine forest (see point 1 in "Results"; see also: Faliński 1980, 1986).

The presence of an alien species in a given end plant community, e.g. in a forest, psammophilous grassland or raised bog community, may relate back to earlier phases of its development by way of succession, and does not have to be the result of a more recent disturbance (degeneration) and subsequent regeneration.

Consider the different relationships which may ensue between incoming plant species which establish themselves in a community and the components present previously (Fig. 9).

(1) the supplementary relationship, well exemplified by *Elodea canadensis*, which easily and permanently found a place in Europe's aquatic plant communities;

(2) the compensatory relationship, when a place vacated by an original component of a community is occupied by individuals of an alien species (*Sambucus racemosa* on the place of bread-leaves shrubs and trees);

(3) the substitutive relationship, involving the displacement of an original component from a community by a newcomer of similar biology and ecological requirements.
Such behaviour may apply to the genus *Impatiens* and specifically to the pairs of species *I. noli-tangere* and *I. parviflora, I. noli-tangere* and *I. glandulifera* or *I. noli-tangere*
and *I. capensis* (Pawlaczyk and Adamowski 1991).

(4) the reducing relationship, best exemplified by the limiting action of *Carex brizoides*
in relation to other species of the herb layer in deciduous forest.

(5) the composing relationship well exemplified by the capacity of *Robinia
pseudacacia* to transform forest and xerothermic grassland communities (Ćwikliński
1972), or by the ability of *Acorus calamus* to transform swamps with *Glyceria maxima*
(*Glycerietum maximae → Acoretum calami*; Faliński 1991b).

These relationships are first expressed in the penetration (pro-neophyte) phase, and
are maintained following the formation of an autochthonous population of an alien species
in a community taken over by it (the eu-neophyte phase and even the post-neophyte
phase; Fig. 10).

Many aspects remain unclear, but I believe that long-term studies of the behaviour
of alien species in communities taken over by them may – if linked with their experimental transplantation to communities in which they have not yet come to dominate –
provide an ever better understanding of the nature of the different interspecific relationships (Figs. 9 and 10), and even shed new light on the classic problem of the "social bonds between species", as well on species preferences for defined conditions.

Finally, I suggest that research into the phytocoenotic role of alien species in plant
communities may solve a   problem of vegetation science which still-unclear defined
by Paczoski (1930) as: concerning "the principle by which plants unite together into
communities", as well helping understand the phenomenon of the formation and maintenance, of repeatable species compositions during succession. Solution of this basic
problem deepens our knowledge not only of the processes of the synanthropisation of
vegetation, but also of the autonomous functioning of plant communities. In these studies,
neophytes can play a role similar to that of the radioisotopes of certain elements in
physiological and environmental studies.

Conclusions

1. Invasion of alien plant species is a particular case of migration. The success of
migration beyond the primary geographical distribution range may be considered from
biogeographical, genetic and biocoenotic points of view.

2. **Biocoenotic (phytocoenotic)** success is observed when a species is permanently
incorporated into existing (natural) plant communities in its new homeland, it partici-
pates in their structure, internal dynamics and functions, when it goes through the entire cycle of reproduction, when it survives in the communities due to the formation of
secondary, autochthonous populations (*neophytism*).

3. **Biocoenotic success** is even more evident when an alien species is able to change
the species composition, structure and dynamics of the existing community transform-
ing it into communities with totally different, new properties (*post-neophytism*).

4. Factors favouring neophytism are:
   – the disturbance of community structure (degeneration of phytocoenoses);

– anthropogenic changes in the environment and habitat (degradation of the habitat);
– the fragmentation of natural complexes of forest, marshland, etc;
– the enhanced migration of plants, animals and people.
5. Factors limiting neophytism are:
   – the persistence of environmental conditions;
   – a good state of preservation of plant communities;
   – contiguity of complexes of forest, marshland, etc.
   – environmental isolation.
6. **Plant communities undergoing various ecological processes**, such as primary and secondary succession, regression, degeneration and regeneration as well as fluctuation are to a varying extent subject to penetration by alien species and their entering neophyte phase:
   – **degeneration** followed by the **regeneration** of a natural community creates particularly beneficial conditions for naturalization and permanent occurrence of alien species in plant communities which stabilized in the course of time;
   – similar significance can be ascribed to **secondary succession** on the settlement ruins, abandoned routes, and, the least seldom, abandoned farmlands;
   – **fluctuation** taking place in natural undisturbed plant communities, as the process deciding about the stability of a given community and about its relations with the biotope and the biochore, in principle, makes the penetration and naturalisation of alien species impossible.

References

Auge, H. 1997. Biologische Invasionen: Das Beispiel *Mahonia aquifolium*. In: Feldmann, R., Henle, K., Auge, H., Flachowsky, J., Klotz, J. and Krönert, R. (eds.), Regeneration und nachhaltige Landnutzung – Konzepte für belastete Regionen, pp. 124-129. Springer Verlag, Berlin, Heidelberg, New York.
Castri, F. di. 1990. On invading species and invaded ecosystems: the interplay of historical chance and biological necessity. In: Di Castri, F., Hansen, A.J. and Debussche, M., eds. 1990. Biological Invasions in Europe and the Mediterranean Basin, pp. 3-16. Kluwer Academic Publishers, Dordrecht / Boston / London.
Ćwikliński, E. 1972. Przenikanie gatunków synantropijnych do zbiorowisk stepowych w rezerwacie Bielinek nad Odrą. Penetration of synanthropic species into steppe communities in the reservation Bielinek on the Oder. Phytocoenosis 1.4: 273-282.
Ćwikliński, E. 1978. Die Einwanderung der Synanthropen-Art *Impatiens parviflora* DC. in die natürlichen Pflanzengesselschaften. Acta botanica slovaca, Acad. Sci slovacae 3: 17-31.
Ćwikliński, E. 1990. *Rumex confertus* Willd. na terenach kolejowych województw siedleckiego i Białopodlaskiego. *Rumex confertus* Willd. on by rail-way grounds od the Sieldce and Biała Podlaska Districts (Central-Eastern Poland). Zeszyty Naukowe Wyższej Szkoły Rolniczo-Pedagogicznej w Siedlach, Nauki Przyrodnicze 24: 187-199.
Faliński, J.B. 1966. Antropogeniczna roślinność Puszczy Białowieskiej jako wynik synantropizacji naturalnego kompleksu leśnego. Végétation anthropogène de la Grande Forêt de Białowieża comme un résultat de la synanthropisation du terrotoire silvestre naturel. Dissertationes Universitatis Varsoviensis 13: 1-256.
Faliński, J.B. 1968. Stadia neofityzmu i stosunek neofitów do innych komponentów zbiorowiska. Stages of neophytism and the reaction of neophytes to other components of the community. In: Synantropizacja szaty roślinnej. I. Neofityzm i apofityzm w szacie roślinnej Polski. Materiały Sympozjum w Nowogrodzie. Mater. Zakł. Fitosoc. Stos. UW 25: 15-31.
Faliński, J.B. 1969. Neofity i neofityzm. Dyskusje fitosocjologiczne (5). Néophytes et néophytisme. Discussion phytosociologiques (5). Ekol. pol. B.15: 337-355.

Faliński, J.B. 1977. Research on vegetation and plant populations dynamics conducted by Białowieża Geobotanical Station of the Warsaw University in the Białowieża Primeval Forest and in the environ 1952-1977. Phytocoenosis 6(1/2): 1-132.

Faliński, J.B. 1980. Vegetation dynamics and sex structure of the populations of pioneer dioecious woody plants. Vegetatio 43: 23-38.

Faliński, J.B. 1986. Vegetation dynamics in temperate lowland primeval forests. Geobotany 8: 1-537. Dr W. Junk Publishers, Dordrecht/Boston/Lancaster.

Faliński, J.B. 1991a. Procesy ekologiczne w zbiorowiskach leśnych. Ecological processes in the forest communities. Phytocoenosis 3(N.S.) Sem. Geobot. 1: 17-41.

Faliński, J.B. 1991b. Kartografia Geobotaniczna. Część 3. [Geobotany cartography. Part 3.]. Państwowe Wydawnictwo Wydawnictw Kartograficznych, Warszawa.

Faliński, J.B. 1995. Les espèces pionnières ligneuses et leur rôle dans la régénération et dans la succession sécondaire. Colloques phytosociologiques 24: 47-76.

Kornaś, J. 1968a. Stadia neofityzmu i stosunek neofitów do innych komponentów zbiorowiska. Stages of neophytism and the relation of neophytes to other components of the community. In: Faliński, J.B. (ed.), Synantropizacja szaty roślinnej. I. Neofityzm i apofityzm w szacie roślinnej Polski. Mater. Zakł. Fitosoc. Stos. UW Warszawa-Białowieża 25: 15-29.

Kornaś, J. 1968b. Geograficzno-historyczna klasyfikacja roślin synantropijnych. A geographical-historical classification of synanthropic plants. In: Faliński, J.B. (ed.), Synantropizacja szaty roślinnej. I. Neofityzm i apofityzm w szacie roślinnej Polski. Mater. Zakł. Fitosoc. Stos. UW Warszawa-Białowieża 25: 33-41.

Kornaś, J. and Medwecka-Kornaś, A. 1968. Występowanie gatunków zawleczonych w naturalnych i na półnaturalnych zespołach roślinnych w Polsce. The occurrence of introduced plants in natural and semi-natural plant communities in Poland. In: Faliński, J.B. (ed.), Synantropizacja szaty roślinnej. I. Neofityzm i apofityzm w szacie roślinnej Polski. Mater. Zakł. Fitosoc. Stos. UW Warszawa-Białowieża 25: 55-63.

Kowarik, I. 1992. Zur Rolle nichteinheimischer Arten bei der Waldbildung auf innerstädtischen Standorten in Berlin. Verh. Ges. Ökol. 21: 207-213.

Kozlovskaya, N.V. and Parfenov, V.I. 1972. Chorologija flory Belorusii. Minsk.

Kujawa-Pawlaczyk, J. 1991. Rozprzestrzenianie się i neofityzm Impatiens parviflora DC. w Puszczy Białowieskiej. Propagation and neophytism of Impatiens parviflora DC. in the Białowieża Forest. Phytocoenosis 3(N.S.) Sem. Geobot. 1: 213-222.

Lohmeyer, W. and Sukopp H. 1992. Agriophyten in der Vegetation Mitteleuropas. Schriftenreihe für Vegetationskunde 25: 1-185.

Łuczaj, Ł. and Adamowski, W. 1991. Dziczenie irgi lśniącej (Cotoneaster lucidus Schlecht.) w Puszczy Białowieskiej. Cotoneaster lucidus Schlecht. turning wild in the Białowieża Forest. Phytocoenosis 3(N.S.) Sem. Geobot. 1: 269-274.

Nikolaeva, V.M. and Zefirov, B.M. 1971. Flora Belovežskoj Pušči. Minsk.

Paczoski, J. 1930. Lasy Białowieży. Die Waldtypen von Białowieża. Monografie Naukowe 1: 1-575.

Pawlaczyk, P. and Adamowski, W. 1991. Impatiens capensis (Balsaminaceae) – nowy gatunek we florze Polski. Impatiens capensis (Balsaminaceae) a new species in the Polish flora. Fragm. Flor. Geobot. 35(1/2): 225-232.

Rejmánek, M. 1989. Invasibility of Plant Communities. In: Drake, J.A., Mooney, H.A., di Castri, F., Groves, R.H., Kruger, F.J., Rejmánek, M. and Williamson, M. (Eds.). Biological Invasions. A Global Perspective, pp. 369-388. John Wiley & Sins, Chchester – New York – Brisbane – Toronto – Singapore.

Rejmánek, M. 1996. A theory of seed plant invasiveness: the first sketch. Biological Conservation 78: 171-181.

Rikli, M. 1903/1904. Die Anthropochoren und der Formenkreis des Nasturtium palustre D.C. Ber. Zürcherich. Bot. Ges. 8: 71-82.

Sokołowski, A.W. 1981. Flora roślin naczyniowych Białowieskiego Parku Narodowego. Flora of the vascular Plants of Białowieża Park. Fragm. Flor. Geobot. 27(1/2): 51-131.

Sokołowski, A.W. 1995. Flora roślin naczyniowych Puszczy Białowieskiej. The flora of vascular plants in the Białowieża Forest. Wyd. Białowieski Park Narodowy, Białowieża.

Starfinger, U. 1990. Die Einbürgerung der Spätblühenden Traubenkirsche (Prunus serotina Ehrh.) in Mitteleuropa. Landschaftsentwicklung und Umweltforschung 69: 1-118 + Tabs.

Sukopp, H. 1962. Neophyten in natürlichen Pflanzengesellschaften Mitteleuropas. Ber. Deutsch. Bot. Ges. 75: 193-205.

Sukopp, H. 1976. Dynamik und Konstanz in der Flora des Bundesrepublik Deutschland. Schr. R. Vegetationkde 10: 9-27.

Sukopp, H. 1995. Neophytie und Neophytismus. In: R. Böcker, H. Gebhardt, W. Konold and S. Schmidt-Fischer (eds.), Gebietsfremde Pflanzenarten, pp. 3-32. Landsberg.

Thellung, A. 1915. Pflanzenwanderungen unter dem Einfluss des Menschen. Beilb. Engl. Bot. Jahrb. 53. Stuttgart.

Thellung, A. 1918-1919. Zur Terminologie der Adventiv- und Ruderalflora. Allq. Bot. Zeistr. 24. Jena.

Trepl, L. 1984. Über *Impatiens parviflora* DC. als Agriophyt in Mitteleuropa. Diss. Bot. 73: 1-400.

Trepl, L. 1990. Zum Problem der Resistenz von Pflanzengesellschaften gegen biologische Invasionen. Verh. Berliner Bot. Ver. 8: 195-230.

Trepl, L. and Sukopp, H. 1993. Zur Bedeutung der Introduktion und Naturalisation von Pflanzen und Tieren für die Zukunft der Artenvielfalt. Rundgespräche d. Kommission für Ökologie 6: 127-142.

Trzcińska-Tacik, H. 1963. Badania nad zasięgami roślin synantroopijnych. 2. *Rumex confertus* Willd. w Polsce. Studies on the distribution of synanthropic plants. 2. *Rumex confertus* Willd. in Poland. Fragm. Flor. Geobot. 9: 73-84.

Wiśniewski, T. 1923. Przyczynek do znajomości flory Puszczy Białowieskiej. Białowieża 2: 34-61.

INVASIONS OF ALIEN PLANTS INTO HABITATS OF CENTRAL EUROPEAN LANDSCAPE: AN HISTORICAL PATTERN

Petr Pyšek[1], Karel Prach[2] and Bohumil Mandák[1]

[1]Institute of Botany, Academy of Sciences of the Czech Republic, CZ-252 43 Průhonice, Czech Republic; e-mail: pysek@ibot.cas.cz; [2]Faculty of Biology, University of South Bohemia, Branišovská 31, CZ-370 01 České Budějovice, Czech Republic; e-mail: Karel.Prach@tix.bf.jcu.cz

Abstract

The localities of 53 species alien to the flora of the Czech Republic were collated from their introduction to the present. The information on the distribution of particular species was obtained by using herbaria (contributing 37.4 % of the total number of localities), floristic periodicals (53.4 %) and unpublished data (9.1 %). In total, 32,277 localities were recorded. If given in the original source, information was recorded on the (i) the year of the record, and (ii) the type of the habitat. The classification of habitats yielded 14 types, divided into two main groups, (1) man-made, heavily disturbed habitats, and (2) less disturbed sites of a more natural origin. Aliens occur most frequently in cities and villages (25.6 % of the total number of localities), and riparian habitats (22.4 %). From the viewpoint of management, these habitats represent further potential sources for their spread into the landscape. The contribution of other habitat types never exceeds 10 %. The representation of forests is rather high (9.2 % including forest margins). The majority of "seminatural" habitats, namely scrub and grasslands, prove to be rather resistant. The representation of particular habitas has been changing remarkably in the course of the 20th century, the most conspicuous trend being a decrease in relative representation of urban habitats and an increasing role of habitats facilitating dispersal (including roads, railways and paths). These trends can be interpreted as a consequence of changes of intensity and type of disturbances affecting the landscape during this period. Every habitat, including forest interiors, has at least some well established invaders.

Introduction

In research on biological invasions, the importance of recipient habitat has been increasingly recognized (Lodge 1993). Although the species-view approach to predicting plant invasions has had some success recently (Rejmánek 1995, 1996) it is clear that attempts to predict the outcomes of invasions have a limited chance of succeeding if they do not take the characteristics of invaded habitats into account (Scott and Panetta 1993).

Nevertheless, scepticism concerning our predictive power with respect to plant invasiveness has its precedents. Recently, Williamson (1996) summarized available data on the role of habitat, carried out a critical assessment and came to similar conclusions to Crawley (1987) who compared the number of alien species in particular habitats in the British Isles, and Usher et al. (1988) who evaluated invasions into nature reserves at the global level. Studies yielding solid quantitative data are rather few and so the generalizations are not well established either (Williamson 1996):

Plant Invasions: Ecological Mechanisms and Human Responses, pp. 23–32
edited by U. Starfinger, K. Edwards, I. Kowarik and M. Williamson
© *1998 Backhuys Publishers, Leiden, The Netherlands*

1. Each plant community is, in principle, invasible.
2. It is probable that some communities are more vulnerable to invasions than others.

In general the invasibility of communities and ecosystems depends much on (a) the position of invaded communities on environmental gradients (moisture, nutrients, disturbance, successional age), and (b) the biotic characteristics of invaded communities (Rejmánek 1989; Hobbs and Huenneke 1992; Hobbs and Humphries 1995; Tilman 1997). Disturbance is another important factor, particularly in man-made habitats where its intensity and frequency is usually high. Not only the intensity and frequency of disturbance but also the change in its regime can increase the vulnerability of plant communities to invasion (Hobbs and Humphries 1995).

Because of the lack of available data, our understanding of why some communities are more prone to invasions by alien plants than others remains, to a large extent, rather intuitive, and more carefully designed studies are needed to verify the hypotheses.

This chapter summarizes 250 years of history of plant invasions into the Czech flora by invaded habitat, and provides a large body of quantitative data to assess the invasibility of particular habitat types, broadly defined within the range of a landscape approach. It also evaluates the changes in the role of particular habitats on an historical time scale.

Data sources and methods

Species selection and collation of the data

Fifty three species alien to the flora of the Czech Republic (area of 78,854 km^2) were selected and information on their localities from the early days up to present was collated. The criteria for species selection were as follows:

(a) Alien status. The species considered were neophytes, i.e. introduced to the territory of the Czech Republic after 1500 A.D. The immigration status was obtained from Czech floras (Hejný and Slavík 1988–1992, Slavík 1995–1997), relevant papers on the topic (Opravil 1980) and databases covering neighbouring Central European countries (Frank and Klotz 1990; Ellenberg *et al.* 1991). The species were introduced between 1738 and 1963 (Pyšek *et al.,* unpublished data) and their area of origin is as follows: America 28, Europe or Eurasia 14, Asia 10, Australia 1.

(b) Degree of naturalization. The aim was to include all the major invaders in the Czech flora so as to get the most complete picture of habitats invaded. Naturalized, commonly occurring invaders were not considered only if taxonomically problematic, so not easily recognizable, and so the reliability of floristic records would be low (e.g. the species of *Aster, Stenactis, Oenothera*).

Species used in the analysis are listed in Appendix 1.

Information on the distribution of particular species was got from the following sources (only the occurrence in the wild was considered, i.e. records in cultivation were not included into the analysis):

(a) Major herbaria (Charles University Praha – PR, National Museum Praha – PRC, Masaryk University Brno – BRNU) as well as various local herbarium collections (CB, CHOM, HR, LIM, LIT, MJ, PL, ROZ, Příbram, Sokolov). In total, these herbaria comprise about 5,100,000 of specimens of vascular plants. Herbaria contributed 37.4 % to the total number of localities analysed in the present study.

(b) Botanical literature: All major floras, periodicals, floristic works and some manu-scripts (dissertations, theses) were checked for the occurrence of the species. The proportion of published records was 53.4 % of the total number of localities.
(c) Unpublished data obtained by personal communications, including our own field research in the last few years. Unpublished data contributed 9.1 % to the total number of localities collated.

The work used the fact that in the Czech Republic there is a high density of floristic records going back a long time (Pyšek 1991; Pyšek and Prach 1993, 1995; Mandák and Pyšek, this volume).

Classification of habitats

If given in the original source, information was recorded on the (i) year of the record, and (ii) type of the habitat. Habitats were classified into 14 habitat types listed in Table 1. This classification might seem rather vague but seems to be the only possible one, considering the detail of habitat description normally given in floristic data.

Particular habitats differ in their abundance in the landscape and in the frequency and character of disturbances. On the basis of their origin and the intensity of distur-bance, they were divided arbitrarily into two main groups: (1) man-made, heavily dis-turbed habitats (6 types), comprising urban sites, roads ditches and banks, railway ar-eas, dumps of various materials, fields and old-fields, paths and their margins; (2) less disturbed sites of a more natural origin (8 types) including water courses and their shores, fishponds, scrub, grasslands, forests and their margins, and also large parks and the surroundings of solitary objects in relatively undisturbed landscape, such as chalets, gamekeeper's lodges etc. (Table 1). We are aware that this division is rather rough and arbitrary, but the data do not allow a more sophisticated delimitation to be used.

Results

The role of particular habitats

In total, 32,277 localities of the alien species considered were recorded, distributed rather unevenly among particular habitats (Table 2). Aliens occur most frequently in human settlements (including both large cities, towns and villages), which contribute 25.6 % to the total number of localities reported. Riparian habitats, i.e. surroundings of both running and still waters, are the second most important, together representing 22.4 %. The contribution of other habitat types never exceeds 10 %. However, the rep-resentation of forests, the relatively least disturbed habitats, is rather high (9.2 % if margins are included). The majority of "seminatural" habitats, mostly scrub and grass-land, are rather resistant (Fig. 1).

Changes in the importance of habitats: temporal trends

The increase in the total number of localities at 50-years intervals is given in Table 2. The representation of particular habitats has changed remarkably in the course of the 20th century (Fig. 2, Table 2). There was a strong decrease in urban habitats, contrib-

Table 1. Classification of habitats given in floristic records. Particular habitats are arranged according to the level of disturbance, i.e. approximately from those disturbed and heavily affected by man to those with more "seminatural" character. The evaluation of the possibilities for the transport of diaspores is relative and based on observational experiences.

Habitat	Characteristics	Frequency in the landscape / Area	Intensity / type of disturbance	Possibilities for transport of diaspores	Status (seminatural vs. man-made
Urban	ruderal habitats and industrial sites in cities, towns and villlages	very frequent, large area	very high / various kinds	very high	man-made
Roads	road ditches and margins in settlements and open landscape	very frequent, small area	high / transport, trampling, salt treatments	very high	man-made
Railways	railway stations and tracks	frequent, small area	high / transport	very high	man-made
Arable land	managed fields, abandoned fields	very frequent, large area	high/agricultural management	moderate	man-made
Dumps	deposits of various character, i.e. dumps, rubbish tips, spoil heaps, dung heaps etc.	frequent, locally more represented, small area	low-high (depending on successional stage)	high (low after dumping)	man-made
Paths	paths in forests, meadows etc. and their margins (not affected by traffic)	very frequent, small area	moderate / various, e.g. trampling	moderate-high	man-made
Ponds	water bodies, mostly ponds and their shores and litoral, wetlands	locally frequent, small area	moderate / eutrofication, fishery	low-moderate	seminatural
Water courses	running waters, i.e. rivers, and brooks, and their shores	very frequent, small area	moderate / floods, eutrofication	high	seminatural
Scrubland	drier habitats dominated by shrubs, including dry xerotherm grasslands	frequent, small area	low	low	seminatural
Grasslands	fresh to moist grasslands subjected to regular management or abandoned	frequent, large area	moderate / mowing, agricultural management	low-moderate	seminatural
Forests	coniferous forests, deciduous woodlands, both seminatural and monocultural	very frequent, large area	low / forestry management practices	low	seminatural
Forest margins	transitions between forests and surrounding habitats	frequent, small area	moderate / various	low	seminatural
Parks	managed parks and gardens, public greenery, chateau gardens	less frequent, small area	low / gardening practices	low-moderate	seminatural
Solitary objects	chalets, gamekeeper's lodges and their surroundings etc.	rather rare, small area	low	low	seminatural

Table 2. Number of localities recorded in particular habitats (see Table 1 for habitat description and characteristics) during 50 year periods. Total number of localities does not correspond exactly to the figure obtained by summing up the numbers for particular periods because some of the records were not dated in the original sources. Similarly, the sum of total numbers in particular habitats exceeds the total of localities analysed (n = 32,277) as some localities were attributed to more than one habitat. Habitats are ranked by decreasing importance.

Habitat	-1850	1850-1900	1900-1950	1950-1995	Total	%
Urban	72	569	1754	5857	8331	25.61
Water courses	7	239	736	3162	4159	12.78
Ponds	5	173	556	2392	3136	9.64
Railways	0	78	389	2099	2581	7.93
Forests	1	44	374	1732	2155	6.62
Roads	2	33	220	1783	2045	6.29
Arable land	6	111	389	1419	1944	5.97
Paths	0	52	270	1582	1911	5.87
Deponies	3	32	249	1415	1716	5.27
Scrubland	2	39	322	1073	1440	4.43
Grasslands	7	72	262	1087	1432	4.40
Forest margins	0	15	163	666	848	2.61
Parks, gardens	11	74	151	349	569	1.75
Solitary objects	0	8	24	237	269	0.83
Total	116	1539	5859	24853	32536	

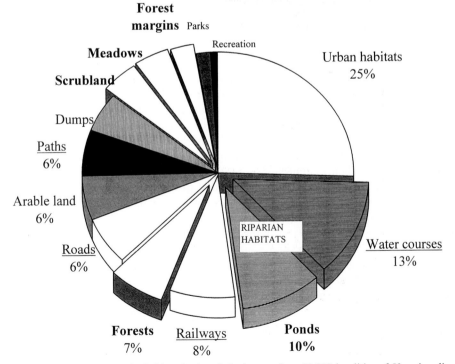

Fig. 1. Frequency of particular habitats in the whole data set from 32,277 localities of 53 major aliens to the Czech flora recorded between 1738–1995. Habitats considered as less heavily disturbed (i.e. "seminatural") are shown in bold letters, "transport" habitats are underlined. Approximate percentage representation is shown for more important habitats (exceeding 5 %).

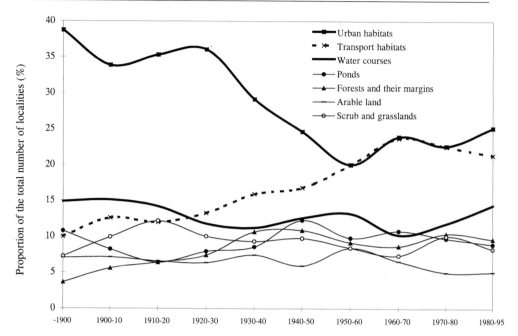

Fig. 2. Changes in representation of habitats of alien species during the 20th century. Proportional contributions to the total are displayed for particular decades. Some habitats were grouped to make the trends obvious, the values for each habitat can be found in Table 2. Transport habitats include road ditches and margins, railway areas and paths, but exclude water courses.

uting 38.7 % in 1900 but only 25.6 % in the last 15 years, and this trend was highly significant (slope of the regression of percentage representation in particular decades on time was significantly different from zero, $r = -0.85$, $F_{1,8} = 21.5$, $P < 0.01$). The most remarkable increase, on the other hand was found in "transport" habitats (including roads, railways and paths), representation of which started at 10 % in 1900 and between 1980–95 exceeded 20 %. This trend was also highly significant ($r = 0.94$, $F_{1,8} = 62.1$, $P < 0.0001$) and the same was true for dumps ($r = 0.81$, $F_{1,8} = 15.9$, $P < 0.01$). Forest habitats and their margins also enhanced their representation, reaching 3.6 % at the beginning of the century but fluctuating around 10 % in the last few decades ($r = 0.78$, $F_{1,8} = 12.4$, $P < 0.01$). Other habitat types do not exhibit interpretable trends in their representation over time ($P > 0.05$): $r = 0.26$, $F_{1,8} = 0.59$ for ponds, $r = 0.42$, $F_{1,8} = 1.8$ for water courses, $r = 0.51$, $F_{1,8} = 2.8$ for arable land, and $r = 0.09$, $F_{1,8} = 0.07$ for scrubland (Fig. 2).

The trends are not consistent between more disturbed (man-made) vs. less disturbed ("seminatural") habitats – e.g. the decrease in urban sites is compensated by an increase in "transport" habitats such as water courses, whereas the increase in forests is balanced by a decrease in parks, etc. so no change in the pattern of representation of both was found (regression slopes not significantly different from zero, $r = 0.09$, $F_{1,8} = 0.07$, $P = 0.79$).

Table 3. Major invaders in particular habitats. For each habitat, the proportional contribution of particular species to the total number of localities recorded in the habitat was calculated and three species with highest representation are listed. Total percentage contribution of these three species is shown in the last colums.

Habitat	Major invaders	Σ %
Urban	*Chamomilla suaveolens* 8.3, *Galinsoga parviflora* 6.5, *Amaranthus retroflexus* 6.0	20.8
Roads	*Cardaria draba* 7.7, *Chamomilla suaveolens* 6.2, *Epilobium ciliatum* 5.5	19.4
Railways	*Cardaria draba* 10.3, *Oenothera biennis* 7.5, *Conyza canadensis* 6.8	24.6
Arable land	*Veronica persica* 25.6, *Galinsoga parviflora* 20.4, *Amaranthus retroflexus* 7.6	53.6
Deponies	*Amaranthus retroflexus* 11.2, *Cardaria draba* 7.5, *Galinsoga parviflora* 6.5	25.2
Paths	*Juncus tenuis* 26.6, *Chamomilla suaveolens* 7.4, *Epilobium ciliatum* 5.8	39.8
Ponds	*Acorus calamus* 23.6, *Elodea canadensis* 18.7, *Potentilla norvegica* 8.5	50.8
Water courses	*Impatiens parviflora* 8.4, *Bidens frondosa* 7.4, *Impatiens glandulifera* 6.6	22.4
Scrubland	*Bryonia alba* 16.1, *Robinia pseudoacacia* 10.1, *Cardaria draba* 5.6	31.8
Meadows	*Trifolium hybridum* 29.9, *Juncus tenuis* 11.0, *Epilobium ciliatum* 7.8	48.7
Forests	*Juncus tenuis* 23.4, *Impatiens parviflora* 17.9, *Digitalis purpurea* 7.1	48.4
Forest margins	*Lupinus polyphyllus* 12.4, *Juncus tenuis* 10.8, *Digitalis purpurea* 10.0	33.2
Parks, gardens	*Geranium pyrenaicum* 16.6, *Impatiens parviflora* 11.7, *Galinsoga ciliata* 5.1	33.4
Recreation areas	*Impatiens parviflora* 8.5, *Juncus tenuis* 8.1, *Lupinus polyphyllus* 6.6	23.2

Species in particular habitats

Table 3 summarizes the major invaders of particular habitats. Some species occur in several habitats, e.g. *Chamomilla suaveolens, Impatiens parviflora, Juncus tenuis, Galinsoga parviflora, Epilobium ciliatum*. The joint contribution of the three most frequent species varies from 19.4 % to 53.6 % with respect to the habitats. Habitats with a high value for the three most frequently encountered invaders are those with rather specialized invasive aliens, notably arable fields, ponds, meadows, and forests. On the other hand, a less specialized alien flora is typical of urban habitats, roads, railways, water courses and recreational areas (Table 3).

Discussion

The data are remarkable for (a) the number of records included, (b) the historical time-scale considered, and (c) the degree of completeness of the alien flora sampled. The pooled data representing more than 32 thousand localities of which 91.2 % were classified by habitat, are a very good measure of importance of particular habitats. The 53 species cover most major invasives of the Czech flora, a reasonably representative sample of the alien flora. This is the first quantitative data set in which it is possible to evaluate changes in the role that particular habitats played in the Central European landscape on the time scale of centuries.

We show that the major habitats of aliens in the Czech Republic are urban and riparian sites (Wittig 1991; de Waal *et al.* 1994; Kowarik 1995; Sukopp *et al.* 1995; Pyšek 1998). For management, these habitats are further potential sources of spread. This is confirmed by the increase in the importance of transport sites over time. Roads and railways are always connected with cities and villages and the steadily increasing rep-

Table 4. Comparison between the proportinal extent of land-use types in the Czech Republic (taken from statistical yearbook – Anonymous 1996) and their invasibility, expressed by the contribution to the total number of localities of alien species recorded in the present study.

Land-use category	Corresponding habitats in the present study	Area in the Czech Republic (%)	Contribution to the number of localities (%)
Arable land	Arable land	39.9	6.0
Meadows and pastures	Grassland, scrubland	14.4	8.8
Forests	Forests, forest margins	33.3	9.2
Surface waters	Ponds, water courses	2.0	22.4
Built-up area and other sites	Urban, railways, roads, paths, deponies, parks and gardens, solitary objects	10.4	53.6

resentation of these sites from the 1920s is an indication that alien species penetrate gradually more and more into the landscape, outside the limits of human settlements. However, almost every habitat has its own, well established invaders whose properties fit its characteristics (Crawley 1987). In quantitative terms, even forest interiors have some important invaders, *Impatiens parviflora* being a typical example (Trepl 1984).

Another limit to the data set is that it does not take into account the extent of the particular habitats in the landscape (Pyšek and Pyšek 1995); these are not known for the whole country. The only way to do this here was to group our habitat types into broader categories corresponding to administrative land use categories, whose area is regularly published for the whole country in statistical yearbooks (Anonymous 1996). The comparison between the proportion of particular land-use types and proportion of localities of alien plants in the corresponding habitat categories is given in Table 4. This is only tentative as some categories could not be related unambiguosly to the land-use types (e.g. paths can occur in many habitats). Scrubland was included into the category "meadows and pastures" as scrub usually develops on agricultural land, often in extensive or abandoned pastures, in abandoned meadows or as hedges between particular parcels. Also, the pattern of land-use has changed. Nevertheless, it still provides us with a rough picture of the importance of broadly defined land-use types.

A remarkable difference between the area of the habitat and its contribution to the abundance of aliens was found (χ^2 goodness-of-fit test with expected values derived from the area of the habitat: $\chi^2 = 85722.5$, df 4, p $< 10^{-9}$). The results indicate a higher invasibility of settlements and surface water areas and a low invasibility of less disturbed habitats, i.e forests, grassland and scrubland (Table 4). The low proportion of records from arable land could be explained by these habitat types harbouring archaeophytes which were not included in the present study.

The decrease in the proportion of urban habitats between 1920s and 1960s reflects changes of intensity and type of disturbance. Early in this century, cities were best places where the influx of aliens was concentrated, and where these species had the best chance of surviving or possibly establishing (Gilbert 1989; Wittig 1991). The surrounding open landscape was disturbed less and in a different way. From the 1930s on, the landscape has become gradually more disturbed which meant easier penetration of aliens. From the 1950s, the extent of urban habitats started to increase again as a consequence of remarkable building activity, e. g. large scale estates at the periphery of big cities, leaving extensive wasteland to be colonized and providing for the initial successional stages of alien species (Mandák and Pyšek, this volume).

Acknowledgments

Helpful comments from Ingo Kowarik and Mark Williamson are greatly acknowledged. Our thanks are due to many colleagues for technical help and providing us with their unpublished data, especially V. Chán, J. Hadinec, M. Štech, A. Pyšek, R. Hlaváček, O. Šída, K. Bímová and I. Ostrý. We thank J. Hladík, Plzeň, for software services. The work was partly supported by grant no. 204/93/2440 from the Grant Agency of the Czech Republic. Last but not least, we thank Uwe Starfinger, Berlin, for bringing us all together once again.

References

Anonymous, 1996. Statistical yearbook of the Czech Republic. Czech Statistical Office.

Crawley, M. J. 1987. What makes a community invasible? In: Gray, A. J., Crawley, M. J. and Edwards, P. J. (eds.), Colonization, succession and stability. pp. 429–543. Blackwell Sci. Publ., Oxford.

de Waal, L.C., Child, E.L., Wade, P. M. and Brock, J. H. (eds.) 1994. Ecology and management of invasive riverside plants. John Wiley and Sons, Chichester.

Ellenberg, H. *et al.* 1991. Zeigerwerte von Pflanzen in Mitteleuropa. Scripta Geobotanica 18: 1–248.

Frank, D. and Klotz, S. 1990. Biologisch-ökologische Daten zur Flora der DDR. Wiss Beitr. Martin Luther Univ. Halle-Wittenberg 32: 1–167.

Gilbert, O. 1989. The ecology of urban habitats. Chapman and Hall, London.

Hejný, S. and Slavík, B. (eds.) 1988–1992. Flora of the Czech Republic. Vol. 1-3. Academia. Praha. [in Czech.]

Hobbs, R. J. and Huenneke, L. F. 1992. Disturbance, diversity and invasion: implications for conservation. Conserv. Biol. 6: 324–337.

Hobbs, R. J. and Humphries, S. E. 1995. An integrated approach to the ecology and management of plant invasions. Conserv. Biol. 9: 761–770.

Kowarik, I. 1995. On the role of alien species in urban flora and vegetation. In: Pyšek, P., Prach, K., Rejmánek, M. and Wade, P. M. (eds.), Plant invasions: general aspects and special problems. pp. 85–103. SPB Academic Publ., Amsterdam.

Lodge, D. M. 1993. Biological invasions: lessons for ecology. Trends Ecol. Evolut. 8: 133–137.

Mandák, B. and Pyšek, P. 1998. History of the spread and habitat preferences of *Atriplex sagittata* (Chenopodiaceae) in the Czech Republic. This volume.

Opravil, E. 1980. On the history of synanthropic vegetation 1–3. Živa 28 (66): 4–5, 53–55, 88–90. [in Czech.]

Pyšek, P. 1991. *Heracleum mantegazzianum* in the Czech Republic: the dynamics of spreading from the historical perspective. Folia Geobot. Phytotax. 26: 439–454.

Pyšek, P. 1998. Alien and native species in Central European urban floras: a quantitative comparison. J. Biogeogr. 25: 155–163.

Pyšek, P. and Prach, K. 1993. Plant invasions and the role of riparian habitats – a comparison of four species alien to central Europe. J. Biogeogr. 20: 413–420.

Pyšek, P. and Prach, K. 1995. Invasion dynamics of *Impatiens glandulifera* – a century of spreading reconstructed. Biol. Conserv. 74: 41–48.

Pyšek, P. and Pyšek, A. 1995. Invasion by *Heracleum mantegazzianum* in different habitats in the Czech Republic. J. Veget. Sci. 6: 711–718.

Rejmánek, M. 1989. Invasibility of plant communities. In: Drake, J. A., Mooney, H. A., di Castri, F., Groves, R. H., Kruger, F. J., Rejmánek, M. and Williamson, M. (eds.), Biological invasions: a global perspective. pp. 369–388, John Wiley and Sons, Chichester.

Rejmánek, M. 1995. What makes a species invasive? In: Pyšek, P., Prach, K., Rejmánek, M. and Wade, P. M. (eds.), Plant invasions: general aspects and special problems. pp. 3–13, SPB Academic Publ., Amsterdam.

Rejmánek, M. 1996. A theory of seed plant invasiveness: the first sketch. Biol. Conserv. 78: 171–181.
Scott, J. K. and Panetta, F. D. 1993. Predicting the Australian weed status of southern African plants. J. Biogeogr. 20: 87–93.
Slavík, B. 1995–1997. Flora of the Czech Republic. Vol. 4-5. Academia, Praha. [in Czech.]
Sukopp, H., Numata, M. and Huber A. (eds.) 1995. Urban ecology as the basis of urban planning. SPB Academic Publ., Amsterdam.
Tilman, D. 1997. Community invasibility, recruitment limitation, and grassland biodiversity. Ecology 78: 81–92.
Trepl, L. 1984. Über *Impatiens parviflora* DC. als Agriophyt in Mitteleuropa. Diss. Bot. 73: 1–400.
Tutin, T. G. *et al.* 1964–80. Flora Europaea. Vol. 1–5. Cambridge University Press.
Usher, M. B. 1988. Biological invasions of nature reserves: a search for generalisation. Biol. Conserv. 44: 119–135.
Williamson, M. 1996. Biological invasions. Chapman and Hall, London.
Wittig, R. 1991. Ökologie der Großstadtflora. G. Fischer, Stuttgart.

Appendix 1. The species alien to the Czech flora which were used for the analysis of habitat preferences. Fifty-three species were considered. Nomenclature follows Tutin et al. (1964–1980).

Acer negundo, Acorus calamus, Ailanthus altissima, Amaranthus albus, Amaranthus powellii, Amaranthus retroflexus, Ambrosia artemisiifolia, Ambrosia trifida, Amorpha fruticosa, Bidens frondosa, Bryonia alba, Bunias orientalis, Cardaria draba[1], Chamomilla suaveolens, Chenopodium botrys, Chenopodium foliosum, Chenopodium pumilio, Conyza canadensis, Corydalis lutea, Cymbalaria muralis, Digitalis purpurea, Echinocystis lobata, Elodea canadensis, Epilobium adenocaulon, Galinsoga ciliata, Galinsoga parviflora, Heracleum mantegazzianum, Hordeum jubatum, Impatiens glandulifera, Impatiens parviflora, Iva xanthiifolia, Juncus tenuis, Lupinus polyphyllus, Lycium barbarum, Mimulus guttatus, Oenothera biennis, Physocarpus opulifolia, Pinus nigra, Pinus strobus, Potentilla norvegica, Reynoutria japonica, Reynoutria sachalinensis, Robinia pseudoacacia, Rudbeckia laciniata, Sisyrinchium angustifolium, Solidago canadensis, Solidago gigantea, Telekia speciosa, Trifolium hybridum, Veronica filiformis, Veronica persica, Xanthium spinosum.

[1] Some sources classify this species as an archaeophyte (Opravil 1980).

ON SUCCESS IN PLANT INVASIONS

Uwe Starfinger
Technische Universität Berlin, Institut für Ökologie, Schmidt-Ott-Str. 1, 12165 Berlin, Germany; e-mail: starfinger@gp.tu-berlin.de

Abstract

The term success is often used in invasion biology, especially in studies that seek the causes and conditions of successful as opposed to failed invasions. Characters of species known to be connected to invasion success are sought in order to use them for the prediction of future invasions. Success is, however, not a simple phenomenon but different categories independent of each other are conceivable: whereas all plant invaders have to go through the step from being imported to growing outside of cultivation in order to be successful, processes like filling out a large secondary range, attaining high frequency, abundance, or dominance, or becoming completely independent of human influence by growing in natural vegetation as agriophytes, can all be seen as leading to (different types of) success. Species can thus have different degrees of success in different categories.

This study uses North American species invading in Germany as an example to show how the knowledge about ecological behavior in the native range can be used to predict invasion success in defined categories only. Plants with a large native range and those that are known as apophytes in their native range, i.e. that occur in vegetation with marked human impact, occur with greater frequency in the secondary range. The success stage of agriophytism (i.e. growing in near-natural vegetation), however, is not correlated to any of the features studied. Apophytism of plant species in the original range as a possible predictor of invasion success deserves more attention.

Introduction

Of the many plant species that are transported by man into areas where they did not occur naturally, only few are able to grow and establish outside of cultivation in the new area. A central question in invasion studies is: What determines whether a species will be an invader or not? The quest for an explanation of invasion success was one of the main issues in the SCOPE program 37 (Drake *et al.* 1989).

The word "success" is consequently used much in invasion studies, a major question being how success is realized. As Pyšek (1995) has shown, many terms in plant invasion studies are used without explicit definitions and consequently with different meanings. The same is true for invasion success.

In this paper I will first look at the varying use of the word success in plant invasion studies. In the second part some aspects of the success of North American species in Germany are studied.

Dimensions of success

Many studies have sought typical characters of successful invaders in order to explain past invasion success and to eventually predict future invasions. In all these studies it

Plant Invasions: Ecological Mechanisms and Human Responses, pp. 33–42
edited by U. Starfinger, K. Edwards, I. Kowarik and M. Williamson
© 1998 Backhuys Publishers, Leiden, The Netherlands

is usually not said explicitly what exactly is meant by success. Generally it is accepted that the first step of a successful invasion that has to follow the import or introduction of propagules or entire plants is the occurrence outside of containment or cultivation as a wild or feral plant, spontaneously, or as a casual. A species that does not make this transition can hardly be called an invader – much less a successful one. From here, however, different roads may lead to (different types of) success: A mere increase in local abundance or the area covered by the species might well be called success, though this is seldom done. However, there are instances where frequency as such is understood as success (Table 1). Mostly, a second step is regarded as a prerequisite for success, the establishment of the species. In the central European terminology of non-native plants (Sukopp 1998), a differentiation of species according to the degree of naturalization success is done: ergasiophytes are plants found only cultivated, ephemerophytes occur irregularly outside of cultivation, epecophytes are established in vegetation or on sites markedly altered by human impact, and agriophytes are fully naturalized and grow also in (near-) natural vegetation (as described by Schroeder 1969). The numbers of species in these categories decrease from the first to the last class and the transition along these is understood as "naturalization success" with the agriophytes as the most successful species.

Partly similar is the treatment of success in Williamson's (1996) tens rule. He gives success rates of ca. 10% for the transition between the following stages: imported, introduced (=casual), established, and pest. The first stages correspond to the above, in this system, however, ultimate success that is attained by only a small percentage of all species consists in becoming a pest.

Several authors use the term invader only for plant species that grow (and expand) in vegetation without marked human impact or for species with detrimental consequences (Mack 1996). This can be seen as an implicit definition of success. More ways of understanding or applying the term success are listed in Table 1.

Table 1. Examples for the (implicit or explicit) use of the term success in the literature

Dimension	Measure	Source
Frequency	No of sites	Hartmann and Konold 1995
Durability	time	Kowarik 1992, Kowarik 1995
Time lag between introduction and spread	time	Kowarik 1992, Kowarik 1995
Influence on vegetation	e.g. cover %, No of species displaced	Falinski 1998
Independence of human influence (agriophytism)	yes/no	Schroeder 1969, Lohmeyer and Sukopp 1992
Rate of spread	area/time	Hartmann and Konold 1995
Pest status	cost	Holdgate 1986, Mack 1996; Williamson 1996; Williamson 1998
Abundance	Absolute number	Kowarik 1992, Pyšek *et al.* 1995
Dominance		
Synanthropic range size	area	Pyšek *et al.* 1995

Prediction of invasion success

The search to explain and predict naturalization success of species is motivated by two aims: it can help to prevent future invasions if potential invaders are identified before they start invading, or to control invading species more effectively at the beginning of the invasion process. It is also connected to general ecological and evolutionary theory and can influence understanding of, e.g., competition, genetics, disturbance and stability of ecosystems, etc.

Explanation and prediction of invasion has been tried with two tools: The main approach consists in the search for biological characters of successful invaders. The most cited is the work by Baker (1965) who has compiled a list of traits of the ideal weed – which is often an invader. Others have expanded the list so that it contains physiological, demographic, genetic, and life-history traits of successful invaders, e.g. Bazzaz (1986), Roy (1990). In spite of the effort invested, prediction of future invasions on the basis of individual features of species is poor (Mack 1996). Perrins *et al.* (1992) showed that prediction of weed status of annuals on the basis of Baker characters is weak.

The other is the use of the performance of the species in other regions: According to Williamson and Fitter (1996) the best descriptor for invasion success is past invasion success, i.e. a plant that has invaded one area, may do so in another. Also, information on the extension of the natural range has been used as a predictor, e.g. Lodge (1993), Rejmánek (1996).

Even though some wide-ranging predictions about invasions are possible, the success of individual species can only be predicted with weak probabilistic correlations (Ehrlich 1989). For an explanation of the success of an individual invasion, features of the invader, the invaded ecosystem, and circumstances of the introduction have to be taken into account. Studies that compare the ecology of a species in the natural and the synanthropic range can help explain the success of individual species in showing how the species copes with abiotic and biotic factors in both regions (e.g. Sukopp and Starfinger 1995; Starfinger 1997).

Case study

One aspect of species' behavior in its home range has not been used in studies of plant invasion success: It is the ability to grow on sites or in vegetation markedly altered by human influence. Native plants occurring on strongly altered sites have been termed apophytes by Rikli (1903/1904), see also Sukopp (1998). As with studying past invasion success (Williamson and Fitter 1996), apophytism may be a useful measure allowing for better predictions of invasion success.

In the following, I use plant species native to eastern North America, naturalized in central Europe, to study two aspects of species' performance in the native range, i.e. range size and apophytism, and two different success measures, i.e. frequency in the new range and the occurrence of species in near-natural vegetation (agriophytism). In particular, correlations between the features concerning the natural range and those concerning the synanthropic range will be sought. The aim is to compare the prediction power of the two features of the home range for invasion success, and to show how different types of success depend on different traits.

Table 2. Descriptors used in the analysis

FREQ
Number of dots in the maps of Haeupler and Schönfelder 1988 and Benkert *et al.* 1996.
7: >1000; 6: >500; 5: >250; 4: >100; 3: >50; 2: >25; 1: <25
AGRIO
Agriophyte in Germany in Lohmeyer and Sukopp 1992.
2: yes; 1: possibly; 0: no
APO
Apophyte in NE – North America.
2: yes; 1: intermediate; 0: no
RANGE
Extension of the range in North America.
5: "whole area";
4: "almost whole area";
3: 20-30 degrees of latitude;
2: 10-20 deg. of lat.;
1: <10 deg. of lat.;

Species from eastern North America occurring in central Europe were chosen because, a) there is general similarity in climate and vegetation, and b) central Europe
has received many species from North America.

Methods

Non-native species of North American origin in the flora of Germany were selected
using the data base by Frank and Klotz (1990). In addition Lohmeyer and Sukopp's
(1992) list of agriophytes was used. Information concerning range size and apophytism
in the native range was collected from floras of eastern North America (Gleason and
Cronquist 1963, Britton and Brown 1970, Strausbaugh and Core n.d.) and supplemented
by my own observations during several trips in North America.

For statistical analysis the information was classified in the following ways (Table
2):

Five classes of range sizes were formed: Species that, according to the floras, occur
in the whole area were rated "5", those in „almost the whole area" "4". Where the
floras describe the range using names of states and provinces or parts thereof, the approximate north-south extension of the range was estimated in degrees of latitude; those
species which have a range of 20–30 degrees were given a ranking of "3", 10–20 degrees a "2", and a "1" for less than 10 degrees.

Apophytism was classified into 3 classes: "0" represents species for which no indication was found that they grow in conditions with marked human impact, e.g., described in the floras as occurring in woods, bogs, etc. The intermediate rating "1" was
given to species apparently not confined to pristine situations. True apophytes were
classified with "2". These occur regularly and/or predominantly under conditions strongly
influenced by man, in the floras this reads as, e.g., road verges, waste places, cultivated grounds, etc.

Frequency in Germany was given as the number of grid squares containing a species in two distribution atlases of Germany (Haeupler and Schönfelder 1988; Benkert
et al. 1996). Grid squares in these atlases are 6′ longitude and 10′ latitude, approxi-

mately 11 by 11 km². In the East German atlas these are divided into quarters. In total there are 2084 mapping units in West, and 3625 in East Germany. In both atlases the number of dots was counted for each species; the number was estimated for more frequent species. The resulting number was classed in one of seven groups (Table 2).

Lohmeyer and Sukopp (1992) and personal communications by H. Sukopp were used for assessing agriophytism in Germany. Whether a species is an agriophyte in Germany is usually a straightforward yes/no decision. However, as there are some doubtful cases, three classes were formed and the doubtful cases treated as intermediate.

The resulting data set was analyzed using Kwikstat. Spearman's rank correlation coefficients were calculated.

Results

Table 3 shows the 78 North American species in Central Europe with their ratings. Of these, 31 are classified as apophytes in the natural range, 28 as not apophytic, and 19 as intermediate. More than half of the species have ranges covering less than 20° of latitude, only two are in the 20–30° bracket (Table 4).

In Germany, most of the species are rare, only 18 species occur in more than 500 grid squares. Many of the most frequent species in Germany are classed as agriophytes here. It must however be noted, that this does not mean, that all or even the majority of their occurrences are agriophytic. On the contrary, most sites where the frequent species occur are more or less strongly disturbed and only a few of the occurrences of the species are agriophytic, e.g. *Conyza canadensis* is present in many waste places and on disturbed grounds and is only rarely found in natural vegetation on river banks and sea shores. There are also examples of very frequent species that never attain the rank of agriophytes, e.g., *Lepidium densiflorum* and *Juncus tenuis*. Among the species that are rare in Germany, some grow in natural vegetation as agriophytes, e.g., *Kalmia latifolia*, *Sarracenia purpurea*, *Physocarpus opulifolius*. Some of these have been planted directly in these bogs. They are, however, considered as fully naturalized agriophytes because they grow and reproduce without direct human influence (Lohmeyer and Sukopp 1992). Agriophytes with limited overall frequency in Germany are often confined to bogs, swamps, or hygrophilous flood plain vegetation, where they are completely naturalized but lack dispersal chances due to the patchy distribution of these habitats (Lohmeyer and Sukopp 1992).

Correlations are given in Table 5. Naturalization success of species expressed as frequency in Germany is higher in those species that have a large natural distribution range. The correlation to apophytism in the home range is even stronger, indicating that apophytes make successful invaders. Apophytes also have a larger distribution range in their home continent than non-apophytes. The success of agriophytism, however, is independent of the two characters of the species in their natural range used here. In conclusion it can be said that ecological characters shown in the natural range can help to predict the probability of invasion success in a new range. However, different dimensions of success are reached independently of each other, and are consequently not correlated with the same characters of species.

Table 3. North American species in Germany Nomenclature follows Frank and Klotz (1990)

NAME	FREQ	AGRIO	APO	RANGE
Acer negundo	7	2	2	3
Agrostis scabra	1	1	1	4
Amaranthus albus	6	1	2	5
Amaranthus blitoides	4	1	2	5
Ambrosia artemisiifolia	5	0	2	5
Ambrosia psilostachya	3	0	2	4
Ambrosia trifida	2	0	1	5
Amelanchier spicata	3	1	0	1
Amorpha fruticosa	1	0	1	4
Artemisia biennis	2	0	2	4
Artemisia dracunculus	4	0	2	4
Asclepias syriaca	1	0	1	2
Aster laevis	1	0	0	2
Aster novae-angliae	4	1	1	2
Aster novi-belgii	6	2	1	2
Aster tradescanti	3	2	1	1
Azolla caroliniana	1	0	0	4
Bidens connata	5	2	0	2
Bidens frondosa	7	2	2	5
Chenopodium ambrosioides	1	0	2	4
Chenopodium capitatum	1	0	2	3
Conyza canadensis	7	2	2	5
Cornus sericea	1	2	0	2
Coronopus didymus	4	0	2	2
Cuscuta campestris	3	1	0	5
Cuscuta gronovii	2	2	0	4
Echinocystis lobata	4	2	2	4
Elodea canadensis	7	2	1	1
Elodea nuttallii	3	2	1	1
Epilobium adenocaulon	7	2	0	2
Erechtites hieracifolia	1	2	2	2
Erigeron annuus	7	2	2	2
Euphorbia maculata	1	0	2	2
Fraxinus pennsylvanica	2	1	0	2
Helianthus tuberosus	6	2	2	5
Hordeum jubatum	4	0	2	5
Hypericum canadense	1	1	0	2
Iris versicolor	1	2	0	2
Iva xanthiifolia	4	0	1	2
Juncus canadensis	1	0	0	1
Juncus tenuis	7	0	2	5
Kalmia angustifolia	1	2	0	2
Lepidium densiflorum	6	0	2	5
Lepidium neglectum	2	0	0	1
Lepidium virginicum	5	0	2	2
Lindernia dubia	1	2	0	2
Mimulus moschatus	1	1	0	1
Myriophyllum heterophyllum	1	2	1	2
Nicandra physalodes	3	0	2	4
Oenothera biennis	7	2	2	4
Oenothera oakesiana	1	2	0	1
Oenothera parviflora	5	0	0	1
Oxalis dillenii	3	0	2	5
Oxycoccus macrocarpa	1	2	0	2

Table 3. Cont.

NAME	FREQ	AGRIO	APO	RANGE
Panicum capillare	3	1	2	4
Parietaria pensylvanica	2	0	0	2
Parthenocissus inserta	3	2	0	2
Physocarpus opulifolius	1	2	1	2
Pieris floribunda	1	2	0	1
Pinus strobus	1	1	0	2
Prunus serotina	7	2	2	2
Quercus rubra	3	0	1	2
Rhus typhina	2	1	1	1
Robinia pseudoacacia	7	2	2	1
Rudbeckia hirta	5	1	1	4
Rudbeckia laciniata	6	2	1	5
Rumex triangulivalvis	2	1	0	1
Sagittaria latifolia	1	2	1	4
Sarracenia purpur	1	2	0	2
Silphium perfoliatum	1	0	0	1
Sisyrinchium montanum	1	0	1	1
Solidago canadensis	7	2	2	5
Solidago gigantea	7	2	2	4
Solidago graminifolia	1	2	0	2
Spiraea tomentosa	1	1	0	2
Symphoricarpos albus	6	0	1	2
Veronica peregrina	4	2	2	5
Xanthium spinosum	2	0	2	5

Table 4. Frequency distribution of the descriptors

	0	1	2	3	4	5	6	7
Range in N. America		15	30	2	15	16		
Frequency in Germany		28	9	10	8	5	6	12
Apophyte in N. America	28	19	31					
Agriophyte in Germany	29	15	34					

Table 5. Matrix of correlation coefficients (n=78)

	Range	Apophyte	Agriophyte
Frequency	0.330	0.523	0.216
p-value	0.00	0.00	0.06
Agriophyte	-0.119	-0.135	
p-value	0.30	0.24	
Apophyte	0.580		
p-value	0.00		

Discussion

Restrictions of the available data introduce some bias in the analysis presented. Floras and distribution maps used for the synanthropic range of the species studied, i.e. Germany, contain mainly established species. One dimension of success, the occurrence of a plant outside of cultivation, can consequently not be dealt with here. Also, the combined number of dots for different sizes of map squares in West and East Germany makes the frequency estimate used here a somewhat rough one. Some inaccuracy is obviously present in the assessment of apophytism and range size in the original range of the species, because the floras used describe the sites, habitats, and range sizes with little detail. These inaccuracies, however, are assumed to merely add noise to the data set, because no deterministic relation between them and the correlations analyzed is assumed. Another problem may lie in the fact that genetic change may have occurred so that the plants in the new range differ from the ancestors that once were transported from the original range.

Some of the agriophytes of Germany are rare, and some of the most abundant non-native plants are never found in (near-) natural vegetation. That these dimensions of success are not related to each other in this study can be seen as a result of the influence of chance and history in biological invasions. Some of the rare agriophytes in bogs have even been planted deliberately. However, they show that the ability to grow in undisturbed near-natural vegetation can be caused by entirely different biological characters than that to expand in a new range. Without the history of the introduction these invasions cannot be explained.

The fact that frequency in the synanthropic range can be predicted with some accuracy on the basis of information about the apophytism of a species in its home range, however, shows that plant traits may cause the ability to grow in vegetation influenced by man and to invade in similar ways (Rejmánek 1996; Williamson 1996).

Germany as a receiving region for plant invaders differs from other regions of the world in the extent of man-made changes to the vegetation. The large human impact on most plant communities can contribute to the success of North American apophytes here, so that this result may not be generalizable.

In conclusion, these findings may add to the growing understanding of which plants are successful invaders and how they can be separated from the less successful ones.

Acknowledgment

I thank Herbert Sukopp and Ingo Kowarik for inspiring discussions. Both of them, and Mark Williamson and Keith Edwards helpfully commented on the manuscript.

References

Baker, H.G. 1965. Characteristics and modes of origin of weeds. In: Baker, H.G. and Stebbins, G.L. (eds.) The genetics of colonizing species. pp. 147-172. Academic Press, New York.
Bazzaz, F.A. 1986. Life history of colonizing plants: some demographic, genetic, and physiological features. In: Mooney, H.A. and Drake, J.A. (eds.) Ecology of biological invasions of North America and Hawaii. pp. 96-110. Springer, New York.
Benkert, D., Fukarek, F. and Korsch, H. (eds.) 1996. Verbreitungsatlas der Farn- und Blütenpflanzen

Ostdeutschlands. Gustav Fischer, Jena, Stuttgart, Lübeck, Ulm.

Britton, N. and Brown, A. 1970. An illustrated flora of the Northern United States and Canada. Dover, New York.

Drake, J.A., Mooney, H.A., di Castri, F., Groves, R.H., Kruger, F.J., Rejmánek, M. and Williamson, M. (eds.) 1989. Biological invasions. A global perspective. SCOPE 37. John Wiley & Sons, Chichester, New York, Brisbane, Toronto, Singapore.

Ehrlich, P.R. 1989. Attributes of invaders and the invading processes: vertebrates. In: Drake, J.A., Mooney, H.A., di Castri, F., Groves, R.H., Kruger, F.J., Rejmánek, M. and Williamson, M. (eds.) Biological invasions. A global perspective. pp. 315-328. John Wiley & Sons, Chichester, New York, Brisbane, Toronto, Singapore.

Falinski, J.B. 1998. Invasive alien plants and vegetation dynamics. In: Starfinger, U., Edwards, K., Kowarik, I. and Williamson, M. (eds.) Plant invasions: ecological mechanisms and human responses. pp. Backhuys, Leiden.

Frank, D. and Klotz, S. 1990. Biologisch-ökologische Daten zur Flora der DDR. Wissenschaftl. Beitr. Martin-Luther-Univ. Halle-Wittenberg 1990/32(P 41): 1-167.

Gleason, H.A. and Cronquist, A. 1963. Manual of vascular plants of Northeastern United States and adjacent Canada. D. van Nostrand, Princeton NJ, Toronto, London, Melbourne.

Haeupler, H. and Schönfelder, P. 1988. Atlas der Farn- und Blütenpflanzen der Bundesrepublik Deutschland. Eugen Ulmer, Stuttgart.

Hartmann, E. and Konold, W. 1995. Späte und Kanadische Goldrute (*Solidago gigantea* et *canadensis*): Ursachen und Problematik ihrer Ausbreitung sowie Möglichkeiten ihrer Zurückdrängung. In: Böcker, R., Gebhardt, H., Konold, W. and Schmidt-Fischer, S. (eds.) Gebietsfremde Pflanzenarten. pp. 93-104. ecomed, Landsberg.

Holdgate, M.W. 1986. Summary and conclusions: characteristics and consequences of biological invasions. Phil. Trans. R. Soc. London B 314: 733-742. The Royal Society, London.

Kowarik, I. 1992. Einführung und Ausbreitung nichteinheimischer Gehölzarten in Berlin und Brandenburg. Verhandlungen des Botanischen Vereins von Berlin und Brandenburg (Berlin) Beih. 3: 1-188.

Kowarik, I. 1995. Time lags in biological invasions with regard to the success and failure of alien species. In: Pyšek, P., Prach, K., Rejmánek, M. and Wade, M. (eds.) Plant invasions – general aspects and special problems. pp. 15-38. SBP Academic Publ., Amsterdam.

Lodge, D.M. 1993. Biological invasions: lessons for ecology. TREE 8(4): 133-137.

Lohmeyer, W. and Sukopp, H. 1992. Agriophyten in der Vegetation MIteleuropas. Schr.Reihe Vegetationskde. 25: 1-185.

Mack, R.N. 1996. Predicting the identity and fate of plant invaders: emergent and emerging approaches. Biol. Conserv. 78: 107-121.

Perrins, J., Williamson, M. and Fitter, A. 1992. Do annual weeds have predictable characters? Acta Oecologica 13(5): 517-533.

Pyšek, P. 1995. On the terminology used in plant invasion studies. In: Pyšek, P., Prach, K., Rejmánek, M. and Wade, M. (eds.) Plant invasions – general aspects and special problems. pp. 71-81. SBP Academic Publ., Amsterdam.

Pyšek, P., Prach, K. and Smilauer, P. 1995. Relating invasion success to plant traits: an analysis of the Czech alien flora. In: Pyšek, P., Prach, K., Rejmánek, M. and Wade, M. (eds.) Plant invasions – general aspects and special problems. pp. 39-60. SBP Academic Publ., Amsterdam.

Rejmánek, M. 1996. A theory of seed plant invasiveness: the first sketch. Biological Conservation 78: 171-181.

Rikli, M. 1903/1904. Die Anthropochoren und der Formenkreis des *Nasturtium palustre* DC. Ber. Zürcherich. Bot. Ges. 8: 71-82.

Roy, J. 1990. In search of the characteristics of plant invaders. In: di Castri, F., Hansen, A.J. and Debussche, M. (eds.) Biological invasions in Europe and the Mediterranean Basin. pp. 335-352. Kluwer, Dordrecht.

Schroeder, F.-G. 1969. Zur Klassifizierung der Anthropochoren. Vegetatio 16: 225-238.

Starfinger, U. 1997. Introduction and naturalization of *Prunus serotina* in central Europe. In: Brock, J.H., Wade, M., Pyšek, P. and Green, D. (eds.) Plant invasions: studies from North America and Europe. pp. 161-171. Backhuys, Leiden.

Strausbaugh, P.D. and Core, E.L. n.d. Flora of West Virginia. Seneca Books, Grantsville, WV.

Sukopp, H. 1998. On the study of anthropogenic plant migrations in central Europe. In: Starfinger, U., Edwards, K., Kowarik, I. and Williamson, M. (eds.) Plant invasions: ecological mechanisms and human responses. pp. 43–56. Backhuys Publishers, Leiden.

Sukopp, H. and Starfinger, U. 1995. *Reynoutria sachalinensis* in Europe and in the Far East: a comparison of the species ecology in its native and adventive distribution ranges. In: Pyšek, P., Prach, K.,

Rejmánek, M. and Wade, M. (eds.) Plant invasions – general aspects and special problems. pp. 151-159. SBP Academic Publ., Amsterdam.

Williamson, M. 1996. Biological invasions. Chapman & Hall, London, Weinheim, New York, Tokyo, Melbourne, Madras.

Williamson, M. and Fitter, A. 1996. The characters of successful invaders. Biol. Conserv. 78: 163-170.

Williamson, M. 1998. Measuring the impact of plant invaders in Britain. In: Starfinger, U., Edwards, K., Kowarik, I. and Williamson, M. (eds.) Plant invasions: ecological mechanisms and human responses. pp. 57–68. Backhuys Publishers, Leiden.

ON THE STUDY OF ANTHROPOGENIC PLANT MIGRATIONS IN CENTRAL EUROPE

Herbert Sukopp
Institut für Ökologie, Technische Universität Berlin, Schmidt-Ott-Str. 1, D 12165 Berlin, Germany

Abstract

The relationships between man and plants have been important in geobotanical research since the middle of the 19th century concentrating on the time of introduction and on the degree of naturalization (Fig. 1). The concepts and terminology of the Swiss botanist Thellung have influenced the Central European approach up to the present day. In his studies he combined natural science and cultural history in a specific way, the Thellungian paradigm.

Three specific traits of Central European research in anthropogenic plant migrations are dealt with: the study of agriophytes, the concept of anecophytes and the concentration on fundamental studies instead of control of invasive species.

The term agriophyte was introduced to describe plants which first reached an area as a result of human activity, but became a permanent constituent of the natural vegetation, no longer relying on human activities for their further existence.

From the beginning of our century, many authors accepted the anthropogenic origin of many agrestals and ruderals. If their present and past distributions are strongly restricted to artificially disturbed sites, they are called anecophytes. Their original habitat is unknown ("homeless plants"), therefore closely resembling many of the cultivated plants which have not been found anywhere in the wild state.

A third specific trait of Central European research may be seen in the fact that basic studies prevail, whereas applied aspects, especially the control of invasive species, were not the starting point.

Introduction

The study of anthropogenic plant migrations started with the investigation of cultivated plants and those which turned wild. Willdenow (1792) published his "Grundriß der Kräuterkunde" (compendium of botany), which went through many editions, was widely read, and had great significance for plant geography, among other things. The chapter "History of plants" can be regarded as the foundation for this branch of science. The first paragraph (in the 5th edition 1810) runs:

"The history of plants involves the influence of climate on vegetation, the changes of climate endured by plants and how they have been preserved by Nature, migration of plants, and finally their distribution over the globe". Concerning plant migrations he states: "But more than wind, weather, seas, rivers and beasts promote the dispersal of plants; man does it. ... Wars which were fought between different nations, the Migration of the Peoples, the Crusades to Palestine, even the trade brought a large amount of plants to us as well as they dispersed our plants in other regions ..." (transl. H.S.).

Plant Invasions: Ecological Mechanisms and Human Responses, pp. 43–56
edited by U. Starfinger, K. Edwards, I. Kowarik and M. Williamson
© 1998 Backhuys Publishers, Leiden, The Netherlands

Willdenow developed the ideas of plant geography together with Alexander v. Humboldt (1769-1859), who had been a close friend since 1788 (Jahn 1966). Humboldt's fundamental work on plant geography appeared in 1805 and 1807 in Paris as "Essai sur la Géographie des Plantes", and 1807 in Tübingen as "Ideen zu einer Geographie der Pflanzen". More significant than the concepts he introduced was his understanding of science. Humboldt "strived ... for an overview of the perceptible world in its entirety and at gaining the broadest possible comprehension of the interaction of the forces" (Schmithüsen 1957). In Humboldt (1807) we find: *In the great linkage of causes and effects, no substance, no activity may be considered in isolation. The equilibrium which rules amongst the perturbations of apparently antagonistic elements results from the free play of dynamic forces, and a complete overview of nature, the final purpose of all physical studies, can only be achieved if no force and no formative process is neglected, so that for the philosophy of nature a broad, promising field of study will be formulated.*"

Here Humboldt adopts an idea of Friedrich Wilhelm Schelling (1775-1854), the head of German "Naturphilosophie", that all natural phenomena are based "on the never ending conflict of the opposing basic forces of material".

Humboldt's conception became the starting point for the (further) development of ecology (Trepl 1987). Humboldt not only set out a definitive basis for the development of ecology, he founded plant geography in an institutional sense, which in the Berlin tradition was continued by Engler and Diels.

In this tradition, Chamisso described in 1827 the effect of human culture after his circumnavigation on the Rurik 1815-1818:

"Wherever a human being settles, the face of nature is changed. His domesticated animals and plants follow him; the woods become sparse; and animals shy away; his plants and seeds spread themselves around his habitation; rats, mice and insects move in under his roof; many kinds of swallow, finch, lark and partridge seek his care and enjoy, as guests, the fruits of his labor. In his gardens and fields a number of plants grow as weeds among the crops he has planted. They mix freely with the crops and share their fate. And where he no longer claims the entire area his tenants estrange themselves from him and even the wild, where he has not set foot, changes their form" (transl. H.S.).

This quotation is taken from an instructional work, "Botany for the non-botanists", which Chamisso wrote for the Culture Ministry: "A survey of the most useful and harmful plants, whether wild or cultivated, which occur in North Germany. Including views of botany and the plant kingdom", a title in the tradition of Natural History.

Hellwig (1886) looked for the origin of agrestal and ruderal plants and distinguished between native plants and old introductions.

The concept of synanthropisation (Falinski 1966, 1986) refers to the totality of the changes produced by humans in the flora, vegetation, communities and sites (ecological change – eurytopisation; geographical – cosmopolitisation; and historical allochtonisation). The essence of synanthropisation, i.e. the transformation process of the vegetation, fauna and abiotic milieu under human influence, is the basic principle of substitution. Thus the following are substituted in the course of synanthropisation: stenotopic by eurytopic, endemic by cosmopolitan and autochtonous by allochthonous

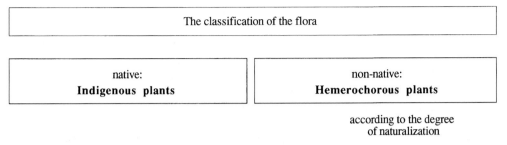

The classification of the flora

| native:
Indigenous plants | non-native:
Hemerochorous plants |

according to the degree
of naturalization

| only in natural
habitats:

**Ahemero-
phytes** | also in
man-made
habitats:

Apophytes | only in
man-made
habitats:

Anecophytes

Indigenophyta
anthropogena | naturalized in
man-made
habitats:

Epecophytes | naturalized in
natural or
semi-natural
habitats:

Agriophytes |

Fig. 1. Classification of the flora of a certain area according to indigenous status or degree of naturalization. After Sukopp and Scholz 1997.

components. Falinski (1969) had in mind particularly the case of displacement when he describes various stages of agriophytism.

The investigations of the Working Group on "Synanthropic Plants" at the Botanical Institute of the Czechoslovakian Academy of Sciences in Pruhonice, covered both epecophytes, in particular prospective quarantine weeds (Hejny *et al.* 1973), as well as agriophytes, especially along river systems (e.g. Kopecky 1967, 1970, 1978), producing monographs and carpo-biological investigations of certain species (see Lohmeyer and Sukopp 1992, p. 12/13). A survey of the agriophytes of Poland (Kornas 1968) and the monographs from Krakow should also be mentioned.

Synopses were published for Central Europe by Sukopp (1962, 1976), Joenje (1987), Weeda (1987), Kornas (1990), Lohmeyer and Sukopp (1992), Weber (1997), Pyšek and Prach (1997) and Eliaš (1997)

The Thellungian paradigm

The relationships between man and plants have been important in geobotanical research since the middle of the 19th century. Plants introduced to a certain region due to direct or indirect human action (hemerochores or anthropochores) are numerous (about 16 per cent in the Pteridophyta and Spermatophyta in some European countries) and lead to vegetation changes. Studies concentrated on the time of introduction and on the degree of naturalization (Fig. 1).

The time of introduction is an extremely important point for historical considerations. Therefore, archaeophytes, which were imported into a region in the time between the neolithic age up to the end of the middle ages, are distinguished from neophytes with expansion since 1500, AD.

Fig. 2. A. Thellung (1881–1928)

The degree of naturalization i.e. different integration of non-native species into the local vegetation, is a second means of classifying hemerochores. This ecologically more interesting point of view is of high relevance also in terms of future synanthrophic changes of the flora (see Jäger 1988). Normally, four groups are distinguished. First the ergasiophytes, cultivated plants, which cannot survive without permanent human care and, consequently are not naturalized. The ephemerophytes are a second group transported with goods such as subtropical fruits (Jauch 1938) or wool (Probst 1949), or as ornamentals, and are occasionally imported in high numbers of species and individuals, but which cannot survive in the new surroundings because they lack the ability to set seeds or to survive cold winters. Only the third and the fourth group are naturalized: "A species is naturalized if it did not exist in a certain area before historical times, but then reaches this area through intentional or unintentional human activities (or by unknown means), and now displays there all the features of a wild native species, i.e. it grows and multiplies by natural reproductive means (seeds, tubers, bulbs etc.) without direct human aid, occurs more or less frequently and regularly at the sites suited to it, and has maintained itself" (Thellung 1912, p. 638; transl. H. S). A third group are the epecophytes, which include nearly all field weeds, especially many archaeophytes. A last, essential but less numerous group, are the agriophytes, which proved to be competitive with natural plant communities. Lohmeyer and Sukopp (1992) give a survey of agriophytes in Middle Europe, considering 230 species.

The term agriophyte (naturalized alien, Lousley 1953) was introduced by Kamychev (1959) to describe plants which first reached an area as a result of human activity, but later became a permanent constituent of the natural vegetation, no longer relying on human activities for their further existence. The concept goes back to Thellung (1918/ 19), who classified species that "settled natural sites, amongst native vegetation, and

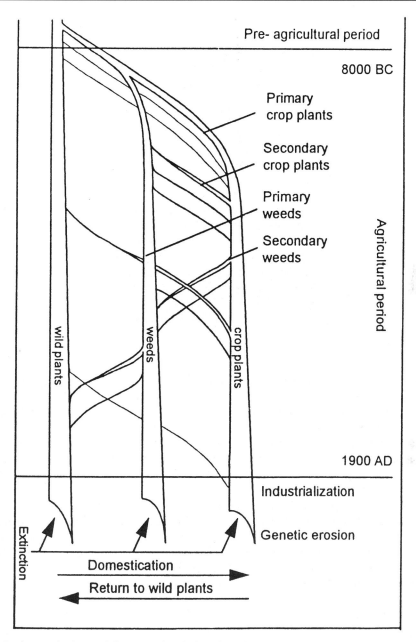

Fig. 3. Plants under human influence and evolution of weeds and crops (after Hammer 1988 and Gladis 1996). Pre-agricultural period = 8000 BC; Industrialization = post-1900. From Hammer et al. 1997.

were able to establish themselves permanently, being independent of human activity for their further existence".

The concepts and terminology of the Swiss botanist Thellung (1912, 1918/19) have influenced the Central European approach up to the present day. A number of earlier and contemporary attempts in the same direction proved less successful (initially A. de Candolle 1855; then e.g. Rikli 1903/04; Linkola 1916).

Thellung's achievement lay not in the development of a new approach, but in the systematic summarization of the accepted, and unquestioned basic concepts and methods which were used, with some variations, in the 'adventive floristics' of his time. As he wrote, he attempted to provide an 'exact definition of these terms according to prevailing usage' (1918/19; he based his work above all on Rikli). Thellung discussed and defined terms like 'native', 'introduced', 'aliens', 'casuals' etc. in French, German and English and created a scientific (Greek) terminology. One is tempted to speak of a paradigm (Trepl 1990). There are difficulties involved with using the paradigm concept for a descriptive, non-theoretical science which is not clearly defined and is without a fairly closed scientific community. Scheuermann (1948) writes that Thellung had 'opened up research in the field of alien plants'. He combined natural science and cultural history in a specific paradigm, later criticized by 'exact' natural scientists.

There are several summaries of the further development of this system of concepts and terminology (particularly Schroeder 1969), one of the aims of which was to achieve a consistent use of terminology. Thellung's terms were still being used, as Schroeder (1969) writes, but usually without adhering to his terminology. Some authors also felt the need for simplification.

A comparison of the Central European terminology with the meaning of terms used in studies on plant invasions was given by Pyšek (1995).

Insertion in natural and man-made vegetation: agriophytes and epecophytes

Given the large differences that exist in the extent to which particular habitats or communities in a region have been invaded, most comparisons between spontaneous areas or synanthropic ones should probably be done within carefully specified habitat and geographical limits (e.g. Sukopp and Sukopp 1988). Thus, the difference in the ratio of introduced to total number of species observed among countries by Heywood (1989) are done at a scale that may often be too coarse. Even small countries typically contain many different habitats as well as biogeographically distinct subregions, each of which may have a unique invasion history and be differently susceptible to invasion. Generalizations and patterns will be more discernible if ecologically and biogeographically homogeneous assemblages are compared. In particular, it is important to treat agricultural, aquacultural, urban, and other human-altered communities differently and separately from communities that have been less overtly affected by humans.

The concept of agriophytes

In Central Europe, vegetation surveys have reached a level which makes it possible to investigate the process of agriophytism for all types of vegetation. In particular the study of nitrophile plant communities in areas beside large rivers that flood regularly have led to new advances. It has become clear that agriophytes play a major role in the floristic structure of the therophyte communities, which develop every summer on initially bare riversides as the water level falls. *Bidens frondosa* and *Xanthium albinum* on the sandy shores of the River Elbe bear striking witness to this (Lohmeyer 1950).

Lists of agriophytes have been published for different parts of the world (Table 1).

Table 1. Lists of agriophytes

1855	Candolle	World
1931	Steffen	East Prussia
1962	Sukopp	Central Europe
1967	Sjörs	Northern Europe
1968	Kornas	Poland
1971	Rousseau	Québec and Ontario
1977	Wittig	Westfalen
1983-1987	Fukarek and Henker	Mecklenburg
1985	Weeda	Netherlands
1986	Falinski	Bialowieza
1987	Masing	Estonia
1989	Rejmánek	World
1991	Burda	Southeast Ukraine
1991	Protopopova	Ukraine
1992	Lohmeyer and Sukopp	Central Europe
1993	Borhidi	Hungary
1994	Clement and Foster	British Isles ("naturalized": amongst native vegetation so as to appear native)
1995	Andersen	Denmark
1995	Johnson	Queensland
1995	Ricardo Napoles et al.	Cuba
1996	Ryves et al.	British Isles
1996	Natali and Jeanmonod	Corsica
1997	Gudzinskas	Lithuania

The role of epecophytes (established aliens, Lousley 1953) in urban flora and vegetation

Comparisons between some European settlements have shown a close relationship between the presence of alien species and the size of settlements (Linkola 1916, Falinski 1971; Sukopp and Werner 1982; Pyšek 1989). This is usually explained by considerable habitat heterogeneity, the role of big cities as centers of species' immigration and the better adaptation of alien species to man-made perturbations (Sukopp and Trepl 1987; Kowarik 1990, 1995; Pyšek 1993).

Due to industrialisation, many introduced species reached Central Europe in the second part of the 19th century (Jäger 1991). Besides changes in species number, there were also shifts in species' frequency. A comparison of species' current frequency in Berlin, with data from the 19th century (Ascherson 1864), reveals differences between native and non-native species. Two-thirds of the species which became more frequent during the last 120 years are neophytes. Similar trends have been found in other cities (Klotz 1984, Sudnik-Wojcikowska 1987, Jackowiak 1989).

The hypothesis that disturbance promotes the establishment and spread of alien species is accepted in most studies on biological invasions. In urban environments, it is mainly the man-made components of disturbance which affect species composition and promote non-natives. Kowarik (1995) showed that naturalized non-natives are enhanced on sites subject to a high level of disturbance. The highest percentages of alien species are found in ruderal vegetation units with annuals as the dominating life form.

Invasion biology has had surprisingly little guidance from, or made contributions to, the larger disciplines of ecology, biogeography, and evolutionary biology. Because of the emphasis on human-introduced species, the importance of invasion throughout the history of life has generally not been appreciated, and few researchers have taken advantage of a systematic approach to the subject (Vermeij 1996).

The flora of cities also comprises a high number of native plants able to settle on man-made sites, the so called apophytes (Rikli 1903/1904): "Native plants, i.e. whose individuals have partly left their original habitats and have settled spontaneously (using natural means of dispersal) man-made habitats (wasteland or cultivated land)" (Thellung 1918/19). Sixty-three percent of all native plants of Berlin's flora are apophytes (Kowarik 1988).

Wittig *et al.* (1985) classified the urban flora according to its behavior to the urban environment: urbanophobe – urbanoneutral – urbanophile.

Migration and evolution

In plant geography, little attention has been paid to ongoing evolutionary processes indirectly promoted by human interference on vegetation. Though several scientists concede the stimulation of man-made evolutionary effects via hybrid contact, the discussions are mainly concerned with the actual destruction of vegetation and species extinction and, moreover, on plant dispersal and species migration induced by man, which is thought to be detrimental. Coupled with concern over increasing biological diversity, a more positive attitude towards invasion is emerging, as occurred during domestication of cultivated plants (Scholz 1997).

From the beginning of our century, many authors accepted the anthropogenic origin of many agrestals and ruderals (Scholz 1996). If their present and past distributions are strongly restricted to artificially disturbed sites, they are called anecophytes (obligatory weeds, not occurring in natural vegetation, Zohary 1962) which in Europe comprise most of the so-called archaeophytes, sensu Thellung (Scholz 1995). Their original habitat is unknown ("homeless plants"), therefore closely resembling many of the cultivated plants which have not been found anywhere in the wild state (Sukopp and Scholz 1997).

Weeds and cultivated plants have developed under similar evolutionary processes (Fig. 3, from Hammer *et al.* 1997). Both of them originate from the beginning of agriculture, when human beings started to create environments favorable for useful plants (Hammer 1988). Over time these areas developed into agricultural ecosystems. Thus, weeds also found the conditions necessary for their existence. While the remains of cultivated plants are often found in archaeological excavation sites, remains of weeds are analyzed by pollen analysis and the analysis of macrofossils (Godwin 1949, 1975, Jensen 1985, 1987, Willerding 1986, Frenzel 1992).

Primary weeds and cultivated plants both originate from wild species, whose diversification extended after the start of domestication by humans (Fig. 3). With selection, cultivated plants developed the characters useful to us. Under this pressure, divergent selection of weeds occurred (Kupzow 1980), for example, forming large numbers of small seeds, fast maturation and low growth habit (e.g. *Capsella bursa-pastoris* and *Stellaria media*).

After agricultural ecosystems were established, weeds also evolved with convergent selection so that weed characteristics have become similar to those of the cultivated plants with which they coexist, e.g. *Agrostemma githago, Bromus secalinus, Lolium remotum*, *Lolium temulentum* and *Silene linicola* (Rothmaler 1947; Kupzow 1980; Scholz 1996). Typical examples of such characters are the loss of specialized seed-spreading mechanisms, the formation of larger diaspores, the synchronous determination of maturity, the loss of mechanical protection and dispersal devices, the reduction of toxic or bitter ingredients, and seed and fruit color different from those of the respective wild progenitor. Because of similar growth habits and other characters, these weeds are extremely well adapted to the cultivated plants with which they grow, producing the phenomenon of crop plant mimicry. The convergent adaptation of the reproductive system of those plants is obvious by the fact that their diaspores go through harvesting, threshing and cleaning (by winnowing) together with the seeds of the cultivars.

Unfortunately, the dynamics of these processes are rarely observed. Weeds of the convergent type are, in most cases, extinct in our agricultural ecosystems due to sophisticated methods of seed separation and herbicide application. Only a few of the many weed species had enough plasticity to adapt to the enormous changes in farming methods. Divergent weeds are also strongly influenced by developments in agriculture. Recent populations of *Papaver rhoeas* show much less genetic variability than they did 30-40 years ago in the same fields (Hammer and Hanelt 1980).

As with primary and secondary crops, we can also define primary and secondary weeds. The latter originate from cultivated plants when characteristics typical of cultivated plants return to characteristics found in the respective wild progenitors. The fragility of the rachis in cereals, causing the spontaneous spread of the seeds, is an example of a fundamental secondary weed trait. Introgressions of the wild characters from closely related wild plants or weeds into crops can have the same result (Hammer 1984, Baker 1991).

Recent weed evolution after species introduction is evident in many cases. For instance, in Scandinavia and Russia, several New World *Epilobium* species have started to evolve like some *Oenothera* species (Skvortzov 1995).

Consequently, "we must realize that most naturalized aliens", being anecophytes or not, "will persist and become part of our more permanent flora" and "create new ecological communities". More important, "we would be wise also to recall that it is the widely dispersed and fairly abundant species that are most likely to survive stresses causing extinction's, and be the founder stocks for new diversification" (Scholz 1997).

The Central European perspective

The Thellungian paradigm

In Central European studies of anthropogenic plant migrations natural science and cultural history were combined in a specific way, the Thellungian paradigm (ch. 2). Research in this field developed relatively independently in Central Europe, particularly when compared with English-speaking countries. In this context it is not possible to arrive at anything like a "natural" system. It is only possible to find different groupings for different purposes (Schroeder 1969)

Urban flora

As a consequence, studies of single populations/plant species are connected with studies of biotic communities/vegetation. Studies of urban flora and vegetation, another specificity of Central Europe, showed the low degree of integration of urban biocoenoses. They are non-equilibrium systems, with stochastic processes being more important than deterministic ones. Succession in urban biocoenoses, which are, compared with non-urban ones, subject to strong and extremely variable anthropogenic influences, is strongly linked to site history. Hence, successional communities are not deterministically directed towards pre-determined states; they are dominated by chance, unpredictable and not repeatable. No reasonable approximation to climax conditions exists. The initial species composition remains relevant for further development.

A major reason for the (relative) unpredictability of succession in urban ecosystems is the high degree to which these systems are subject to invasions of "aliens"; the biogeographical spectrum of species composition of cities is very different from that of the surrounding countryside. The main cause may lie (a) in the conditions of naturalization, i.e. in the high invasibility of the biocoenoses, and (b) in the conditions of dispersal (introduction, transportation). Disturbances generally increase invasibility and urban ecosystems are disturbed ones. Towns are open to invasions of alien species, their number is unforeseeable (Trepl 1994).

Urban biocoenoses are an extreme example of communities produced by successive invasions and not by co-evolutionary development. In principle, the historic uniqueness of urban situations, i.e. the combinations of environmental factors and organisms, differentiates urban ecosystems from most natural ones, even those subject to strong disturbance.

Basic vs. applied studies

A third specific trait of Central European research may be seen in the fact, that fundamental (basic) studies prevail, whereas applied aspects, especially the control of invasive species were not the starting point. In nature conservation, sometimes a unidimensional view of organisms is taken, viewing on "exotics" or "invaders" in an anthropomorphic way. Such words were used with a specific connotation. Depending on the "Zeitgeist", this may be colored positively or negatively. Central European experience is that gradation of non-natives is studied in many examples, but could not up to now be operationalized for prognostic purposes (Breckling 1993). Lohmeyer and Sukopp (1992) have shown that only long-term studies could give reliable results.

Acknowledgments

I thank Keith Edwards, Bogdan J. Falinski, Slavomil Hejny, Jan Kornás +, Ingo Kowarik, Wilhelm Lohmeyer, Hildemar Scholz, Uwe Starfinger, Ludwig Trepl and Mark Williamson for helpful discussions.

References

Andersen, U. V. 1995. Comparison of dispersal strategies of alien and native species in the Danish flora. In: Pyšek, P., Prach, K., Rejmánek, M. and Wade, M. (eds.). Plant Invasions – General Aspects and Special Problems: 61-70.

Ascherson, P. 1864. Flora der Provinz Brandenburg, der Altmark und des Herzogthums Magdeburg, 2. Abt. Specialflora von Berlin, Berlin 1864.

Baker, H.C. 1991. The continuing evolution of weeds. Econ. Bot. 45: 445-449.

Borhidi, A. 1993. Social behaviour types of the Hungarian flora, its naturalness and relative ecological indicator values. Pecs. 93 p.

Breckling, B. 1993. Naturkonzepte und Paradigmen in der Ökologie. Einige Entwicklungen. Wissenschaftszentrum Berlin für Sozialforschung FS II 93-304. Berlin. 53 S.

Burda, R.J. 1991. Anthropogenic transformation of flora. – [Russ.] Academy Sciences Ukrainian SSR. 169 S. Kiew.

Candolle, A. de 1855. Géographie botanique raisonnée Paris.

Chamisso, A. v. 1827. Übersicht der nutzbarsten und der schädlichsten Gewächse, welche wild oder angebaut in Norddeutschland vorkommen. Nebst Ansichten von der Pflanzenkunde und dem Pflanzenreiche. Ferdinand Dümmler: Berlin. VIII + 526 S.

Clement, E.J. and Foster, M.C. 1994. Alien plants of the British Isles. Botanical Society of the British Isles. London. 590 p.

Eliaš, P. (ed.) 1997. Invázie a invázne organizmy. Slovenský Narodný Komitét SCOPE, Nitra. 213 pp.

Falinski, J.B. (ed.) 1971. Synanthropisation of plant cover. II. Synanthropic flora and vegetation of towns connected with their natural conditions, history and function. (Polish with English summary) Mater. Zakl. Fitosoc. Stos. U.W. Warszawa-Bialowieza 27: 1-317.

Falinski, J.B. 1966. Antropogeniczna roslinnosc Puszczezy Bialowieskiej. Rozpr. Uniw. Warsawskiego 13: 1-256.

Falinski, J.B. 1986. Vegetation dynamics in temperate lowland primeval forests. – Geobotany 8. 537 S. Dordrecht.

Frenzel, B. 1992. The history of flora and vegetation during the Quarternary. – Progress in Botany 53: 361-400.

Fukarek, F. and Henker, H. 1983-1987. Neue kritische Flora von Mecklenburg. T.1-5. – Arch. Freunde Naturgesch. Mecklenburg 23: 28-133. 24: 11-93. 25: 5-79. 26: 13-85. 27: 5-41.

Godwin, H. 1949. The spreading of the British flora considered in relation to conditions of the late-glacial period. – J. Ecol. 37: 140-147.

Godwin, H. 1975. The History of the British Flora, ed. 2. – The University Press, Cambridge.

Gudzinskas, Z. 1997. Conspectus of alien plant species of Lithuania. 1-4. Botanica Lithuanica 3: 3-23; 335-366.

Hammer, K. 1984. Das Domestikationssyndrom. Kulturpflanze 32: 11-34.

Hammer, K. 1988. Präadaptationen und die Domestikation von Kulturpflanzen und Unkräutern. – Biol. Zentralbl. 107: 631-636.

Hammer, K. and Hanelt, P. 1980. Variabilitätsindices von *Papaver rhoeas*-Populationen und ihre Beziehung zum Entwicklungsstand der Landwirtschaft. Biol. Zentralbl. 99: 325-343.

Hammer, K., Gladis, T. and Diederichsen, A. 1997. Weeds as genetic resources. Plant Genetic Resources Newsletter 111: 33-39.

Hejný, S., Jehlík, V., Kopecký, K., Kropác, Z. and Lhotská, M. 1973. Karanténni plevele Ceskoslovenska. Studie CSAV 8. Praha

Hellwig, F. 1886. Über den Ursprung der Ackerunkräuter und Ruderalflora Deutschlands I-II. I. Diss. Breslau. II. Englers Bot. Jahrb. 7,5.

Heywood, V.H. 1989. Patterns, extents and modes of invasions by terrestrial plants. p. 31-55. In: Drake, J.A., Mooney, H.A., Di Castri, F., Groves, R.H., Kruger, F.J., Rejmanek, M. and Williamson, M. (eds.) Biological invasions: a global perspective. Wiley, Chichester.

Humboldt, A. v. 1807. Ideen zu einer Geographie der Pflanzen nebst einem Naturgemälde der Tropenländer. XII + 120 S. Tübingen. Also in: Ostwalds Klassiker exakter Wiss. 248. 1960.

Jackowiak, B. 1989. Dynamik der Gefäßpflanzenflora einer Großstadt am Beispiel von Poznan (Polen). In. Ubrizsy Savoia, A. (ed.) Spontaneous vegetation in settlements (Proc. IAVS Symposium 1988) Part one, pp 89-98. Camerino.

Jahn, I. 1966. Carl Ludwig Willdenow und die Biologie seiner Zeit. Wiss.Z.Humboldt-Univ. Berlin. Math.-nat.R. 15: 803-812.

Jalas, J. 1955. Hemerobe und hemerochore Pflanzenarten. Ein terminologischer Reformversuch. – Acta Soc. Fauna Flora Fenn. 72 (11): 1-15.

Jauch, F. 1938. Fremdpflanzen auf den Karlsruher Güterbahnhöfen. – Beitr.Naturkundl. Forsch. Südwestdeutschl., Karlsruhe, 3: 76-147.

Jäger, E.J. 1988. Möglichkeiten der Prognose synanthroper Pflanzenausbreitungen. – Flora (Jena) 180: 101-131.

Jäger, E.J. 1991. Grundlagen der Pflanzenverbreitung. – In: Schubert, R. (ed.): Lehrbuch der Ökologie. 3. Aufl.: 167-173. Jena.

Jensen, H.A. 1985. Catalogue of late- and post-glacial macrofossils of Spermatophyta from Denmark, Schleswig, Scania, Halland and Blekinge dated 13.000 B.P to 1536 A.D. – Danm. Geol. Unders.Ser. A, No. 6: 1-95.

Jensen, H.A. 1987. Macrofossils and their Contributions to the Spermatophyte Flora of Southern Scandinavia from 13.000 BP to 1536 AD. – Biol. Skrift. 29. 76 pp. – Munksgaard: Copenhagen.

Joenje, W. 1987. The SCOPE programme on the ecology of biological invasions: an account of the Dutch contribution. – Proced. Kon. Ned. Akad. Wetensch. C 90: 3-13.

Johnson, R.W. 1995. The Aliens have landed: An Account of the Development of the Naturalised Flora of Queensland. Proc. Royal Soc. Queensland 105 (1): 5-17.

Kamychev, N.S. 1959. A contribution to the classification of anthropochores. – [Russ.] Bot. Zurn. 44: 1613-1616.

Klotz, S. 1984. Phytoökologische Beiträge zur Charakterisierung und Gliederung urbaner Ökosysteme, dargestellt am Beispiel der Städte Halle und Halle-Neustadt. – Diss. Halle-Wittenberg 283, XIX S.

Kopecky, K. 1967. Die flußbegleitende Neophytengesellschaft Impatienti-Solidaginetum in Mittelmähren. – Preslia 39: 151-166.

Kopecky, K. 1970. Neophyten in den Uferzönosen der Wilden und "Vereinigten" Adler in Nordostböhmen. – [Tschech., dt. Zusammenfass.] Studie CSAV 7: 97-106.

Kopecky, K. 1978. Impact of human habitation on varying species composition of tall-herb communities along brooks on NE slope of the Orlicke hory Mountains. – Preslia 50: 321-340.

Kornas, J. 1968. A tentative list of recently introduced synanthropic plants (kenophytes) established in Poland. – [Poln., engl. Zusammenfass.] Mater. Zakl. Fitosoc. Stos. U.W. 25: 43-53.

Kornas, J. 1990. Plant invasions in Central Europe: historical and ecological aspects. – In: F. di Castri, Hansen, A.J. and Debussche, M. (eds.): Biological invasions in Europe and the Mediterranean Basin. Monogr. Biol. 65: 19-36. Dordrecht.

Kowarik, I. 1988. Zum menschlichen Einfluß auf Flora und Vegetation. Theoretische Konzepte und ein Quantifizierungsansatz am Beispiel von Berlin (West). Landschaftsentwicklung und Umweltforschung 56: 1-280.

Kowarik, I. 1990. Zur Einführung und Ausbreitung der Robinie (*Robinia pseudoacacia* L.) in Brandenburg und zur Gehölzsukzession ruderaler Robinienbestände in Berlin. – Verh. Berl. Bot. Vereins 8: 33-67.

Kowarik, I. 1995. On the role of alien species in urban flora and vegetation. In: Pyšek, P., Prach, K., Rejmánek, M. and Wade, M. (eds.). Plants Invasions – General Aspects and Special Problems: 85-103.

Kupzow, A.J. 1980. Theoretical basis of plant domestication. Theor. Appl. Genet. 57: 65-74.

Linkola, K. 1916. Studien über den Einfluß der Kultur auf die Flora in den Gegenden nördlich vom Ladogasee. I. Allgemeiner Teil. Act. Soc. Faun. Flor. Fenn. 45, No 1: 1-429.

Lohmeyer, W. 1950. Das Polygoneto Brittingeri-Chenopodietum rubri und das Xanthieto riparii-Chenopodietum rubri, zwei flußbegleitende Bidention-Gesellschaften. – Mitt. Florist.-Soziol. Arbeitsgem. Niedersachsen N.F. 2: 12-20.

Lohmeyer, W. and Sukopp, H. 1992. Agriophyten in der Vegetation Mitteleuropas. Schriftenreihe für Vegetationskunde 25: 1-185.

Lousley, J.E. 1953. The Recent Influx of Aliens into the British Flora. In: Lousley, J.E. (ed.). The Changing Flora of Britain. Conference 1952. Oxford BSBI. 203 p.

MacArthur, R.H. and Wilson, E.O. 1967. The Theory of Island Biogeography. Princeton. N.J.

Masing, V. 1987. Naturalized aliens in the vegetation of the Baltic region. – 14. Intern. Bot. Congr. Berlin. 24. Juli – 1. Aug. 1987. 12 p. Lecture.

Natali, A. and Jeanmonod, D. 1996. Flore analytique des plantes introduites en Corse. Annexe n° 4 des "Compléments au Prodrome de la flore de Corse". Ed. des Conservatoire et Jardin botaniques de la Ville de Genève, 211 pp.

Probst, R. 1949. Wolladventivflora Mitteleuropas. – Naturhist.Mus.Solothurn.

Protopopova, V.V. 1991. Sinantropiaja flora Ukraini. Kiev, Naukova dymka.

Pyšek, P. 1989. Archaeophytes and neophytes in the ruderal flora of some Czech settlements. Preslia 61: 209-226 (in Czech).

Pyšek, P. 1993. Factors affecting the diversity of flora and vegetation in central European settlements. Vegetatio 106: 89-100.

Pyšek, P. 1995. On the terminology used in plant invasion studies.In: Pyšek, P., Prach, K., Rejmánek, M. and Wade, M. (eds.). Plants Invasions – General Aspects and Special Problems: 71-81.

Pyšek, P. and Prach, K. (eds.) 1997. Alien plants in the Czech flora. Zprávy Ces. Bot. Spolec., Praha 32: Mater. 14: 1-138.

Rejmánek, M. 1989. Invasibility of plant communities. – In: Drake, J.A., Mooney, H.A., di Castri, F., Groves, R.H., Kruger, F.J., Rejmánek, M. and Williamson, M. (eds.): Biological invasions. Scope 37. S. 369-388. Chichester.

Ricardo Napoles, N.E., Pouyu Rojas, E. and Herrera Oliver, P.P. 1995. The synanthropic flora of Cuba. Fontqueria 42: 367-430.

Rikli, M. 1903/1904. Die Anthropochoren und der Formenkreis des *Nasturtium palustre* DC. – Ber. Zürcherich. Bot. Ges. 8: 71-82. In: Ber. Schweiz. Bot. Ges. 13.

Rothmaler, W. 1947. Artentstehung in historischer Zeit, am Beispiel der Unkräuter des Kulturleins (*Linum usitatissimum*). – Züchter 17/18: 89-92.

Rousseau, C. 1971. Une classification de la flore synantropique du Québec et de l'Ontario. I. Caractère généraux: II. Liste des especes. Naturaliste Canad. 98, 529-533, 697-730.

Ryves, T.B., Clement, E.J. and Foster, M.C. 1996. Alien grasses of the British Isles. Botanical Society of the British Isles. London. 181 p.

Scheuermann, R. 1948. Zur Einteilung der Adventiv- und Ruderalflora. Ber.Schweiz.Bot.Ges. 58: 268-276.

Schmithüsen, J. 1957. Anfänge und Ziele der Vegetationsgeographie. Mitt. Flor.-soz. Arbeitsgem. N.F. 6/7: 52-78.

Scholz, H. 1995. Das Archäophytenproblem in neuer Sicht. – Schr.-R.. Vegetationsk., (Sukopp-Festschrift) 27: 431-439.

Scholz, H. 1996. Ursprung und Evolution obligatorischer Unkräuter. Origin and evolution of obligatory weeds. – In: Fritsch, R. and Hammer, K. (eds.), Evolution und Taxonomie von pflanzengenetischen Ressourcen – Festschrift für Peter Hanelt – . 286. S. – Schrift. Genet. Ressourcen 4: 109-129. – ZADI: Bonn.

Scholz, H. 1997. Plant evolution under the impact of man. Scripta Bot. Belg. 15: 144.

Schroeder, F.-G. 1969. Zur Klassifizierung der Anthropochoren. – Vegetatio 16: 225-238.

Sjörs, H. 1967. Nordisk växtgeografi. – 2. Aufl. 240 S. Stockholm.

Skvortzov, A.K. 1995. Taxonomy and nomenclature of adventive Epilobium species in Russia. Bull. Soc. Nat. Moscou 100: 74-78.

Steffen, H. 1931. Vegetationskunde von Ostpreußen. – Pflanzensoziologie (Jena) 1: 406 S.

Sudnik-Wojcikowska, B. 1987. Die Flora der Stadt Warschau und ihre Veränderungen im 19. und 20. Jahrhundert. – [Pol. with German summary] Bd 1. 242 S.; Bd. 2. 435 S. Warschau.

Sukopp, H. 1962. Neophyten in natürlichen Pflanzengesellschaften Mitteleuropas. – Ber. Deutsch. Bot. Ges. 75: 193-205.

Sukopp, H. 1972. Wandel von Flora und Vegetation in Mitteleuropa unter dem Einfluß des Menschen. – Ber.Landw. 50 (1): 112-139.

Sukopp, H. 1976. Dynamik und Konstanz in der Flora der Bundesrepublik Deutschland. – Schriftenreihe Vegetationsk. 10: 9-26.

Sukopp, H. 1987. On the history of plant geography and plant ecology in Berlin. Englera 7: 85-103.

Sukopp, H. and Scholz, H. 1997. Herkunft der Unkräuter. Osnabrücker Naturwissenschaftliche Mitteilungen 23: 327-333.

Sukopp, H. and Sukopp, U. 1988. *Reynoutria japonica* Houtt. in Japan und in Europe. Veröff.Geobot.Inst.ETH, Stiftung Rübel, Zürich 98: 354-372.

Sukopp, H. and Trepl, L. 1987. Extinction and Naturalization of Plant Species as Related to Ecosystem Structure and Function. Ecological Studies, 61. Schulze, E.-D. and Zwölfer, H. (eds.): 245-276. Springer-Verlag Berlin Heidelberg.

Sukopp, H. and Werner, P. 1982. Nature in cities. Nature and environment series 28. Council of Europe. Strasbourg. 94 p.

Sukopp, H. and Werner, P. 1983. Urban environments and vegetation. In: Holzner, W., Werger, M.J.A. and Ikusima, I. (eds.), Man's impact on vegetation, Geobotany 5: 247-260, Junk, Den Haag, Boston, London.

Thellung, A. 1912. La flore adventice de Montpellier. – Mem. Soc. Nation. Sci. Nat. Math. 38: 57-728.

Thellung, A. 1918/19. Zur Terminologie der Adventiv- und Ruderalfloristik. – Allg. Bot. Z. Syst. 24/25: 36-42. Ausgegeben am 1. August 1922.

Trepl, L. 1987. Geschichte der Ökologie. Vom 17. Jahrhundert bis zur Gegenwart. Zehn Vorlesungen. – Frankfurt a.M. 1987.

Trepl, L. 1990. Research on the anthropogenic migration of plants and naturalisation. Its history and current state of development. 75-97. In: Sukopp, H., Hejny, S. and Kowarik, I. (eds.). Urban Ecology. The Hague. 282 p.

Trepl, L. 1994. Towards a theory of urban biocoenoses. Some hypotheses and research questions. – In. Barker, G.M.B., Luniak, M., Trojan, P. and Zimny, H. (eds.): Urban ecological studies in Europe. Warsaw (Memorabilia Zoologica Nr. 49), 15-19.

Vermeij, G.J. 1996. An agenda for invasion biology. Biological Conservation 78: 3-9.

Watson, H.C. 1847. Cybele Britannica. Vol. 1. London.

Weber, E.F. 1997. The alien flora of Europe: a taxonomic and biogeographic review. J. Veg. Sc. 8: 565-572.

Weeda, E.J. 1985. Veranderingen in het voorkomen van vaatplanten in Nederland. – [Dutch with English summary.] In: Mennema, J., Quene-Boterenbrood, A.J. and Plate, C.L. (eds.): Atlas van de Nederlandse flora 2. S. 9-47. Utrecht.

Weeda, E.J. 1987. Invasions of vascular plants and mosses into the Netherlands. – Proc. Kon. Ned. Akad. Wetensch. C 90: 19-29.

Willdenow, C.L. 1792. Grundriß der Kräuterkunde zu Vorlesungen entworfen. – Berlin, 5.ed. 1810.

Willerding, U. (1986): Zur Geschichte der Unkräuter Mitteleuropas. 382 S. – K. Wachtholtz, Neumünster.

Wittig, R. 1977. Agriophyten in Westfalen. Natur & Heimat (Münster) 37: 13-23.

Wittig, R., Diesing, D., Gödde, M. 1985. Urbanophob – Urbanoneutral – Urbanophil. Das Verhalten der Arten gegenüber dem Lebensraum Stadt. Flora 177: 265-282.

Zohary, M. 1962. Plant Life of Palestine, Israel and Jordan. – The Ronald Press, New York.

MEASURING THE IMPACT OF PLANT INVADERS IN BRITAIN

Mark Williamson
Department of Biology, University of York, York YO10 5DD, England
e-mail: mw1@york.ac.uk

Abstract

The tens rule compares the performance of different imported plants amongst themselves, and indicates that most alien plants in Britain have little impact. Now the performance of aliens and natives are compared. While post-medieval introductions are adequately documented, it is difficult to decide which other species are truly native and which ancient introductions. Various approaches are used. Six measures of impact are considered: the impact on nature reserves, three measures of weediness, and measures of abundance and range. They form a set which, on ordination, gives a series from pure ecological to pure agricultural measure.

Most of these measures compare established, naturalized, aliens with natives. Aliens in general are not significantly different from natives in measured impact. Only in range, where all casual aliens are also considered, is there a marked difference. Aliens have both smaller ranges and a different distribution of range sizes.

Although aliens have an equal or lesser impact than natives, they have an additional impact and some are troublesome in certain situations. Identifying the small proportion of new invaders that will become troublesome is still not feasible.

(Nomenclature follows Stace 1991).

Introduction

Article 8(h) of the Convention on Biological Diversity requires nations to "Prevent the introduction of, control or eradicate those alien species which threaten ecosystems, habitats or species" and this is partly because alien species are a serious threat to biodiversity, "second only to habitat loss" (Glowka *et al.* 1994). Although some aliens are terrible in their effects, not only on biodiversity but also in other ways, it is well known that most alien plant species have little effect (Williamson 1996). It makes sense to try and quantify the variation in effect of alien plants and to measure their impact.

Over ten years ago, I proposed the ten-ten rule for British invaders (Williamson and Brown 1986) which, with further work, became the tens rule (Williamson and Fitter 1996a) (Table 1). The tens rule is quite rough, with approximate limits of 5 and 20 % (there is nothing magic about ten as such) and divides invasions into four qualitative stages, imported, casual, established (or naturalized) and pest. These are defined in Williamson (1996) and in Williamson and Fitter (1996a,b), but the categories are to some extent subjective ones. It is well known that there is a gradation between the stages of casual and naturalized (Williamson 1993) and that pest is a category open to much disagreement and in a different dimension. Nevertheless, the rule indicates clearly that pest invaders are unusual. The exceptions to the rule are also interesting.

Plant Invasions: Ecological Mechanisms and Human Responses, pp. 57–68
edited by U. Starfinger, K. Edwards, I. Kowarik and M. Williamson
© *1998 Backhuys Publishers, Leiden, The Netherlands*

Table 1. The tens rule (Williamson 1996).

5 – **10** – 20 % of imported plants	become casuals
5 – **10** – 20 % of casuals	become established
5 – **10** – 20 % of established	become pests (weeds)

For instance, crop plants do not obey the first 'ten'; in fact almost 95% become casual, far above the 20 % maximum indicated by the rule (Williamson 1994).

Is it possible to find more quantitative measures of the impact of aliens? Can costs be ascribed to the ecological and other damage done by invasive plants?

Measures of impact

There are many possible measures of environmental and economic impact, but many are difficult to estimate. A variety of economic measures, such as contingent valuation, are discussed by Perrings (1995) in the related context of biodiversity. I have been able to find six usable measures (Table 2) of different aspects of the impact of invasive plants in Britain. Of these, only the third (Prus 1997) is a strict economic measure. The remainder are environmental or a mixture of environmental and economic. All relate to the impact of invaders present in Britain, not to the cost of preventing their entry. For some species and some parts of the world such costs are important. In Britain, Colorado beetle *Leptinotarsa decemlineata*, a pest of potatoes, has been eradicated 163 times (Bartlett 1979). There are no comparable costs for British plants.

The first measure of impact, the one most directly related to what most ecologists are concerned about, is the work to control plant species in Wildlife Trust nature reserves in three parts of northern England. The Wildlife Trusts are voluntary bodies, though some of their members are professionals, and they seek and receive advice from statutory bodies such as English Nature. I asked the Cumbria, Northumberland and Yorkshire Trusts for information on which plant species they sought to control in their reserves and the cost and effort of doing so. All the reserves are quite small. Many are only a few hectares, and even the largest are of the order of a hundred hectares. There are problems in comparing costs and effort, and I will deal with that elsewhere. Here I use the list (Table 4) of 33 plants that are, or should be, in the Trusts' view, controlled.

Most of the British countryside is agricultural land, and so the major economic effects of alien plants are found in agriculture. The second and third measures (Table 2) are of the impact of agricultural weeds in the UK. Schering Agriculture (1986) in a guide to both grass and dicotyledonous weeds, gives the rank order of incidence of 44 dicots in its surveys. The surveys, on internal evidence, were done in central and southern England, the economically dominant agricultural area in Britain. Schering was, of course, not interested in where the weeds came from. A mere three taxa are invaders since the beginning of the nineteenth century, *Matricaria discoidea, Veronica persica* and weed beet. The latter is an invasive taxon but not an invasive species, an artifact of agriculture that has arisen in France and Italy. Its species, *Beta vulgaris*, includes also both wild (or sea) beet and numerous agricultural varieties.

However, many of the weeds reported by Schering may be ancient introductions.

Table 2. The data sets used in this study.

1. Records of plants controlled in nature reserves
   Yorkshire, Cumbia & Northumberland Wildlife Trusts
2. Incidence of arable weeds in parts of England
   Schering Agriculture (1986).
3. Herbicide related costs of weed control in UK
   Prus (1997).
4. Average weediness of 49 annuals as scored by 65 scientists
   Perrins, Williamson & Fitter (1992).
5. Incidence in 1 m² quadrats, in the Sheffield region
   Grime, Hodgson & Hunt (1988).
6a. 10 kilometre square records (10^8 m² records)
   Ecological Flora Database (Fitter and Peat 1994) (*ex* Biological Records Centre).
6b. Locality records for aliens
   Clement & Foster (1994).

British floras have tended to regard as native any species established before the time of the first systematic botanic records, from the time of John Ray in the seventeenth century (Webb 1985). Although Webb is rather dismissive of the efforts of quaternary botanists, Godwin (1975) is a remarkable systematic survey of the quaternary record. Using Godwin's information, supported by the archaeobotanical database (Tomlinson 1993; Tomlinson and Hall 1996) in the Environmental Archaeology Unit at York, I divided the species in Schering's list into those for which there was a pre-neolithic record, which I call natives, those which first appear in neolithic times or later, but pre-Ray, which I call ancient introductions, those without a record in the deposits, which I call unknown, and the three introduced post-Ray. Bearing Webb's strictures on taxonomic standards in mind, and noting that the number of records increases markedly from the neolithic, I would not claim that this division is more than a rough approximation.

The third measure (Table 2), the second to concentrate on weeds, is a surprising study (Prus 1997) of the cost and effort of applying herbicides and other weed controls in the UK. I say surprising because, when I first heard of it, I did not think the details of the method could give sensible answers, but they do. The cost of control is a frequent economic measure of the impact of invasives and is near some standard approaches to the economic evaluation of species differences (Pearce and Moran 1994). Prus (1997) has taken three main measures from published statistics by companies and the government: the annual sales of herbicides, the annual costs of application and the annual costs of cultivation to suppress weeds. The Ministry of Agriculture (MAFF) conduct a pesticide usage survey. Companies have to provide information on the individual species controlled as part of the registration process. From such information, and using a variety of subsidiary information, Prus was eventually able to estimate the annual cost (in 1992 pounds) of individual species for the UK as a whole. The most expensive is *Alopecurus myosuroides,* black-grass, at a little over eighty-six million pounds, or, in natural logarithms, 18.27 (Fig. 4). The rank order fits with common perception and has, as Prus (1997) notes, a high correlation with the fourth measure, a survey of perceived weediness (Table 3).

In his Table 7.4, Prus lists all weeds with a weed cost of more than £ 22026 that are said to be possibly introduced (not definitely native) in Clapham *et al.* (1987) or Stace (1991) or are listed as probably not native by Webb (1985) (22026 is e^{10}, or

$\log_e 22026 = 10$). From a variety of sources, Prus (1997) lists the first record of each of these species. This allows a distinction between ancient invaders and modern ones. The ancient invaders mostly have dates between 1900 and 3500 bp, the modern ones date from 500 bp (*Acer pseudoplatanus*, probably introduced into Scotland in the fifteenth century (Jones 1944)), and are mostly of nineteenth and twentieth century date. These data give 218 species with both weed cost and probable alien status.

The fourth measure (Table 2) is a survey of the opinions of 65 scientists about 49 species of British annual plant (Perrins *et al.* 1992). Each scientist was asked to score each species as a weed or not. The average results were scaled from plus two (everyone thought it was weed) to minus two (nobody thought it was a weed). As with the weeds in the Schering survey (set two) there were only three modern weeds (two the same, the third *Impatiens glandulifera* instead of weed beet). The remaining 46 species were classified as native, ancient introduction or unknown using Godwin (1975), exactly as for the Schering weeds. That is, plants were regarded as native or introduced long ago when there was some information to indicate whether or not they occurred in Britain before neolithic times.

The fifth and sixth measures (Table 2) are measures of abundance and distribution. As such they are measures of the success of plants, their ecological impact, rather than of their economic impact. It would be desirable to find some measure of the effect of plants on ecosystems or on ecosystem functioning, but none is available at present.

Abundance and distribution are usually positively correlated, but not always, and the correlation can be quite weak (Gaston 1996; Gaston *et al.* 1997). It is natural to think of abundance as a number and of distribution as a convexly bounded area on a map, but neither is so simple (Gaston 1994). The individuals counted are dispersed in space, and there are several possible measures of range (Gaston 1991). The relation between them is brought out by the modern habit of recording distribution by incidence, by presence or absence, in recording units of a fixed size. National grid squares of 10 km x 10 km (10km squares) are the standard unit in Britain as in some other European countries (with UTM grids). Abundance is the distribution of the species in sampling units covering just one individual. There is clearly a continuous series of possible sizes of sampling units from 1m² quadrats or smaller (for measuring abundance) up to and beyond the 50 km squares (used for mapping distributions at a European scale); abundance and distribution are just two, possibly rather arbitrary, points on a continuous scale. Pearman (1997) discusses how the scale of the recording unit affects perceptions of scarcity and rarity.

There are no widespread surveys known to me that try to measure the abundance of plants in Britain. The closest I have been able to find is (Table 2) survey II of Grime *et al.* (1988). This was of 2748 1m² quadrats chosen to give "a full range of examples of each of the main herbaceous vegetation types" in the Sheffield region of England. The incidence at this scale is given for 281 species; trees, naturally enough, are only recorded as seedlings and saplings. For some species information is also given, as "gregariousness", on incidence in all the 10 cm x 10cm quadrats within the 1m². Neither is strictly a measure of abundance; the 10 cm squares as subdivisions are very aggregated. Allowing for the variation in plant size no quadrat size is ideal; some measure of biomass might be more appropriate. I have used the 1m² data because it was the standard unit of the survey, is not aggregated and is distributed in what might be loosely described as a stratified random sampling scheme. As 1m² is eight orders of magnitude smaller than 10km squares, it is, I hope, reasonable to think of one as abundance and the other as distribution.

The final measure of impact, by distribution, comes from two data sets covering the British Isles. The first data set (6a, Table 2) is for spermatophytes and comes from the Biological Records Centre via the Ecological Flora Database (Fitter and Peat 1994). There are two lists: 197 introduced species and 1419 native (in the conventional flora sense). Each list gives the number of 10 km square records and refers to plants that are regarded as fully naturalized.

The maximum recorded number of squares is 3534 (*Plantago lanceolata*), with 3533 records for *Trifolium repens*, out of a possible total of 3630. There are only five intro-duced plant species in the highest category, the right hand bin, of Fig 1: *Acer pseudo-platanus* 3287, *Aegopodium podagraria* 2864, *Cymbalaria muralis* 2041, *Matricaria discoidea* 3314 and *Veronica persica* 2338. Only *A. podagraria* is an ancient (proba-bly Roman) introduction, while only *A. pseudoplatanus* occurs at all widely amongst native plants.

The second data set for distribution (6b, Table 2) is for alien vascular plants except grasses (Clement and Foster 1994). They record by locality, using place names on herbarium labels and similar information, and do not attempt to distinguish casual and established locality records. For many species, they give information by species under the categories casual (persisting less than two years), persistent (longer but probably not permanent), established (likely permanent), naturalized (established in native veg-etation) and introduced (planted), but I have not attempted to use that information. The established category seems to correspond to locally naturalized in the usage of Williamson (1993) rather than to established as used in the tens rule (Table 1).

Clement and Foster label the number of locality records by bullets (as shown on Fig. 1). The categories are 1-4 (one bullet), 5-14 (two bullets), 15-49, 50-499 and over 500 localities; only the first two are from strict counts, the others are estimates. To convert these to 10km squares I used the detailed comparisons of records at different scales in Pearman (1997) and my own study of the number of 10km square records for five bullet species (those with 500 or more localities) in the Ecological Flora Da-tabase. The answers fortunately correspond. On average, three localities equal one 10km square. That is, there is a marked cut off of 10km square records of five bullet species at around 167 squares. There are exceptions, notably *Allium triquetrum* which is very abundant in Cornwall and is rightly a five bullet species, though with only 86 10km square records. Nevertheless, this 3:1 rule allows a satisfactory plotting of locality records to match 10km square records on the logarithmic scales used in Figure 1. I have not attempted to smooth within bullet categories, though the actual number of species (the five listed above) is used for the highest category in the five bullet set.

None of the six measures has had a phylogenetic correction applied. This is the correction that can be used to correct for the fact that closely related species will be more like each other in their ecology, because of characters inherited from a common ancestor, than more distantly related ones (Harvey 1996). There is disagreement about how it should be used in ecology (Westoby *et al.* 1995) and it seems in any case to be less important with quantitative characters, as used here, than with qualitative ones (Fitter 1995). It is also impossible to use with the correlations and multivariate analy-sis of the next section. As will be seen, the conclusions are so striking (for set six of Table 1) or so insignificant (all the other sets) that it is unlikely that a phylogenetic correction would affect the conclusions.

The correlations of the impact measures

All six data sets are intended to measure impact; they should be positively correlated if they do. For all, except the records from nature reserves, there is at least a rank order of impact. The Schering data are only as ranks, the others are all distinctly non-normal in distribution as can be seen in Figures 1–4. Species recorded in one survey are not necessarily recorded in others; there is only a partial match in any pair and no simple data matrix for the set. For each pair I have taken the species that do match, ranked them in that match and calculated a rank correlation coefficient (Table 3). The number of pairs used is different for each value of Table 3, but it is, nevertheless, a positive definite matrix. A Principal Coordinate Analysis shows that about 50% of the variance is associated with one axis, which has the five surveys strung out more or less evenly in the order shown in Table 3. This is also the order used in Table 2. The validity of that axis is shown by the way the size of the correlations diminishes steadily from the principal diagonal to the top right and bottom left corners.

The order of the five surveys in Table 3 runs from pure agricultural weediness to a measure of range. The Prus survey and, even more, the Perrins *et al.* survey, include a measure of abundance; Prus because that comes into the cost, Perrins because rare plants are generally not perceived as weeds whatever their other ecological characteristics. The largest correlation is between Prus and Perrins. The Grime *et al.* survey, which was the nearest I could find to a direct measure of abundance, is well correlated with perceived weediness (Perrins *et al.*) and with range (number of 10km square records), though the Perrins *et al.* survey is almost as well correlated with range.

With only five measures considered, it would be unlikely that there would be more than one important axis. It is satisfactory that there is one is so readily intelligible, and gives confidence that, collectively, these measures span a useful series of impacts.

Impacts measured

Of the six measures only one shows a striking difference in the impact of native and alien species, and that is the measure of range. I will therefore consider the results of the measures listed in Table 2 from the last to the first.

Figure 1 shows that most native species are widespread and most aliens have a very restricted distribution; natives are skewed to the left, aliens to the right. Gaston *et al.* (in press) have studied the distribution of 10 km square records in various groups in

Table 3. The matrix of rank correlations between data sets 2-6 of Table 2. The labels on the right identify the source, the labels at the bottom indicate the type of data.

1	.52	.32	.12	.00	Schering Agriculture
.52	1	.88	.39	.09	Prus
.32	.88	1	.53	.51	Perrins *et al.*
.12	.39	.53	1	.53	Grime *et al.*
.00	.09	.51	.53	1	Ecological. Flora Database

weed incidence	weed cost	perceived weediness	1 m² records	(10 km)² records	

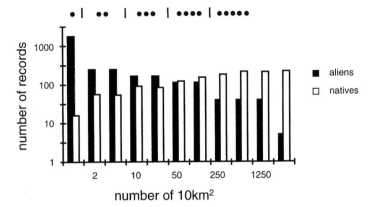

Fig. 1. Histograms of the number of records of individual species of plants in different numbers of 10 km x 10 km grid squares. Natives are those so-labelled in the Ecological Flora Database (Fitter and Peat 1994), aliens from Clement and Foster (1994). Note the logarithmic scale on both axes. The bullets at the top are the Clement and Foster categories; the method of translating to 10km squares is described in the text. The histogram interval is approximately times-the-square-root-of-five (0.35 on a $\log_{10}$ scale). The divisions between the bins come at 1.3, 3.03, 6.8, 15.1, 33.9, 75.9, 169.8, 380.2, 851.1 and 1905.5 10km squares.

Britain under power transformations. They find that the cube root transformation is the closest to making as many as possible symmetrical, but that even so the distributions are typically not normal on any power transformation.

Abundance has no clear upper limit, but distribution cannot be greater than the area studied. Power transformations are appropriate for data limited only by zero, as abundance is, but folded transformations (to use Tukey's term) are to be preferred when the data has both an upper and lower limit, as distributions do. Atkinson (1985) gives three families of folded transformations, but all have the logit as the analogue of the logarithm. Tukey (1977) calls the logit a flog (short for folded logarithm); it is $\log\{p/(1-p)\}$ where p is the proportion of the maximum for each data point.

Both natives and aliens in Fig. 1 are made reasonably symmetrical by the logit transformation. It seems that in the same way that the logarithm is the natural and normal transformation for abundance (Williamson 1972) the logit is for distributions. This point will be elaborated, with more data sets, in another paper.

Whatever the best mathematical form for the two distributions shown in Figure 1, there is no doubt that they are markedly different, and the difference consistently favours natives whatever the state of the alien. For instance, there are 31 native species more widespread than the most widely distributed alien *Matricaria discoidea*. The difference applies to established aliens as well as casuals, bringing out once again that that distinction is more a quantitative than a qualitative one.

Data sets 2-5 deal largely with established aliens, though Prus' study of weed costs finds that many casual species, particularly so-called volunteers from crops, are costly to control. Volunteers are plants persisting, re-emerging or growing from seed, after the crop has been harvested. The results for the Grime *et al.* survey of 1m², the Perrins *et al.* survey of perceived weediness, the Prus study of weed cost and the Schering listing of weed incidence are shown in Figs 2 – 5. As there are several more or less coincident points, the number of points in these figures appears to be, but is not, less than those listed in Table 5.

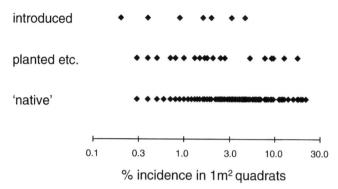

Fig. 2. Incidence in 2748 1m² quadrats chosen to give "a full range of examples of each of the main herbaceous vegetation types" in the Sheffield region of northern England (Grime *et al.* 1988 p. 9). 'Introduced' are all modern introductions except *Myrrhis odorata*, which is normally listed as introduced, although "Accepted, with reservations, as native" by Clement and Foster (1994). 'Planted etc.' are "doubtfully native [in the Sheffield region] or have an increased geographical or ecological range as a result of escaping from cultivation" (Grime *et al.* 1988 p. 20). 'Native' are all other species in the survey, and so include those regarded as ancient introductions in Figs. 3-5. Note the logarithmic scale of the abscissa.

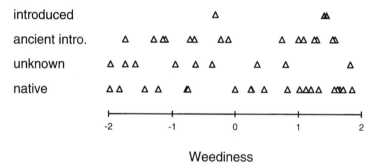

Fig. 3. Distribution of the average weediness as perceived by 65 scientists, from Perrins *et al.* (1992). '2' is always a weed, '-2' is never a weed. Introduced are modern (post-medieval) introductions; the other three categories are based on the records in Godwin (1975).

The differences in the distributions shown in the four sets can be tested by one way Analyses of Variances or by their non-parametric equivalents. Biologists tend to feel more comfortable with the latter, statisticians point to the remarkable robustness of the analysis of variance. However; the weed incidence (Schering 1986), perceived weediness (Perrins *et al.* 1992.) and abundance (Grime *et al* 1988.) sets all have no significant differences between their categories on any test, while the weed cost (Prus 1997) set (Fig. 4) is significant on all. The significance levels in weed cost are 0.003 for the Analysis of Variance, 0.016 for Kruskal-Wallis, 0.026 for Mood (all from Minitab). The reason why the Prus weed cost data are statistically significant is that ancient introductions are associated with a higher cost, largely because of the cluster of points at an abscissa value of just over 15 (Fig. 4). Even with the weed cost data, there is no difference between post-medieval introductions and natives.

Even the significant difference, between the weed cost of ancient introductions and other plants, could be an artifact of the way these are distinguished. This was by que-

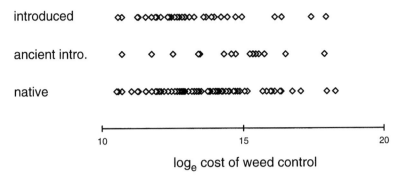

Fig. 4. Distribution of weed cost, on a log$_e$ scale, of all plants with a cost of over 10 on this scale. Introduced are modern introductions, ancient introductions are those doubtfully native in Clapham *et al.* (1987) or Stace (1991) or regarded as not native by Webb (1985).

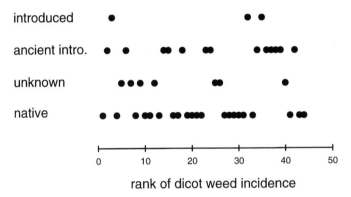

Fig. 5. Rank of the incidence of dicotyledonous weeds from Schering Agriculture (1986). Classification as in Fig. 4. Commoner weeds have a higher incidence rank.

ries about native status in Clapham *et al.* (1987), Stace (1991) or Webb (1985). The principal reason for making such queries seems to have been an association with arable land. On the other hand, as the Prus weed cost survey includes many more species than either the Schering weed incidence or the Perrins *et al* perceived weediness surveys, Figures 3 and 5 might support the view that those latter surveys would have produced the weed cost result if more species had been included. The view that ancient invaders are more readily identified if they are weedy, and that their appearance of being more weedy is an artifact, seems preferable.

For established aliens and widespread casuals, I conclude there is no good evidence that they are either more or less weedy, species for species, than natives.

The same conclusion seems to apply to the final data set (1, Table 2), those species that require control of nature reserves in parts of northern England (Table 4). Most of the species controlled are native. But then, most widespread species are also native (Fig. 1). Are there fewer or more introduced plants in Table 4 than would be expected? Examine Table 5.

One difficulty in interpreting Table 5 is that different criteria are used for assessing native status in different columns. A chi-squared test of modern introductions against other categories combined shows two apparently strongly significant results. The Grime

Table 4. Plants controlled, or which the Trusts consider should be controlled, in nature reserves in northern England managed by the Cumbria Wildlife Trust, the Northumberland Wildlife Trust and the Yorkshire Wildlife Trust. The number after each species is the number of reserves in which it is, or should be, controlled.

Native	Planted
Alnus glutinosa, 1	*Fagus sylvatica ,* 1
Arctium minus, 1	*Pinus sylvestris,* 1
Betula spp., 7	
Brachypodium pinnatum, 3	
Chamerion angustifolium, 4	Total 2 species, 2 cases
Cirsium arvense, 7	
Cirsium vulgare, 1	
Crataegus monogyna, 3	
Cytisus scoparius, 1	
Elytrigia repens, 1	Introduced
Epilobium hirsutum, 1	*Acer pseudoplatanus,* 5
Fraxinus excelsior, 2	*Fallopia japonica,* 1
Glyceria fluitans, 2	*Impatiens glandulifera,* 5
Hippophaë rhamnoides, 2	*Lysichiton americanum,* 1
Ilex aquifolium, 1	*Rhododendron ponticum,* 7
Juncus spp., 3	5 species, 19 cases
Phragmites australis, 2	
Prunus spinosa, 1	
Pteridium aquilinum, 12	
Rosa pimpinellifolia, 1	
Rubus fruticosus, 3	
Salix cinerea, 1	
Salix spp., 1	
Senecio jacobaea, 6	
Typha latifolia, 1	
Urtica dioica, 3	
Total 26 species, 71 cases	

et al. (1988) 1m^2 data are lacking introduced species, while the Prus (1997) weed cost data have more than expected. Both results, and all the variation in the bottom line of Table 5, may have a simple explanation.

From Fig. 1, it is evident that sets that consist predominantly of widespread species will be poor in introduced species. Sets containing less widespread species will have a higher frequency of introduced species. The Grime *et al.* (1988) survey concentrated on common species, those that are well represented in a sample of 1m^2 quadrats. Correspondingly, it has a low proportion of introduced species. The weed incidence (Schering 1986) and perceived weediness (Perrins *et al* 1992) surveys concentrated on widespread species. The 10 km square and nature reserve results are quite similar, indicating that both are dealing with effectively random samples of the flora. The weed cost (Prus 1997) survey brings in volunteers and other casuals and uses a criterion weakly related to distribution, and so has a higher proportion of introduced species.

If that explanation is right there is no reason to suppose that there is anything unusual about the proportion of controlled species on nature reserves.

Table 5. The numbers in different categories recorded in the six surveys.

Data set	Weed incidence	Weed cost	Perceived weediness	1 m² quadrat	10 km square	Nature reserves
native	21	147	21	245	1419	26
planted				27		2
unknown	7		9			
ancient introductions	13	19	16			
(modern) introduced	3	52	3	9	197	5
(modern) introduced, as %	7	24	6	3	12	15

Conclusion

The six data sets point to a single and clear conclusion, that established aliens in Britain have about the same impact, species for species, as native ones. Add to that that casual aliens have little impact except in special situations such as crop volunteers, and that even established aliens are markedly less widespread than natives, it might be thought that aliens do not constitute a problem to worry about. That would be a bad mistake. Although aliens are, in Britain, no more undesirable than many natives, they are additional to natives, an incremental cost. For instance, *Rhododendron ponticum* and *Heracleum mantegazzianum*, not in Grime *et al.* (1988) and only recorded in 1495 and 522 10 km squares respectively, are still costly problems in many places where they occur. We can just be glad that Britain is still free of the really serious alien problems that have occurred elsewhere. Nevertheless, almost the only good indicator that a new species will be a problem is its behaviour elsewhere (Williamson 1996). Prediction is still not feasible.

Acknowledgements

I am grateful to Alastair Fitter, Kevin Gaston, Richard Law, Charles Perrings and Charlotte Williamson for their helpful comments. I am obliged to Susan Brudenell and Alastair Fitter for extracting the 10 km square records, which are part of the NERC Biological Records Centre data, from the Ecological Flora Database and to Meg Stark for the final versions of the figures. Allan Hall was exceedingly helpful on the subject of ancient invaders, and searched the Environmental Archaeology Unit's archaeobotanical database for me. For the Wildlife Trust records I am indebted to Kerry Milligan (Cumbria), Ian Douglas and David Stewart (Northumberland), David Hargreaves and Helen Jackson (Yorkshire) and the following reserve chairmen or wardens: C. Alder, Alan Ball, Katy Bottrell, J. Carr, John Drewett, Robert Evison, Keith Gittens, G.E. Griffith, E.W. Hartley, Rob Knight, Peter Larner, Colin Marsden, Dennis Moffatt, Christopher Needham, Joyce Payne, B.R. Spence, Andrew Thompson, D. Wales and to several who did not wish to be named.

References

Atkinson, A.C. 1985. Plots, Transformations and Regression. Oxford University Press, Oxford.

Bartlett, P.W. 1979 Preventing the establishment of Colorado beetle in England and Wales. In: Ebbels, D.L. and King, J.E. (eds.), Plant Health and the European Single Market. pp. 247-257. Blackwell Scientific Publications, Oxford.

Clapham, A.R., Tutin, T.G. and Moore, D.M. 1987. Flora of the British Isles. 3rd edition. Cambridge University Press, Cambridge.

Clement, E. J. and Foster, M. C. 1994. Alien Plants of the British Isles. Botanical Society of the British Isles, London.

Fitter, A. H. 1995. Interpreting quantitative and qualitative characteristics in comparative analysis. J. Ecol. 83: 730.

Fitter, A. H. and Peat H. J. 1994. The ecological flora database. J. Ecol. 82: 415-425.

Gaston, K. J. 1991. How large is a species' geographic range? Oikos 61: 434-438.

Gaston, K. J. 1994. Rarity. Chapman and Hall, London and New York.

Gaston, K. J. 1996. The multiple forms of the interspecific abundance-distribution relationship. Oikos 76: 211-220.

Gaston, K. J., Blackburn, T. M. and Lawton, J. H. 1997. Interspecific abundance-range size relationships: an appraisal of mechanisms. J. Anim. Ecol. 66: 579-601.

Gaston, K. J., Quinn, R. M., Blackburn, T.M. and Eversham, B.C. in press. Species-range size distributions in Britain. Ecography:

Glowka, L., Burhenne-Guilmin, F. and Synge, H. 1994. A Guide to the Convention on Biological Diversity. IUCN, Gland and Cambridge.

Grime, J. P., Hodgson, J. G., and Hunt, R. 1988. Comparative Plant Ecology. Unwin Hyman, London.

Godwin, H. 1975. History of the British Flora. 2nd edition. Cambridge University Press, Cambridge.

Havey, P. H. 1996. Phylogenies for ecologists. J. Anim. Ecol. 65: 255-263.

Jones, E. W. 1944. *Acer*, Biological Flora of the British Isles. J. Ecol. 32: 215-252.

Pearce, D., and Moran, D. 1994. The Economic Value of Biodiversity. Earthscan, London.

Pearman, D. 1997 Towards a new definition of rare and scarce plants. Watsonia 21: 231-251.

Perrings, C. (co-ordinator) 1995. Economic values of biodiversity. In: Heywood, V.H. (ed.), Global Biodiversity Assessment. pp. 823-914. Cambridge University Press, Cambridge for the United Nations Environment Programme.

Perrins, J., Williamson, M. and Fitter, A. 1992. A survey of differing views of weed classification: implications for regulation of introductions. Biol. Conserv. 60: 47-56.

Prus, J. L. 1997. New Methods of Risk assessment for the Release of Transgenic Plants. Ph.D. thesis, Cranfield University.

Schering Agriculture 1986. Weed Guide. Revised edition. Schering Agriculture, Nottingham.

Stace, C. A. 1991. New Flora of the British Isles. Cambridge University Press, Cambridge.

Tomlinson, P.T. 1993. Design and implementation of a relational database for archaeological records for Great Britain and Ireland. Circaea 10: 1-30.

Tomlinson, P. T. and Hall, A. R. 1996. A review of the archaeological evidence for food plants for the British Isles: an example of the use of the Archaeobotanical Computer Database (ABCD). Instant Archaeology 1 (http://intarch.ac.uk/journal/issue1/tomlinson_index.html).

Tukey, J.W. 1977. Exploratory Data analysis. Addison-Wesley, Reading, Massachusetts.

Webb, D. A. 1985. What are the criteria for presuming native status? Watsonia 15: 231-236.

Westoby, M., Leishman, M. and Lord, J. 1995. Issues of interpretation after relating comparative datasets to phylogeny. J. Ecol. 83: 892-893.

Williamson, M. 1972. The Analysis of Biological Populations. Edward Arnold, London.

Williamson, M. 1993. Invaders, weeds and the risk from genetically manipulated organisms. Experientia 49: 219-224.

Williamson, M. 1994. Community response to transgenic plant release: predictions from British experience of invasive plants and feral crop plants. Molecular Ecology 3: 75-79.

Williamson, M. 1996. Biological Invasions. Chapman & Hall, London

Williamson, M. and Brown, K. C. 1986. The analysis and modelling of British invasions. Phil. Trans. R. Soc. Lond. B 314: 505-522.

Williamson, M. and Fitter, A. 1996a. The varying success of invaders. Ecology 77: 1661-1666.

Williamson, M. and Fitter, A. 1996b. The characters of successful invaders. Biol. Conserv. 78: 163-170.

PERCEPTION AND LEGISLATION

CURRENT LEGAL STATUS REGARDING RELEASE OF NON-NATIVE PLANTS AND ANIMALS IN GERMANY

Ulrike Doyle[1], Andreas Fisahn[2], Harald Ginzky[2] and Gerd Winter[2]
[1]Fachgebiet II 1.3, Umweltbundesamt, Postfach 33 00 22, 14191 Berlin, Germany;
E-mail: ulrike.doyle@uba.de; [2]Forschungsstelle für Europäisches Umweltrecht,
Universität Bremen, Universitätsallee GW I, 28359 Bremen, Germany

Abstract

In this paper the law and practice in Germany of the release of non-native species into the wild is discussed. Major results are the following:

Applications for and authorization of releases are rare in Germany. It appears that the legal framing of releases in Germany does not adequately grasp the case of unintended release. In this context the privilege of agriculture and forestry needs to be discussed.

The German licensing criteria for releases including the methodology of risk assessment need to be further developed. Risk assessment methods as developed for releasing genetically modified organisms may be consulted. On the long run a joint regulation of both non-native and genetically modified organisms may be envisaged.

Introduction

The extinction of species is everywhere in the public eye, but alongside this process native flora and fauna are increasingly being influenced and modified by the proliferation of alien plant and animal species. Biological invasions are becoming recognized as a world-wide problem for nature conservation, and there is a growing demand for some form of regulatory response to this problem (Sjöberg and Hokkanen 1996; Office of Technology Assessment (U.S. Congress) 1993). The question we shall address here is whether current legislation in Germany can respond adequately to the constellation of problems which arise from this. We shall begin by examining the various ways in which they can be regulated at national level. The international community has addressed the protection of species in a range of conventions. The present paper will examine whether existing national regulations reflect international requirements, to the extent that the European Community has adopted such requirements on behalf of Member States. After that the licensing practice will be discussed followed by a comparison of the regulation of genetically modified organisms with non-native organisms.

The concept of release (in German: *Ausbringung*) is not used in any of the relevant regulations. In this study, release is used as a generic term to convey the various legal concepts applied to the fact. It embraces all ways in which non-native animals and plants are introduced into the environment.

Plant Invasions: Ecological Mechanisms and Human Responses, pp. 71–83
edited by U. Starfinger, K. Edwards, I. Kowarik and M. Williamson
© *1998 Backhuys Publishers, Leiden, The Netherlands*

National law

The framework legislation in Germany for central protective regulation covering the release of alien animals and plants is defined in the Federal Nature Conservation Act, which applies nation-wide, and in the Conservation Acts adopted by each of the *Länder*, or federal states. The release of non-native fish is covered by specific fisheries legislation. In Germany, it is a federal responsibility to regulate marine and coastal fishing. The *Länder* are responsible for inland waters. Relevant provisions are also found in hunting, forestry law, agricultural law, and law relating to pesticides, the protection of animals, and epidemics.

The Federal Nature Conservation Act

§ 20 d para. 2 of this Act sets out framework provisions for the release of non-native animals and plants into the environment: "Alien wild and non-wild species of animals and plants may only be set free or introduced into the wild if permission is granted by the authority responsible under state law. This does not apply to the cultivation of plants in agriculture and forestry. Permission shall be refused if the danger cannot be ruled out that the native community of flora and fauna will be contaminated (in German: *Verfälschung*) or that the survival or propagation of native species of wild animals or plants or of populations of such species will be placed at risk." The Federal Nature Conservation Act also includes import regulations and rules with regard to control.

We shall begin below by describing and discussing the field of application of § 20 d para. 2 of the Act and the licensing conditions. We shall then look at the relevant provisions in the amendment to this Act.

Field of application of § 20 d para. 2 / Federal Nature Conservation Act
§ 20 d para. 2 of the Federal Nature Conservation Act requires a licence for the release of alien wild and non-wild species of animals and plants (Fig. 1). The rule covers two types of release of alien animals and plants into the environment: one is *setting free* (German: *Aussetzen)* and the other is *introduction* (German: *Ansiedeln)* in the wild. "Setting free" is defined unanimously in the literature as "leaving to its own devices", i.e. the party responsible for the release does not thereafter tend to the specimens or provide in any way for their survival.[1] "Introduction", on the other hand, is taken to mean the planned input of animals with a view to establishing a local population, or the sowing, cultivation or other release of plants, both of these usually in conjunction with some form of care.[2]

The terms animals and plants as applied by § 20 d para. 2 are broadly defined. In line with § 20 a para. 1 no. 1 lit. b and no. 2 lit. b of the same Act, the "eggs, larvae, pupae and other developmental forms" of animals and the "seeds, fruits and other developmental forms" of plants must also be covered, as their release can also induce the establishment of alien species of flora and fauna. The definition of animals and plants reflects the scientific terminology.[3] However, it has not yet been possible to clarify whether micro-organisms fall under the field of application, and if so, which.

---

[1]Gassner, § 20 d marginal 9, Kolodziejcok, § 20 d marginal 21, Ebersbach, p. 198, Müller-Boge, p. 17, Apfelbacher, p. 251. See also BT-Drs. 10/5064, p. 19.

[2]Gassner, § 20 d marginal 9, Kolodziejcok, § 20 d marginal 21, Ebersbach, p. 198, Müller-Boge, p. 17, Apfelbacher, p. 251.

Federal Nature Conservation Act § 20 d para. 2	The Wild	Settled Areas
Plants:		
setting free (negligent conduct)	no licensing required	no licensing required
introduction (activity conforms to a plan)	**licensing required**	**no licence required** (cultivating alien plants in agriculture and forestry)
Animals:		
setting free (negligent conduct)	**licensing required**	**licensing required ***
introduction (activity conforms to a plan)	**licensing required**	**licensing required**

*but no longer when the amendment to the Federal Nature Conservation Act is implemented

Fig. 1. Field of application of § 20 d para. 2 of the Federal Nature Conservation Act (Germany)

"Alien" in the sense of § 20 d para. 2 applies to any species not or no longer encountered in the locality in which it is to be released. The geographical reference here is, therefore, to the area in which the specific release is to take place.[4] A non-indigenous sub-species of a native species may also be alien (§ 20 a para. 3 of the Federal Nature Conservation Act). The purpose of this distinction is to prevent alien sub-species from displacing native related species by interbreeding.[5]

In terms of time, the decision depends on the natural state of a specific area at the moment of licensing. It is of no significance whether the species or sub-species has ever been established in that area or anywhere else in Germany in the past. In other words, the re-introduction of species which were once established in Germany but had been displaced at the time of application would fall under § 20 d para. 2 of the Federal Nature Conservation Act.

The release of *non-wild species* of animals and plants also calls for a licence. This means, in particular, that the setting free of alien domesticated animals also falls under the licensing requirement in § 20 d para. 2.

Whereas introduction assumes that the activity conforms to a plan, setting free could conceivably imply negligent conduct.[6]

According to the text of the law, *introduction of plants only requires licensing if it occurs in the wild*. If we follow the wording, this qualification does not apply to setting free. The wild is equated with unpopulated areas as defined in § 1 para. 1 of the Federal Nature Conservation Act.[7]

[3]Cf. § 20 a para. 3 Federal Nature Conservation Act.
[4]Ebersbach, p. 197 ff., Gassner, § 20 d marginal 9, Kolodziejcok, § 20 d marginal 19, Müller-Boge, p. 18.
[5]Battefeld, § 25 marginal 8.
[6]Kolodziejcok, § 20 d marginal 21, Ebersbach, p. 198, Müller-Boge, p. 18, Apfelbacher, p. 251.
[7]Kolodziejcok, § 20 a marginal 29. Similarly Ebersbach, p. 198.

The cultivation of ornamental or useful plants in gardens, allotments, parks, cemeteries, other green spaces created within settlements and comparable spaces serving human use is not seen as falling within the sphere of "the wild", given that the vegetation on such land is essentially artificial, having been placed there by human hand.[8] Whereas sowing seeds on a bank dividing arable farm land would fall within the sphere of "the wild" .

No licence is required under § 20 d para. 2 clause 2 of the Federal Nature Conservation Act for cultivating alien plants within the framework of agriculture or forestry. This exemption applies primarily to useful crops and cultivated plants.[9] The release of alien animals, however, must under the terms of § 20 d para. 2 be licensed, even if it occurs within the framework of agriculture or forestry.

Conditions of licence
According to § 20 d para. 2 clause 3 of the Act, permission "shall be refused if the danger cannot be ruled out that the native community of flora and fauna will be contaminated or that the survival or propagation of native species of wild animals or plants or of populations of such species will be placed at risk." There is no case law to date to clarify the interpretation of these conditions of licence, since no court has yet passed down a verdict either on § 20 d para. 2 of the Federal Nature Conservation Act or on the regulations for implementation adopted by the *Länder*. A universally understood interpretation of the terms "contamination", "danger to survival" etc. is still also under discussion from the scientific point-of-view;[10] all we have at present are a number of case studies.

The first factor to bear in mind is that the purpose of both conditions is to protect *native* flora and fauna. In this respect, § 20 a para. 4 clause 1 of the Federal Nature Conservation Act contains a legal definition, according to which an animal or plant living in the wild is native if "the area of occurrence or regular migration lies or has in the course of history lain fully or partly within the area covered by this Act or is spreading by natural means into the area to which this Act applies". According to § 20 a para. 4 clause 2 of the Federal Nature Conservation Act, a wild species of animal or plant should also be regarded as native if "animals and plants of the relevant species which have turned wild or were introduced under human influence survive as a population in the wild over several generations without human aid in the area covered by this Act". An alien species under § 20 d para. 2 of the Federal Nature Conservation Act may, therefore, nevertheless be "native" in the sense of § 20 a para. 4 of the same Act if it is established in or regularly migrates to another part of Germany (which is the area covered by this Act) or if it has done so at an earlier date (in the course of history).[11]

Under § 20 d para. 2 clause 3 of the Act, a danger to native species must be ruled out in order to obtain a licence. It follows from this that the danger does not have to be demonstrated positively. To refuse a licence, it is sufficient for the authority to have some indication that the native community of flora and fauna is jeopardised.[12] The applicant must then prove that the danger definitively does not exist. The upshot of §

[8]Kolodziejcok, § 20 d marginal 28, Meßerschmidt, § 20 d marginal 6.
[9]Apfelbacher, p. 252.
[10]Auhagen, p. 15 ff.
[11]Gassner, § 20 d marginal 9, Kolodziejcok, § 20 d marginal 19.
[12]See also Schink, p. 452, as well as Battefeld § 25 marginal 14.

20 d para. 2 clause 3, therefore, is that the burden of proof lies with the applicant. However, as the decision is founded on a forecast, exaggerated expectations cannot be levelled at the furnishing of evidence, especially as proof of the contrary cannot be provided with absolute certainty.[13]

Assessment

It is recommended that negligent behaviour be incorporated into the restrictions of § 20 d para. 2 of the Act, because this provides the basis for monitoring provisions.[14]

There is a problem here in that introduction of plants only requires licensing when it takes place in the wild. No licence is required for introducing alien plants in settlement areas or artificially maintained spaces such as parks. This granting of privilege is not, however, convincing, as alien plants can spread beyond these sites.[15] In addition, they can cause on site damage. Finally, exempting cultivated landscapes contradicts the aim of the legislation, as it is above all on sites such as these that the planned introduction of alien plants occurs. Most neophytes began by being cultivated for ornamental purposes, especially in parks and gardens, only to spread from there "into the wild".[16]

The final criticism refers to the exemption of agriculture and forestry. In the forestry sector at least, the release of alien woody species has already caused ecological damage.[17] So far ecological risks have primarily been induced by the release of alien plants, both ornamental and crop plants.[18] After all, plants cultivated for agriculture and forestry can run wild, generating ecological changes.

Another aspect which merits discussion is that § 20 d para. 2 of the Federal Nature Conservation Act describes two different constellations, yet stipulates the same conditions for both. This is due to the fact that alien species of animals and plants can also be native if they formerly occurred elsewhere in Germany as mentioned in § 20 a para. 4 of the Act. This means that § 20 d para. 2 refers on the one hand to the release of alien species of animals and plants not previously encountered in Germany (henceforth: first introduction) and on the other to the re-introduction of animal and plant species which were formerly native.

One argument in favour of a legal distinction between first and re-introduction is that, whereas the introduction of previously unknown alien species would lead to increasingly similar communities the world over,[19] re-introduction can stabilise and consolidate regional specificity. In other words, re-introduction can in itself be an asset. This should be taken into consideration in the legal regulation of the issue. This fundamental assessment is not refuted by the fact that re-introductions are not in themselves accompanied by lesser ecological risks than first introductions.[20]

[13]Gassner, § 20 d marginal 12.

[14]As opposed to extending the duty to obtain a licence in Blum, § 44 marginal 1, which would mean a substantial proliferation of duties to monitor.

[15]Cf. Kowarik, p. 44 on the spread of non-native woody plants beyond the borders of settlement areas.

[16]Kübler, p. 89 ff. and Sukopp, p. 14 ff. See also Battefeld, § 25 marginal 9, who cites as examples the angelica tree, Canadian goldenrod and the giant varieties of knotgrass, etc. These species, he maintains, have in some instances displaced native ones.

[17]Cf. Knoerzer, p. 67 ff. on the proliferation of the Douglas fir and Kowarik, p. 44 ff. on the spread of Black Locust and the Black Cherry.

[18]Sukopp, p. 14.

[19]Sukopp's reasoning, p. 3 ff.

[20]Reichholf, p. 37 ff.

Both constellations confront a problem in that § 20 a para. 4 clause 2 of the Federal Nature Conservation Act also regards species as native, and therefore worthy of protection, when they have been released by humans and survived in the wild for several generations. The consequence of this provision is that formerly alien species which have become established are covered by the protective aim of § 20 d para. 2 of the Act, regardless of whether they themselves cause ecological damage. In this way, the late black cherry from North America would be deemed a native of Berlin's forests, according to the legal definition, deserving protection from any new alien species and in accordance with the general rules. In this respect, the legislature is called upon to offer clarification.

Import regulations
Native flora and fauna can also be protected by regulating the import of alien animals and plants. Under § 21 a para. 1 clause 1 no. 3 of the Federal Nature Conservation Act, the Federal Ministry for the Environment, Nature Conservation and Nuclear Safety is empowered to issue ordinances prohibiting the import of certain animal and plant species not covered by the EC's CITES Regulation[21] or else to make such import dependent on a licence under § 21 b of the Federal Nature Conservation Act if this is necessary because of a danger "that the native community of flora and fauna will be contaminated or that the survival or propagation of native species of wild animals or plants or of populations of such species will be placed at risk."

To date no ban on imports has been issued under § 21 a para. 1 clause 1 no. 3 of the Act. What has been introduced, however, is the requirement through § 6 para. 1 of the Federal Protection of Species Ordinance to acquire an import licence for the species named in Annex 3 column 1. The risk that native flora and fauna might be contaminated is to be seen in the cases, for example, of the dwarf gull, the grey heron, the American bullfrog, the fire-bellied salamander and so on.[22]

The Federal Nature Conservation Act does not provide adequate scope for monitoring and imposing penalties. Stipulations refer only to a universal right to information (§ 23) and a general duty to tolerate (§ 10).

Draft Amendment to the Federal Nature Conservation Act
The draft amendment to the Federal Nature Conservation Act[23] includes a modified text on the release of alien animals and plants. The new version envisages two major changes with regard to the issue which concerns us:

First, as opposed to setting free, only the introduction of alien animals into the wild will require licensing, as the former is covered by legislation on the protection of animals. This means that a licence will no longer be necessary, as it was, for the release of alien animals in settlement areas. § 3 no. 3 of the Protection of Animals Act prohibits the setting free of domestic animals in general.

---

[21]Council Regulation (EEC) no. 3236/82 of 3 December 1982 on Implementation in the Community of the Convention on International Trade in Endangered Species of Wild Fauna and Flora, OJ L 384/1.
   The Federal Nature Conservation Act has essentially transposed the provisions of the EC's CITES Regulation. On the basis of art. 15 of the Regulation, however, it was left to Member States to decide whether they wished to uphold or adopt stricter measures for various reasons, including the protection of native species.
[22]This suggests that the risk of contamination (§ 21 b para. 1 clause 1 no. 3 of the Federal Nature Conservation Act) should be assessed. Annex 3, however, consists mainly of species for which the risk of species extinction has priority.
[23]BT-Drs. 13/6441.

Secondly, the duty to obtain a licence will not apply where alien animals are used as a biological form of plant protection. The reason given for this is that licensing a plant protection method requires an assessment of its impact on the natural balance.[24]

Criticism must be levelled at the fact that in future the setting free of alien animals will not call for a licence. The reference to animal protection legislation is not acceptable because the purpose of this legislation is to protect individual animals and not nature or landscape.[25] Besides, limiting the duty to obtain a licence to release in the wild is not convincing.[26]

The time aspect in the amendment does not seem devoid of problems. Formerly native plant species which have been displaced within the last 100 years are not alien according to the legal definition in § 8 para. 2 no. 6 of the draft, so that under § 37 para. 3 clause 2 no. 2 of the draft they do not require a licence for release. This would mean that re-introduction could take place without a licence, but it does not take into account the discussion about local genetic diversity.

The criticisms outlined above also apply to the new draft. Here, again, agriculture and forestry enjoy a privileged status, and introduction of plants only requires licensing when it is to take place in the wild.

The power to issue legal ordinances granted under § 47 para. 2 no. 3 of the draft enables the Federal Ministry for the Environment, Nature Conservation and Nuclear Safety to prohibit the possession, safekeeping and marketing – as defined in § 39 para. 1 of the draft – of non-native animal and plant species which are not specially protected if this is necessary due to the danger of contaminating the local community of fauna and flora. This means that the Federal level can even prohibit the possession and marketing of non-native species when they are not specially protected, if this serves the purpose described.

The relationship between legislation on nature conservation and other statutes

Under § 20 para. 2 of the Federal Nature Conservation Act, the Act's rules with regard to the protection of species and any legal provisions enacted on the basis of those rules shall be without prejudice to "the provisions of legislation governing plant protection, animal protection, epidemics, and forestry, hunting and fishery". The provisions encountered in the hunting and fishery laws broadly reflect the provisions in the Federal Nature Conservation Act. The other laws are only indirectly pertinent, given that they pursue a different purpose, and will, therefore, not be discussed here.

Standards established under EC law

EC legislation on the protection of species

The EC CITES Regulation,[27] which transposed the Washington Convention on the Protection of Species, empowers Member States under art. 15 to uphold or adopt stricter

---

[24]Ibid. See § 15 of the Plant Protection Act.
[25]BT-Drs. 13/6441, p. 64.
[26]See above.
[27]Council Regulation (EEC) no. 3236/82 of 3 December 1982, see Fn 23.

measures in order, amongst other things, to preserve native species. The Regulation applies directly in Germany. The new EC Regulation on the Protection of Species,[28] which came into force on 1 June 1997, transfers any regulation of imports to protect native species to the European level. Pursuant to art. 4 para. 6 lit. b of the Regulation, the Commission can restrict the import – in general or in relation to specific countries of origin – of live specimens of species whose introduction into the natural habitat of the Community has been proven to pose an ecological threat to indigenous species of animals and plants living in the Community in a wild state.

Bird Conservation Directive and Flora Fauna Habitat Directive

The Bird Conservation Directive[29] and Flora Fauna Habitat Directive[30] do not enter force directly in Germany and must, as a result, be properly transposed in the new Federal Nature Conservation Act. The draft does not keep pace with this, especially in comparison with the particular procedural provisions laid down in the Flora Fauna Habitat Directive.

With specific regard to the introduction of wild birds, art. 11 of the Bird Conservation Directive stipulates: "Member States shall see that any introduction of species of bird which do not occur naturally in the wild state in the European territory of the Member States does not prejudice the local flora or fauna. In this connection they shall consult the Commission."

A similar stipulation is contained in art. 22 of the Flora Fauna Habitat Directive, this time with regard to the introduction of additional animal species and also plants: "In implementing the provisions of this Directive, Member States shall: a) study the desirability of re-introducing species in Annex IV that are native to their territory where this might contribute to their conservation, provided that an investigation, also taking into account experience in other Member States or elsewhere, has established that such re-introduction contributes effectively to re-establishing these species at a favourable conservation status and that it takes place only after proper consultation of the public concerned; b) ensure that the deliberate introduction into the wild of any species which is not native to their territory is regulated so as not to prejudice natural habitats within their natural range or the wild native fauna and flora and, if they consider it necessary, prohibit such introduction. The results of the assessment undertaken shall be forwarded to the committee for information; ..."

Licensing practice

This outline of licensing practice is based on our survey of higher to top-tier authorities responsible for hunting, fishery and forestry in the federal states of Niedersachsen,

---

[28]The EC CITES Regulation is to be replaced on 01/06/1997 by the new EC Regulation on the Protection of Species (Text in Common Position (EC) no. 26/96 adopted by the Council on 26 February 1996 with regard to issuing Council Regulation (EC) no. 338/97 of 9 December 1996 on protecting specimens of wild animal and plant species by monitoring trade, OJ C 196/58).

[29]Council Directive of 2 April 1979 on the Conservation of Wild Birds, OJ L 103/1.

[30]Council Directive 92/43/EEC of 21 May 1992 on the Preservation of Natural Habitats and of Wild Fauna and Flora.

Nordrhein-Westfalen, Baden-Württemberg and Hessen. The findings can be summarised as follows:

Urban population centres reported the same problems as more rural districts with regard to the typical problems associated with the release and spread of garden and ornamental plants and domestic pets.

A distinction between deliberate and negligent release is rarely drawn in practice, as the procedure is usually only confronted with the result, the occurrence of an alien species, and it is difficult to ascertain as a rule how this species actually entered the wild. It is not easy for the authorities to establish, for example, whether the tortoises they discover ran away from their owners or whether their owners had been trying to get rid of them. The problem is similar with plants.

As far as *animals* are concerned, most cases of release related to alien fish or amphibians. Top of the list were various species of tortoise or turtle. A number of authorities identified the appearance of alien birds. Among the mammals, the most frequent reports were of alien species of game. The picture for *plants* shows that most of the species named were garden plants.

As a general rule the authorities surveyed did not react to reported releases of non-native organisms by imposing penalties or ordering the perpetrators to remove the offending organism, the main reason being that they were unable to ascertain the perpetrator's identity. Only one instance was reported in which the offender was ordered to eliminate the deed and placed under prohibition.

Applications for licences to release alien animals and plants are only received in comparatively modest quantities (e.g. 20 applications in those four federal states over the last ten years). There is also a definite north-south divide. Most of the applications (11) were lodged in Baden-Württemberg. Most applications are made with the aim of re-introducing formerly native species of game (6) or for research purposes (4). At the same time, however, no mention was made of any legal dispute over the issue of a licence, nor of any claims for compensation due to the release of alien species.

In processing applications to release alien species, the authorities rely as a rule on internal knowledge, including consultation with other specialist authorities, and on scientific material which is universally accessible. The authorities do not either carry out preparatory experiments themselves or ask the applicant to do this (if the latter have not already carried out such experiments already as part of their scientific research, e.g. for university projects). None of the authorities surveyed had any table or scheme for evaluating the species or the potential dangers beyond the text of the law. Where uncertainty prevails, this (more or less overtly) weighs against the applicant.

Comparison with the release of genetically modified organisms

As both non-native and genetically modified organisms (GMO) introduce new genetic material into native populations licensing procedures and risk assessment methods as developed for releasing genetically modified organisms may be consulted.

As far as the wording of the law is concerned, if we compare the risk assessment applied under the Federal Nature Conservation Act to the release of alien species with that applied by the Genetic Engineering Act to the release of GMOs, the former proves more restrictive. Harmful consequences may not be compensated by any advantages to be gained from release, which they may under § 16 para. 1 of the Genetic Engineering Act. Nor does the applicant have any entitlement in the licensing procedure, which is at the discretion of the authority. The burden of proof lies with the applicant, who also bears the burden of doubt. It should be borne in mind, however, that the narrow interpretation of constituent facts (especially the "wild" element) pursuant to the Nature Conservation Act means that there are relatively few instances of the release of alien species into the environment which require licensing.

The history of neophyte proliferation shows that the period which elapses between release and proliferation can be expected to be of the magnitude of decades to centuries. Due to the effects of time lag, new species will continue to spread in future with sometimes unexpected consequences, even if no further new species are added (Kowarik 1996). This means that "the danger ... that the native community of flora and fauna will be contaminated" has already materialised (Böcker *et al.* 1995, Doyle 1996, Gebhardt *et al.* 1996). It would be desirable, in order to keep the further proliferation of alien species to a minimum, at least to make sufficient use of the scope offered by existing legislation.

Nevertheless, any impression that the release of alien species is subject to stricter regulation will be corrected by comparing licensing practice for the release of alien species with that for the release of GMOs (compare Fisahn 1998). The salient difference between the two procedures relates to the effort invested around risk assessment. For the release of alien species, risk assessment involves a not particularly elaborate method which draws on a relatively limited, certain base of empirical and theoretical knowledge about how native and non-native species relate. This risk assessment is made easier by the unambiguous allocation of the burden of proof, which is borne by the applicant. Licences for the release of GMOs, on the other hand, are founded on far more extensive scientific research into the properties of the GMO and its potential environmental impact. Given the reference to current scientific understanding, it is hardly likely that any decision will rest on a burden of proof. The difference can be illustrated in quantitative terms, too. Whereas the report and decision of the licensing authority in the case of alien species covered two pages at most, in the case of GMOs this documentation has easily amounted (so far, anyway) to fifty or a hundred pages. This indicates that the supervision of GMO release is more precise than for species release.

This cannot be explained in terms of the substantive provisions of the law, as the conditions for species release are more restrictive, as we have seen. One factor towards an explanation is the social context. The public regards GMO release as a problem, whereas little concern is shown about problems arising from the release of alien species. Secondly, the way the procedure is organised evidently leads to different intensities of investigation. Responsibility for GMO licensing rests in Germany with the Robert Koch Institute, a centralised authority with a considerable staff of experts, whereas species licensing is usually "only" the task of the higher-level nature conservation authorities, who see it as one job among many. Finally, GMO licensing entails a certain public participation, if limited, whereas species licensing does not involve the public at all, with not even specialist organisations being consulted.

For the foreseeable future, however, we can probably expect little more than modest steps towards a more differentiated regulatory regime for alien organisms. This includes a need for Germany, like others, to devise methods of risk assessment. These could be designed to deal variously with, in particular, the above-mentioned differences between natural, cultivated and genetically modified properties, perhaps by making more allowance for the risks incurred by re-introducing formerly native organisms than for those induced by the first introduction of organisms with artificially modified characteristics or organisms drawn from other habitats.

Recommendations

For a better success in practice of stopping unwanted releases of non-native plants and animals the following recommendations are suggested.

Field of application:
It would be advisable to formulate the elements of prohibition separately from the elements of permission. This would make it possible to state more clearly that negligent release constitutes an offence. According to the general definition of terms, no licence is required for introducing alien animals and plants in settlement areas or in artificially cultivated spaces. Given the real risks, this does not serve the purpose of the law.

The distinction between first introduction and re-introduction:
A differentiated approach is advisable. This need can also be derived from the international standards and EC law (Convention on the Protection of Biological Diversity, Alpine Convention, Flora Fauna Habitat Directive). With regard to the release of alien species which have never been native to Germany, it is worth considering whether it would not be appropriate to ask for proof of benefits, e. g. as in art. 17 clause 2 of the Alpine Protocol.

Imports:
The power granted in the new EC Diversity Regulation to control imports in order to protect nature and landscapes from non-native species seems to have been established at the wrong level. The draft amendment to the Federal Nature Conservation Act compensates by allowing for the introduction of a ban on possession by means of a legal ordinance.

The privileged status of agriculture and forestry:
The privilege which agriculture and forestry enjoy does not serve the purpose of the law. The relevant legal provisions do not adequately cover the dangers and risks associated with the release of alien species of plants.

Animal protection regulations:
The amendment to the Federal Nature Conservation Act removes any duty to license the setting free of alien animals. This means that aspects of nature conservation need to be incorporated in the Animal Protection Act.

Unintended release:
A federal framework regulation should be adopted, analogous to the provisions in various Nature Conservation Acts passed by the *Länder*, requiring a licence to open and operate establishments in which alien species of flora and fauna are kept. The safety standards should be stipulated in non-statutory regulations.

Monitoring:
To ensure appropriate enforcement, Germany would have first of all to ascertain whether federal and state provisions grant the authorities enough powers to monitor developments. An explicit regulation would make the competence of the monitoring authorities clear.

International co-operation:
The relevant EC provisions do prescribe some duties to report and consult. These must be taken into account. The Federal Nature Conservation Act does not make such provision. Consultation would above all be necessary and meaningful when release might have cross-border consequences.

Acknowledgement

The investigations were conducted and funded within the framework of the Environmental Research Plan of the Federal Ministry of the Environment under project number 108 02 901/02.

References

Apfelbacher, D. 1987. Naturschutzrecht nach der ersten Novelle zum Bundesnaturschutzgesetz. NuR: 9 (6): 241-255.

Auhagen, A. 1991. Vorschlag für eine Präzisierung der Definition der in Roten Listen verwendeten Gefährdungsgrade. Landschaftsentwicklung und Umweltforschung Sonderheft S 6: 15-23.

Battefeld, K. U. 1996. Hessisches Naturschutzrecht, Kommentar, Loseblattsammlung. C.F. Müller, Heidelberg.

Blum, P., Agenda, C. A. and Franke, J. 1990. Niedersächsisches Naturschutzgesetz, Kommentar, Loseblattsammlung. Kommunal- und Schulverlag, Wiesbaden.

Böcker, R., Gebhardt, H., Konold, W. and Schmidt-Fischer, S. (eds.) 1995. Gebietsfremde Pflanzenarten – Auswirkungen auf einheimische Arten, Lebensgemeinschaften und Biotope. Ecomed, Landsberg.

Doyle, U. 1996. Seminar: „Changes in fauna and flora as a result of watercourse development – neozoans and neophytes" – summary. In: Umweltbundesamt (eds.), Texte des Umweltbundesamtes 74/96. pp. 113-119. Umweltbundesamt, Berlin.

Ebersbach, H. 1981. Tieraussetzung im Natur-, Jagd- und Fischereirecht, NuR 3 (6):195-201.

Fisahn, A. 1998. Die Genehmigung der Freisetzung gentechnisch veränderter Organismen eine Fallstudie. In: Winter, G. (ed.), Die Prüfung der Freisetzung gentechnisch veränderter Organismen. Recht und Genehmigungspraxis. Forthcoming. E. Schmidt, Berlin, Bielefeld, München.

Gassner, E., Bendomir-Kahlo, G., Schmidt-Rätsch, A. and Schmidt-Rätsch, J. 1996. Bundesnaturschutzgesetz, Kommentar. C. H. Beck, München.

Gebhardt, H., Kinzelbach, R. and Schmidt-Fischer, S. (eds.). 1996. Neozoen – neue Tierarten in der Natur. Ecomed, Landsberg.

Knoerzer, D., Kühnel, U., Theodoropoulos, K. and Reif, A. 1995. Zur Aus- und Verbreitung neophytischer Gehölze in Südwestdeutschland mit besonderer Berücksichtigung der Douglasie (*Pseudotsuga menziesii*). In: Böcker, R., Gebhardt, H., Konold, W. and Schmidt-Fischer, S. (eds.), Gebietsfremde Pflanzenarten. pp. 67-81. Ecomed, Landsberg.

Kolodziejcok, K.-G., Recken, J., Apfelbacher, D. and Bendomir-Kahlo, G. 1991. Naturschutz, Landschaftspflege und einschlägige Regelungen des Jagd- und Forstrechts, Kommentar Loseblattsammlung (23. Lfg. – Stand: September 1991). E. Schmidt, Berlin, Bielefeld, München.

Kowarik, I. 1995. Ausbreitung nichteinheimischer Gehölzarten als Problem des Naturschutzes? In: Böcker, R., Gebhardt, H., Konold, W. and Schmidt-Fischer, S. (eds.), Gebietsfremde Pflanzenarten. pp. 33-56. Ecomed, Landsberg.

Kowarik, I. 1996. Auswirkungen von Neophyten auf Ökosysteme und deren Bewertung. In:

Langzeitmonitoring von Umwelteffekten transgener Organismen. Umweltbundesamt (eds.), Texte des Umweltbundesamtes 58/96. pp. 119-155. Umweltbundesamt, Berlin.

Kübler, R. 1995. Versuche zur Regulierung des Riesenbärenklaus. In: Böcker, R., Gebhardt, H., Konold, W. and Schmidt-Fischer, S. (eds.), Gebietsfremde Pflanzenarten. pp. 89-92. Ecomed, Landsberg.

Meßerschmidt, K. 1996. Bundesnaturschutzrecht: Kommentar zum Gesetz über Naturschutz und Landschaftspflege (German federal nature conservation law – commentaries) Loseblattausgabe. C. F. Müller, Karlsruhe, Heidelberg.

Müller-Boge, M. 1996. Die Neozoen im aktuellen Recht – Aussetzung und Einfuhr. In: Gebhardt, H., Kinzelbach, R. and Schmidt-Fischer, S. (eds.), Neozoen – neue Tierarten in der Natur. pp. 15-23. Ecomed, Landsberg.

Reichholf, J. H. 1996. Wie problematisch sind die Neozoen wirklich? In: Gebhardt, H., Kinzelbach, R. and Schmidt-Fischer, S. (eds.), Gebietsfremde Tierarten. pp. 37-48. Ecomed, Landsberg.

Sjöberg, G. and Hokkanen, H. M. T. 1996. Conclusions and recommendations of the OECD workshop on the ecology of introduced, exotic wildlife: Fundamental and economic aspects. Wildl. Biol. 2: 131-133.

Schink, A. 1989. Naturschutz- und Landschaftspflegerecht Nordrhein-Westfalen, Schriftenreihe des Freiherr-vom-Stein-Institutes Bd. 12, XXI. Deutscher Gemeindeverlag, Köln, Berlin, Hannover, Kiel, Mainz, München.

Sukopp, H. 1995. Neophytie und Neophytismus. In: Böcker, R., Gebhardt, H., Konold, W. and Schmidt-Fischer, S. (eds.), Gebietsfremde Pflanzenarten – Auswirkungen auf einheimische Arten, Lebensgemeinschaften und Biotope. pp. 3-32. Ecomed, Landsberg.

U.S. Congress, Office of Technology Assessment (eds.) 1993. Harmful Non-Indigenous Species in the United States, OTA-F-565. U. S. Government Printing Office. Washington, DC.

A CRITIQUE OF THE GENERAL APPROACH TO INVASIVE PLANT SPECIES

Keith R. Edwards
Institute of Botany, Czech Republic Academy of Sciences, 145 Dukelská, 37982 Třeboň, Czech Republic; e-mail: keith@butbn.cas.cz

Abstract

It is assumed that the establishment and spread of non-native, invasive species results in decreased diversity of native flora and fauna, possibly threatening rare and/or endangered species, and may change ecosystem processes. Accidental and deliberate introductions of these species also is leading to homogenization of the world's communities, blurring the boundaries between biogeographic realms. While many examples exist for invasive animal species in support of the first two arguments, such data are much less numerous for plant species. Usually, management actions are initiated against invasive plant species before ecological studies are conducted to determine the ecological values and problems associated with these species.

Management actions should be based on sound ecological information; the lack of such information for most invasive plant species means that such studies should become a priority. In addition, management focuses mostly on controlling a particular invasive species, and not on the disturbance which allowed it to establish originally; such management considers the effects of environmental degradation, but not the causes. A more holistic approach is recommended, placing the invasive species within an ecosystem context and incorporating larger spatial and temporal scales than are normally considered. Such an approach requires obtaining data about these upper level processes before formulating and implementing a management plan.

Introduction

The establishment of plants and animals into areas far from their native range has increased markedly in the last 300 years, due to increased human travel and trade, which allows for the possibility of any species from one corner of the globe being introduced anywhere else on Earth (Lodge 1993). This movement of species may represent a major component of global environmental change (Vitousek *et al.* 1997), by blurring the boundaries between biogeographic regions (Elton 1958). This degradation of the distinction between biogeographical realms is leading to the homogenization of the world's communities (Lodge 1993; Vitousek *et al.* 1997), presumably to be dominated by generalist species which are better able to survive and grow in human-impacted ecosystems. It has been hypothesized that the introduction of European plants, animals, and pathogens was instrumental in the success of European colonization of the Americas and Australia (Crosby 1986).

Colonization and change in a species' range have been, and still are, natural processes (Hengeveld 1989). However, it is this increased rate of invasion that alarms ecologists and natural resource managers. The rapid movement of plants and animals takes them out of their evolutionary context, releasing them from the constraints im-

Plant Invasions: Ecological Mechanisms and Human Responses, pp. 85–94
edited by U. Starfinger, K. Edwards, I. Kowarik and M. Williamson
© *1998 Backhuys Publishers, Leiden, The Netherlands*

posed by their natural control agents, with which the particular species evolved (Crawley 1989). This is thought to be a major reason for the success of certain introduced species and why invasive plants, in particular, are larger and more vigorous in their new area of secondary distribution compared to plants of the same species growing in their original native habitat (Crawley 1987; Blossey and Nötzold 1995).

There is an increasing body of literature concerned with invasions and invasive species (*e.g.* Mooney and Drake 1986; Drake *et al.* 1989; Di Castri *et al.* 1990; Pyšek *et al.* 1995; Williamson 1996). Numerous studies in these volumes describe the characteristics of invasive species or of the communities which are susceptible to invasion. Unfortunately, most of this literature is anecdotal, descriptive, or irrelevant to generating a greater understanding of, or allowing for better predictions about, invasive species (Mack 1996). Most of this literature is also focused on the single species of concern, and, thus, it is not comprehensive in its scope (Hobbs and Humphries 1995). Yet, it is these information sources which form the basis for management decisions and actions against invasive organisms.

The objective of this paper is to review the assumptions behind the current single-species focus of active management in regards to invasive plant species and to identify the weaknesses of this approach. A more holistic, ecosystem-level approach is then offered as being a more comprehensive and effective long-term strategy, by placing the invasive plant species into a larger ecological context, both spatially and temporally.

Reasons for management actions

Generally, ecologists and natural resource managers give several reasons for their concerns about invasive species. One reason often stated is that successful invasions lead to reductions in species diversity, with subsequent threats to rare and endangered species, perhaps leading to local extinction of native species; of course, this is considered to be a major ecological and environmental problem. Diversity may result in greater community stability, making the communities more resistant to colonization. One of Elton's (1958) four arguments in support of the diversity/stability hypothesis was the ease by which species-poor islands are successfully invaded by non-native species. There has been much theoretical support for this principle (MacArthur 1955; Bray 1958; Elton 1958; Margalef 1963; Pimm 1984), although there have been few empirical studies to validate it (McNaughton 1977; Tilman and Downing 1994). According to its proponents, any action or event that decreases diversity is considered to be deleterious to community structure and development.

The collected evidence indicates that invasions rarely lead to extinction of native species, especially in continental settings; colonization by invasive species more commonly leads to a restriction of the ecological range of the native species (Vermeij 1996). These observations usually consider only one group of life form, such as plants or birds, and fail to clearly state the spatial scale of observation, thus limiting the usefulness of this information. Daehler and Strong (1996) showed that the invasion of *Spartina* spp., especially *S. alterniflora*, in Pacific coastal estuaries of the western U.S. differentially impacted the diversity and density of different life forms. *Spartina* invasion appears to have a distinct negative impact on native plant species, but the effect is less clear on animal species. There was increased diversity and density of invertebrate species in *Spartina* patches, compared to areas with native plants. The effect on bird life was also not clear, as the impact appeared to be dependent on the state of the patch prior to invasion; previously bare patches had reduced bird use, while there was little impact in already vegetated areas.

A second reason put forth is that invasion by non-native species can lead to changes in ecosystem functions, which include resource use and acquisition, alterations to the disturbance regime, and alteration of the trophic structure (Williamson 1996). Such impacts to ecosystem functions are much better documented for animal and fungal invaders, especially in North America. An exception is the situation on islands, such as Hawaii, where the impacts of invasive plant species, including *Myrica faya*, are well-known. The establishment of *M. faya* leads to increased N content of the soil through its symbiotic relationship with a nitrogen-fixing actinomycete *Frankia* (Loope and Medeiros 1994; Williamson 1996). Such effects are not as well documented for continental regions (Vitousek *et al.* 1997). Certain plant species may affect ecosystem processes, such as *Bromus tectorum*, whose presence leads to increased fire frequency in western U.S. grasslands (Brandt and Rickard 1994). Daehler and Strong (1996) found that *Spartina* spp. invasion of Pacific coast estuaries resulted in increased sediment accumulation, which could restrict water flow and lead to a widening of the floodplain area. Also, Lesica and DeLuca (1996) showed that the non-native *Agropyron desertorum* had greater nutrient and water uptake and retention rates than the native *Agropyron* spp., leading to decreased species diversity and drier, more nutrient-poor habitats. The effects of *A. desertorum* invasions were noticed several decades after it was introduced as a grazing plant.

The above-stated reasons are provided as justifications for implementing management actions quickly, before there is much information collected concerning the ecological impacts and values of the particular invasive plant species (Usher 1988; Luken 1994). For example, the invasive species *Lythrum salicaria* (purple loosestrife) is a well-known invasive plant of temperate, North American wetlands (Rawinski 1982; Thompson *et al.* 1987; Malecki *et al.* 1993; Edwards *et al.* 1995; Edwards 1996). *Lythrum salicaria* was first introduced into eastern North America near the end of the eighteenth century, but only began to spread rapidly along riverways and ditches in the 1930's (Stuckey 1980). Control measures were first attempted in the 1940's and have continued uninterrupted since (Thompson *et al.* 1987). However, the first ecological studies were not done until the late 1960's (Shamsi and Whitehead, 1974), with most being performed since the late 1970's (Rawinski 1982; Thompson *et al.* 1987; Malecki *et al.* 1993; Edwards 1996). The effect of *L. salicaria* on wetland ecosystem processes has only recently begun to be investigated (Welling and Becker 1992). Currently, information concerning the impact of *L. salicaria* on rare and endangered species is mostly anecdotal, while there is a great lack of information regarding its effects on ecosystem processes.

Classical biological control

The situation described above appears to be improving in recent years, as witnessed by the contributions to this conference. One possible reason for the increase in ecological studies of invasive plant species may be the emergence of classical biological control as a main management technique for controlling successful non-native species.

The purpose of classical biological control is to introduce known enemies of the target species in its native range into the invaded community, for which ecological information is needed about the target species and the proposed control agents. Biological control assumes that a main reason that invasive species become successful is because they have been able to move ahead of their natural enemies, thereby providing the invasive species with a competitive advantage over native species. Also, biological control assumes a simple food web, which may not be valid (Strong 1997). The deliberate introduction and establishment of insect herbivores and pathogens are thought to provide greater and longer-term control, with fewer environmental risks, than chemical treatments (Malecki *et al.* 1993; Hobbs and Humphries 1995).

There are noted success stories with biological control, especially in agricultural settings. Other noteworthy successes include the control of Klamath weed (*Hypericum perforatum*) and alligator weed (*Alternanthera philoxeroides*) in the U.S. (Louda *et al.* 1997). However, concerns and criticisms of biological control have been raised:

1. Introducing non-native species as control agents removes them from their evolutionary and ecological contexts (Crawley 1989). The control agents may be released from the constraints imposed by their natural enemies. Rigorous testing of the effects of these agents on native species in the target area, prior to release, partially helps to alleviate the concerns (Howarth 1991). However, these tests may not be sufficient to illuminate all potential problems, as has been illustrated recently (Simberloff and Stiling 1996 a; Louda *et al.* 1997). Past attempts in the U.S. to strengthen regulations on the release of biological control organisms were defeated due to opposition from proponents for biological control (Beardsley 1997).

2. Further homogenization of the world's communities. The deliberate introduction of biological control agents into areas outside of their native range furthers the process of homogenization of the world's communities. Such introductions appear to run counter to goals of preserving native biodiversity (Luken 1994). The deliberate movement of species, for whatever reason, may be viewed as another manifestation of the mechanistic, Cartesian view that has been the prevailing dogma in Western societies for the last 300 years, when what is needed is a more holistic world view (Zohar 1991).

3. Effect of biological control agents on non-target species and communities. There is little information about the post-release effects of biological control agents on non-target native plants and communities. While many advocates for biological control state that there is no evidence of significant adverse ecological effects, critics counter that the lack of evidence does not mean that there are no effects, only that few studies have looked for these effects (Simberloff and Stiling 1996 b; Louda *et al.* 1997; Strong 1997). Recent reports (see Simberloff and Stiling 1996 b; Louda *et al.* 1997) suggest that adverse impacts may be more prevalent than thought previously. In a well researched study, Louda *et al* (1997) show the negative impact on native species of thistle by *Rhinocyllus conicus*, a weevil which was introduced into North America in 1968 as a biological control agent against non-native thistles. *R. conicus* has increased its range and is now attacking native thistle species, with the concern that it may soon attack an endangered thistle, *Cirsium pitcheri*, which is an endemic of the western Great Lakes region (Louda *et al.* 1997; Strong 1997). Another example is that of the snail *Euglandina rosea*, which was introduced to several Pacific islands to control the giant African snail, *Achatina fulica*, a successful invader of crops and gardens. Instead of attacking the targeted snail species, *E. rosea* attacked native tree snails of the genera *Partula* and *Achatinella*, in some cases leading to the extinction of the native snails on several islands (Williamson 1996).

4. Mutations. Biological control agents can evolve so to acquire new hosts or to tolerate a greater range of physical factors (Hopper *et al.* 1993; Simberloff and Stiling 1996 b).

A holistic approach

Management plans for invasive species focus on reducing the density of a particular species, but without addressing the underlying disturbance phenomena which allowed it to invade initially. In this way, management actions attack an effect of environmental degradation, but not the causes (Hobbs and Humphries 1995). This could lead managers to control for one invasive species, only to have another one colonize and establish in the affected community. A new control strategy would need to be developed specifically to combat the new invader. Single species approaches lock the manager into a cycle of controlling one species only to be faced with invasion of other, unwanted species (Luken 1994; Hobbs and Humphries 1995).

It makes more sense to manage for larger scale processes, because these constrain more local factors which reside within the system; the invasive species is placed within the larger context of managing for the ecosystem (Allen and Starr 1982; Allen and Hoekstra 1992; Denny 1992; Hobbs and Humphries 1995). This type of management approach would integrate natural and human aspects in formulating a comprehensive management plan. Such an approach has been evolving for some time, pushed by the knowledge that past, more narrowly-focused management approaches led to ineffectual or even undesired outcomes (Born and Sonzogni 1995). Concepts, such as ecosystem management, integrated environmental management, and watershed management, are similar attempts at devising a management approach that is more environmentally and ecologically aware, and stresses the interactive and interconnected natures of systems (Margerum and Born 1995). All of these concepts emphasize a need for: 1) coordinated control and direction of human activities towards achieving certain goals; 2) a course of action that emphasizes both natural and human resources in the system; and 3) an inclusive approach which emphasizes the scale and interconnectedness of ecological and human factors (Born and Sonzogni 1995).

These holistic concepts require that, first, the spatial and temporal boundaries encompassed in the management plan are clearly delineated and sufficiently broad to include all of the relevant activities in the system. Managers and other interested parties need to ensure that the scope of the management plan includes all of the critical components of the ecosystem, as well as socioeconomic and political components. Second, the connections linking these different components must be identified; this is usually done using analytical tools, such as geographic information systems (GIS) (Born and Sonzogni 1995; Margerum and Born 1995).

The results of these two steps is to produce a very complex picture of the particular issue. If left at this stage, the results would be too complicated to produce a pragmatic

and workable management plan; two other steps are required. First is the need to reduce all of these different and interconnected factors to only those which are critical
for producing a successful management plan. Once these factors are identified, efforts
are concentrated on the tasks which are aimed at these factors and lead to achieving
the stated goals (Margerum and Born 1995). The second step is to include all of the
interested and affected parties in the development and implementation of the plan. This
requires interaction among many entities, some of which may have conflicting views
and values. Coordination among these disparate entities involves exchanging information
through discussions, and official and informal meetings, and conflict resolution to mediate
differences between groups and to keep the process moving (Born and Sonzogni 1995).
This last step, which is an exercise in cooperative decision making, is an on-going activity
throughout the entire process and is very necessary to achieving the desired goals of
the project.

When faced with a successful invasive species, such an integrated approach will
require information not only about the autecology of the particular species, but about
how the invasive species fits into the larger community and landscape contexts. The
disturbance regime of the system will need to be determined, as well as the ecological
impact and value of the invasive species. These, and other, factors will need to be
determined prior to implementation of any control measures. In concert with the more
conceptual ideas stated above, such plans require that: 1) there are clearly defined goals;
2) there is a realistic assessment of what may be achieved (integrated management plans
are ideal situations which should be strived for but are rarely attained; Margerum and
Born 1995); 3) managers obtain the cooperation and involvement of affected landowners
and political entities; and 4) a long-term monitoring program be established (Jordan
et al. 1987; Denny 1992; Kentula *et al.* 1992; Grumbine 1994).

There are several examples of a more comprehensive management approach leading to effective control, especially of invasive plant species, where the initial disturbance was removed and natural ecosystem functions were restored.

1. Many portions of the Kissimmee River in Florida (USA) were channelized, os-
tensibly for flood protection. This led to the establishment of upland vegetation in former
wetlands adjacent to the river, and to the invasion of *Eichhornia crassipes* and *Pistia
stratiotes* in several of the now cut-off oxbows. Removal of the straightened channels
allowed for the resumption of a more natural water flow, flooding the previously drained
wetlands, and substantially decreasing the density of *E. crassipes* and *P. stratiotes* in
the original riverbed (Berger 1993).

2. *Melilotus alba* (white sweet clover) invades oak savanna communities in the upper
midwest USA, when the natural fire regime is disrupted. Resumption of a more natural fire regime led to control of *M. alba*, in addition to restoring the affected oak savanna communities (Kline 1983).

Of course, such an approach may not be as successful where the invasive species
affects ecosystem functions. A good example is the case of *Bromus tectorum* in the
grasslands of the western U.S. This annual grass became dominant following the start
of cattle grazing. The trampling of the native grass species by the cattle, and increased
soil nutrient loading, allowed *B. tectorum* to invade and dominate these grasslands. It
was apparent that intensive cattle grazing was the disturbance which led to the establishment and dominance of *B. tectorum* in these communities; thus, the logical management plan was to remove the cattle and allow the native species to re-colonize these
communities. However, removal of cattle did not result in decreased density of *B. tec-*

torum; in one case, *B. tectorum* is still the dominant 40 years after cattle removal (Brandt and Rickard 1994). When confronted with this type of situation, the manager needs to focus on removing the invasive species directly, and not only any overlying disturbance.

All of the above approaches emphasize active management strategies. However, sometimes it may be better to allow the invasion to run its course naturally. As long as there are no further disturbance phenomena, it would be expected that, over time, the interactions between the invasive species and the native flora and fauna would approach a condition of equilibrium. This places the invasion into a historical perspective (Usher 1988; Allen and Hoekstra 1992). Prowse (1998) provides a good example of the evolution of such interactions between the invasive *Impatiens glandulifera* in Great Britain and native British species. Usher (1988) reported that there was evidence suggesting that such a relationship developed between the introduced mink (*Mustela vison*) and the common eider (*Somateria mollisima*) in southern Sweden. *Elodea canadensis* successfully invaded European waterways in the nineteenth century, resulting in choked river channels and impeding navigation. No control methods successfully reduced *E. canadensis* density; however, *E. canadensis* declined beginning in the 1920's. It is still uncertain what factors caused this decline (Elton 1958).

Similarly, *Myriophyllum spicatum* successfully invaded many lakes and waterways in North America (Patten 1956; Nichols 1975). In the 1970's, *M. spicatum* populations declined in several areas. Carpenter (1980), studying *M. spicatum* in Lake Wingra, Wisconsin, USA, noted that the invasion followed a wave pattern, with invasion and dominance followed by a sustained decline; the process took ten years approximately. The decline appeared to be the result of several interacting factors, including nutrient supply, epiphyte abundance, competition from native macrophytes, and the actions of parasites and/or pathogens (Carpenter 1980).

Conclusion

The purpose of this paper was to show the limitations of the traditional approach to invasive species. This does not mean that I view cavalierly the successful introduction of non-native species, with the possible decrease in diversity and density of native species. It is just that I believe that we need to be less emotional in our approach to invasive species and place them within larger spatio-temporal, as well as historical, contexts than has been done previously. A comprehensive, integrative, holistic management approach offers a better means of obtaining long-term control of invasive plant species, with less chance of deleterious side effects, than the single-species approach, which is current management practice. An integrative approach places the invasion within larger spatial and temporal scales than is the case normally, and requires obtaining data about the ecological impact and value of the particular invasive species, as well as the upper level processes that constrain the invasive population, prior to formulating and implementing a management plan. Clear-cut and achievable goals are set, with the requisite amount of resources needed to successfully achieve them designated; more disturbed areas will probably require greater resource inputs (Christensen *et al.* 1996). In the end, it is a question of the temporal and spatial scale at which one is observing the processes of colonization and establishment, and the value judgements of the observer (Allen and Hoekstra 1992; Eser 1998), which will determine whether a particular invasion is considered to be part of a natural extension of a species range, and allowed to proceed, or a 'malignancy' which must be controlled and managed (Lodge 1993).

Acknowledgements

I thank all of the people with whom I discussed and debated these ideas, thereby resulting in more focused and reasoned arguments (I hope). Special thanks goes to Uwe Starfinger, for not only encouraging me to present this contribution at the Fourth Conference on the Ecology of Invasive Plant Species, but who listened patiently to my rantings on the subject, and to those people in Tim Allen's lab at the University of Wisconsin for the intellectual stimulation that is a constant source of enjoyment. Additional thanks go to Uwe Starfinger and Ingo Kowarik for their comments on the draft, making it a better and more thorough paper.

References

Allen, T.F.H. and Hoekstra, T.W. 1992. Toward a Unified Ecology. Columbia University Press, New York.

Allen, T.F.H. and Starr, T.B. 1982. Hierarchy: Perspectives for Ecological Complexity. University of Chicago Press, Chicago.

Beardsley, T. 1997. Biological noncontrol. Sci. Amer. 277: 17-19.

Berger, J.J. 1993. Ecological restoration and non-indigenous plant species: A review. Restor. Ecol. 1: 74-82.

Blossey, B. and Nötzold, R. 1995. Evolution of increased competitive ability in invasive nonindigenous plants: A hypothesis. J. Ecol. 83: 887-889.

Born, S.M. and Sonzogni, W.C. 1995. Integrated environmental management: Strengthening the conceptualization. Environ. Manag. 19: 167-181.

Brandt, C.A. and Rickard, W.H. 1994. Alien taxa in the North American shrub-steppe four decades after cessation of livestock grazing and cultivation agriculture. Biol. Conserv. 68: 95-105.

Bray, J.R. 1958. Notes toward an ecological theory. Ecology 39: 770-776.

Carpenter, S.R. 1980. The decline of *Myriophyllum spicatum* in a eutrophic Wisconsin lake. Can. J. Bot. 58: 527-535.

Christensen, N.L., Bartuska, A.M., Brown, J.H., Carpenter, S., D'Antonia, C., Francis, R., Franklin, J.F., MacMahon, J.A., Noss, R.F., Parsons, D.J., Peterson, C.H., Turner, M.G. and Woodmansee, R.G. 1996. The report of the Ecological Society of America committee on the scientific basis for ecosystem management. Ecol. Appl. 6: 665-691.

Crawley, M.J. 1987. What makes a community invasible? In: Gray, A.J., Crawley, M.J. and Edwards, P.J. (eds.), Colonization, Succession, and Stability. pp. 429-453. Blackwell Scientific Publications, Oxford.

Crawley, M.J. 1989. Herbivores and plant population dynamics. In: Davy, A.J., Hutchings, M.J., and Watkinson, A.R. (eds.), Plant Population Ecology. pp. 367-392. Blackwell Scientific Publications, Oxford.

Crosby, A.W. 1986. Ecological Imperialism: The Biological Expansion of Europe. Cambridge University Press, Cambridge, 900-1900 pp.

Daehler, C.C. and Strong, D.R. 1996. Status, prediction and prevention of introduced cordgrass *Spartina* spp. invasions in Pacific estuaries, USA. Biol. Conserv. 78: 51-58.

Denny, P. 1992. An approach to the development of environmentally sensitive action plans for river floodplain management in central and eastern Europe. In: Finlayson, M. (ed.), Integrated Management and Conservation of Wetlands in Agricultural and Forested Landscapes. pp. 46-49. IWRB Special Publication, Number 22, Slimbridge, Gloucester, England.

Di Castri, F., Hansen, A.J. and Debussche, M. 1990. Biological Invasions in Europe and the Mediterranean Basin. Kluwer Academic Publishers, Dordrecht.

Drake, J.A., Mooney, H.A., di Castri, F., Groves, R.H., Kruger, F.J., Rejmánek, M. and Williamson, M. 1989. Biological Invasions: A Global Perspective. SCOPE 37. John Wiley and Sons, Chichester.

Edwards, K.R. 1996. Comparative Study of Native European and Invasive North American Populations of *Lythrum salicaria* L. Ph.D. Dissertation, University of Wisconsin. Madison, Wisconsin.

Edwards, K.R., Adams, M.S. and Květ, J. 1995. Invasion history and ecology of *Lythrum salicaria* in North America. In: Pyšek, P. Prach, K., Rejmánek, M., and Wade, M. (eds.), Plant Invasions: General Aspects and Special Problems. pp. 161-180. SPB Academic Publishing, Amsterdam.

Elton, C.S. 1958. The Ecology of Invasions by Animals and Plants. Methuen, London.

Eser, U. 1998. Assessment of plant invasions: Theoretical and philosophical fundamentals. In: Starfinger, U., Edwards, K., Kowarik, I., and Williamson, M. (eds.), Plant Invasions: Ecological Consequences and Human Responses. pp. 95-107. Backhuys Publishers, Leiden.

Grumbine, R.E. 1994. What is ecosystem management? Conserv. Biol. 8: 27-38.

Hengeveld, R. 1989. Dynamics of Biological Invasions. Chapman and Hall, London.

Hobbs, R.J. and Humphries, S.E. 1995. An integrated approach to the ecology and management of plant invasions. Conserv. Biol. 9: 761-770.

Hopper, K.R., Roush, R.T. and Powell, W. 1993. Management of genetics of biological-control introductions. Ann. Rev. Entom. 38: 27-51.

Howarth, F.G. 1991. Environmental impacts of classical biological control. Ann. Rev. Entom. 36: 485-509.

Jordan, W.R., Gilpin, M.E. and Aber, J.D. 1987. Restoration ecology: Ecological restoration as a technique for basic research. In: Jordan, W.R., Gilpin, M.E. and Aber, J.D. (eds.), Restoration Ecology: A Synthetic Approach to Ecological Research. pp. 3-21. Cambridge University Press, Cambridge.

Kentula, M.E., Brooks, R.P., Gwin, S.E., Holland, C.C., Sherman, A.D. and Sifneos, J.C. 1992. An Approach to Improving Decision Making in Wetland Restoration and Creation. Island Press, Washington, D.C.

Kline, V. 1983. Control of sweet clover in a restored prairie (Wisconsin). Restor. Manag. Notes 1: 30-31.

Lesica, P. and DeLuca, T.H. 1996. Long-term harmful effects of crested wheatgrass on Great Plains grassland ecosystems. J. Soil Water Conserv. 51: 408-409.

Lodge, D. 1993. Biological invasions: Lessons for ecology. TREE 8: 133-137.

Loope, L.L. and Medeiros, A.C. 1994. Impacts of biological invasions on the management and recovery of rare plants in Haleakala National Park, Maui, Hawaiian Islands. In: Bowles, M.L. and Whelan, C.J. (eds.), Restoration of Endangered Species. pp. 143-158. Cambridge University Press, Cambridge.

Louda, S.M., Kendall, D., Connor, J. and Simberloff, D. 1997. Ecological effects of an insect introduced for the biological control of weeds. Science 277: 1088-1090.

Luken, J.O. 1994. Valuing plants in natural areas. Nat. Areas J. 14: 295-299.

MacArthur, R.H. 1955. Fluctuations of animal populations, and a measure of community stability. Ecology 36: 533-536.

Mack, R.N. 1996. Predicting the identity and fate of plant invaders: Emergent and emerging approaches. Biol. Conserv. 78: 107-121.

Malecki, R.A., Blossey, B., Hight, S., Schroeder, D., Kok, L. and Coulson, J. 1993. Biological control of purple loosestrife. Bioscience 43: 680-686.

Margalef, R. 1963. On certain unifying principles in ecology. Am. Nat. 97: 357-374.

Margerum, R.D. and Born, S.M. 1995. Integrated environmental management: Moving from theory to practice. J. Environ. Plan. Manag. 38: 371-391.

McNaughton, S.J. 1977. Diversity and stability of ecological communities: A comment on the role of empiricism in ecology. Am. Nat. 111: 515-525.

Mooney, H.A. and Drake, J.A. 1986. Ecology of Biological Invasions of North America and Hawaii. Springer-Verlag, New York.

Nichols, S.A. 1975. Identification and management of Eurasian watermilfoil in Wisconsin. Trans. Wis. Acad. Sci. Arts Let. 63: 116-128.

Patten, B.C. 1956. Notes on the biology of *Myriophyllum spicatum* L. in a New Jersey lake. Bull. Torrey Bot. Club 83: 5-18.

Pimm, S. 1984. The complexity and stability of ecosystems. Nature 307: 321-326.

Prowse, A. 1998. Patterns of early growth and mortality in *Impatiens glandulifera*. In: Starfinger, U., Edwards, K., Kowarik, I. and Williamson, M. (eds.), Plant Invasions: Ecological Consequences and Human Responses. pp. 245-252. Backhuys Publishers, Leiden.

Pyšek, P., Prach, K., Rejmánek, M. and Wade, M. 1995. Plant Invasions: General Aspects and Special Problems. SPB Academic Publishing, Amsterdam.

Rawinski, T.J. 1982. The Ecology and Management of Purple Loosestrife (*Lythrum salicaria* L.) in Central New York. M.S. Thesis, Cornell University. Ithaca, New York.

Shamsi, S.R.A. and Whitehead, F.H. 1974. Comparative eco-physiology of *Epilobium hirsutum* L. and *Lythrum salicaria* L. I. General biology, distribution, and germination. J. Ecol. 62: 279-290.

Simberloff, D. and Stiling, P. 1996 a. How risky is biological control? Ecology 77: 1965-1974.

Simberloff, D. and Stiling, P. 1996 b. Risks of species introduced for biological control. Biol. Conserv. 78: 185-192.

Strong, D.R. 1997. Fear no weevil? Science 277: 1058-1059.

Stuckey, R.L. 1980. Distributional history of *Lythrum salicaria* (purple loosestrife) in North America. Bartonia 47: 3-20.

Thompson, D.Q., Stuckey, R.L. and Thompson, E.B. 1987. Spread, Impact, and Control of Purple Loosestrife (*Lythrum salicaria*) in North American Wetlands. Fish and Wildlife Research Report 2, U.S. Department of Interior, Fish and Wildlife Service, Washington, D.C.

Tilman, D. and Downing, J.A. 1994. Biodiversity and stability in grasslands. Nature 367: 363-365.

Usher, M.B. 1988. Biological invasions of nature reserves: A search for generalisations. Biol. Conserv. 44: 119-135.

Vermeij, G.J. 1996. An agenda for invasion biology. Biol. Conserv. 78: 3-9.

Vitousek, P.M., D'Antonio, C.M., Loope, L.L., Rejmánek, M. and Westbrooks, R. 1997. Introduced species: A significant component of human-caused global change. N. Z. J. Ecol. 21: 1-16.

Welling, C.H. and Becker, R.L. 1992. Life History and Taxonomic Status of Purple Loosestrife in Minnesota: Implications for Management and Regulation of this Exotic Plant. Special Publication Number 146, Minnesota Department of Natural Resources.

Williamson, M. 1996. Biological Invasions. Chapman and Hall, London.

Zohar, D. 1991. The Quantum Self. Flamingo, London.

ASSESSMENT OF PLANT INVASIONS: THEORETICAL AND PHILOSOPHICAL FUNDAMENTALS

Uta Eser

Universität Tübingen, Zentrum für Ethik in den Wissenschaften, Keplerstr. 17, D-72074 Tübingen, Germany; e-mail: uta.eser@uni-tuebingen.de

Abstract

There is a tendency among ecologists to regard biological invasions as obviously negative. To question value-judgements based on 'ecological evidence' I present an analysis of the impact of non-scientific values and norms on the scientific writing about introduced plants. Special concern is given to the suspicion of a xenophobic bias. Scientific terms like 'alien' or 'invasion' bear negative connotations that might influence perception and evaluation. Furthermore, the terminology often reflects a conservation bias. Like the term 'weed' the terms 'invasive' or 'neophyte' denote interference with conservation or management goals. The concept of the plant community also influences assessments. In an organism-like community, intruders necessarily are seen as afflicting the health or integrity of the whole. The concept of disturbance or the hypothesis of natural stability refer to an ideal of nature as a harmonic cosmos. From this perspective, human interventions necessarily are conceived as destructive. Conservationists tend to idealize pristine nature as intrinsically good, harmonic and stable. The opposite image of a nature 'red in teeth and claws' suggests that nature must be controlled and subjected by humans. As spreading introduced plants are neither pristine nor controllable, they are conceived as negative from the perspective of conservation. I claim that the replacement of original vegetation by a new species may not *per se* be assessed negatively, but needs further reasons. The conservation of biodiversity or endangered species is a reasonable argument, the preference for natives is not.

Introduction

Undoubtedly, the introduction of non-native plants can have significant consequences for ecosystems. Ecology can study these effects, but, as a natural science, is limited to their description and explanation. Management decisions, however, require not only mere descriptions of environmental impacts, but their assessment. Whether impacts have to be interpreted as damages depends on scientific research as well as on values and criteria that are less obvious. How to assess plant invasions and how to deal with them is, therefore, still highly controversial.

Whereas invasions of agricultural or pasture land can simply be assessed in terms of economic losses, it is much more difficult to evaluate invasions into natural habitats. Just to adduce the costs of eradication measures is unsatisfactory because they already take for granted what had to be shown: that some kind of ecological damage has happened. The criteria according to which one might determine definitively "ecological damage" are hard to define.

Nevertheless, in some publications the assessment of plant introductions seems to require no further explanation. The mere fact of reduction of species diversity, and replacement of "natives" by "non-natives", is assumed to be undesirable.

Plant Invasions: Ecological Mechanisms and Human Responses, pp. 95–107
edited by U. Starfinger, K. Edwards, I. Kowarik and M. Williamson
© 1998 Backhuys Publishers, Leiden, The Netherlands

"The replacement of a native wetland plant community by a monospecific stand of an exotic weed does not need refined assessment to demonstrate that a local ecological disaster has occurred" (Thompson *et al.* 1987).

If this statement would be right, any further debate about the involved criteria would be unnecessary. But there are cases where even scientists do not agree whether or not certain species are damaging. In July 1997, a controversy took place in the e-mail-server list 'Aliens' about the question *"How many preliminary work must be conducted on a given plant species before it can be deemed 'threatening' and programs initiated for the control or eradication of this species?"* While some participants endorsed preventive measures for the sake of general preservation goals, others asked for more detailed empirical research and a case-by-case approach due to the uniqueness of every single event. But they all seemed to agree that the appropriate evaluation of new species depended on good science. Against this opinion, I will argue in this paper that assessments of ecological impacts of human actions, like species introductions, can never be a matter of science solely. Empirical quantitative studies are a necessary but not a sufficient condition for assessments. These require not only scientific evaluations but also value-judgements[1]. Therefore, they involve criteria that cannot be found empirically. Well-founded assessments should make these criteria explicit.

Accordingly, I start with some remarks on facts and values in ecology and the concept of scientific freedom from values. Then, I explain the method of textanalysis that I used to find explicit or implicit value statements in papers about species introductions. Presenting some results, I will highlight terms and concepts that bear normative implications. These will be discussed with regard to the aims of nature conservation. Last, I consider ethical questions as the responsibilty of scientists and the characteristics of human actions.

Facts and values in ecology

According to the German sociologist Max Weber, science can but describe what is, not what should be or ought to be done. Value judgements, meaning "practical evaluation of facts as desirable or undesirable" are not a question of science (Weber 1917: 499). This statement is known as Weber's thesis of 'science's freedom from values' (*Wertfreiheitsthese*). This epistemological claim is the first argument against the idea that a better scientific description of their consequences would *per se* facilitate the assessment of species introductions. Certainly, better knowledge will and should influence evaluation as well as legislation. But still, the evaluation of any empirical fact depends on more or less subjective preferences and values.

Although the separation of facts and values is a reasonable norm for science it is contrary to scientific practice. Historians and philosophers of science have shown that science is deeply rooted in, and dependent on, the historical context (Fleck 1993, Kuhn 1976). As a social enterprise, science is necessarily biased. Besides its own constitu-

---

[1] Assessment procedures should discriminate between 'evaluation' and 'value-judgement'. Whereas *evaluation* should concern (empirical) judgements of 'factually true or false', *value-judgement* should be limited to (moral) judgements of 'right or wrong' (Eser and Potthast 1997).

tional values, it also implies decisions about contextual values and, thus, takes normative stances (Steen 1995). Taking stances is not a bad thing in itself. A conservationist bias, for example, might be all right for an ecologist. But, following Weber, such a personal conviction has to be made explicit, so that others can decide whether or not they agree with the conclusions.

The topic of non-indigenous species tends to be biased in a particular and, in my opinion, somewhat irritating way. Let me give two examples: the German popular magazine 'natur' titled in 1991:

> "Green invaders. Foreigners on their way to success. Alien plants override German herbs" (Finck 1991, translation by UE).

A newsletter by the Park Service of the Grand Canyon (Arizona, USA) informs the visitor:

> "Alien Invaders: The Grand Canyon is under attack from alien plants! You can help protect our native plants by joining the habitat restoration team in removing these nasty invaders" (Grand Canyon Visitor Information, April 1996).

Sure enough, these statements are not scientific, but want to raise consiousness for problems caused by introduced species or motivate people to help prevent their further spread. Nevertheless, the language that is used appeals to xenophobic, nationalist and racist feelings that have to be rejected. If scientific assessments are based on such prejudices as well, they ought to be questioned in public.

Fortunately, there is a growing awareness among scientists about this problematic tendency in talking about 'natives' and 'aliens'. For example, James Brown stated in the global volume of the SCOPE-Programme on Biological Invasions:

> "There is a kind of a irrational xenophobia about invading animals and plants that resembles the inherent fear and intolerance of foreign races, cultures, and religions. [...] This xenophobia needs to be replaced by a rational, scientifically justifiable view of the ecological roles of exotic species" (Brown 1989: 105).

Other ecologists try to eliminate offensive terms from the debate (Garthwaite 1993, Binggeli 1994) and explicitly reject unreflected patriotism (Reichholf 1996). But still, many scientists do not see that scientific concepts and theories themselves reflect certain worldviews and ideas and are therefore value-laden.

Analyzing texts about plant invasions

Unlike most of the papers that were given in the '4th Conference on the Ecology of Invasive Alien Plants', my paper is not directly concerned with introduced plant species. Rather, it deals with texts about these species. This is an important differentiation: I'm not writing about things "out there", in nature, but about something existing in the texts of scientists and/or environmentalists. Therefore, my method is mainly one of linguistics. I searched through scientific publications about non-native plants in order to find hints at non-scientific influences on the scientific writing. My material are specific papers about species that are considered problematic in the European context (Schwabe and Kratochwil 1991, De Waal *et al.* 1994, Hartmann *et al.* 1994, Cronk and Fuller 1995, Pyšek *et al.* 1995), as well as more general and international publications, especially resulting from the SCOPE-Programme on Biological Invasions (Groves and Burdon 1986, Kornberg and Williamson 1986, MacDonald *et al.* 1986; Mooney and Drake 1986; Joenje 1987; Drake *et al.* 1989; Mooney and Drake 1989; Di Castri *et al.* 1990).

I used the following methods:
1. Characterisation of the semantic field of relevant concepts related to species introductions by using polar opposites, according to Hard (1969). I tried to figure out how different opposites, like nature vs. culture, natural vs. unnatural, alien vs. native and wild vs. cultivated, form part of the concept of 'biological conservation' on one hand and of 'alien invasives' on the other.
2. I interpreted signs that are used in the texts as tracks, supposing that also scientific terms not only denote one precise meaning, but also bear many connotations. Using these connotations, the signs in a text can be read as tracks of a hidden notion (Eco 1977, Ginzburg 1980, Hard 1995). The method of following tracks is a kind of unconventional reading. Associating scientific terms with different contexts, I tried to find unintended messages of intentional phrases.

Hidden values in scientific papers and concepts

Biased terminology

Against the ideal of value-free science, the terminology concerning introduced species reflects a conservation bias. Cronk and Fuller (1995) frankly admit this bias:

> "This work approaches the subject in the context of conservation and concentrates on plant invasions as a threat to wild biodiversity. The definition of invasive plants here [...] reflects this conservation bias [...] (Cronk and Fuller 1995: xiii).

Although many different definitions of the term 'invasive' are in use, most of them include a tendency for the species to expand and this expansion is seen as bearing negative consequences for the native flora and fauna (Pyšek 1995). The same is true for the term *'neophyte'*, which is prevalent in Central Europe. In contrast to its scientific defintion, the term *'neophyte'* in its common use not only means that the introduction occurred less than 500 years ago, but also that the species tends to invade natural areas and to replace native vegetation (e.g. Strohschneider 1991). Sukopp (1995) found that 90 % of the participants of a conference on introduced plant species used the term *'neophyte'* in association to 'naturalization'. This is exactly how Thellung had defined the term at the beginning of the century (Thellung 1918/19). Thus, to call a plant species 'invasive' or *'neophyte'* already means to evaluate it as non-desirable from the perspective of nature conservation.

Every-day-language and scientific language

Besides the conservation bias of the ecology of "invasives", ecological papers in general comprise lots of expressions that also have non-scientific meanings and bear many connotations. To talk about balance of nature, about circles and interconnectedness, about diversity and stability, or about disturbance and catastrophes, does not leave the scientist emotionally unaffected. Terms from different contexts are imported into science, and they still carry their original meaning as connotation. As some of these connotations also concern moral feelings and values, they can influence the assessment.

Elton's term 'invasion' has been criticized as militaristic from the very beginning of the SCOPE programme (Groves and Burdon 1986). Nevertheless, it is still the most common term in articles explicitly dealing with non-indigenous plants (Pyšek 1995). The term 'invasion' hardly can be regarded as value-free. According to Webster's New Encyclopedic Dictionary (1993), 'invasion' means:

> "1. [...] entrance of an army into a country for conquest; 2. [...] the entrance or spread of some usually harmful thing".

To term a plant species 'invasive', due to this common meaning of the term, evokes the idea that its expansion has to do with aggression and destruction. The ability of an introduced plant to successfully colonize new areas, and outcompete other species, still is characterised as "aggressiveness" in many texts (e.g. Harper 1965). Schwabe and Kratochwil (1991) describe *Solidago gigantea* as growing with "intolerant" polycormones and explicitly call its tendency to expand "aggressiv". These notions of aggressiveness and harm necessarily lead to the assumption that invasive species pose a threat to others.

Implications of the concept of community

Not only single terms, but also whole scientific concepts have normative consequences. For example, the concept of the plant community has impacts on the assessment of new plants. The species composition of a given plant community can either be explained by its physical environment or by the internal structure of the community. Although community and environment influence each other, different concepts tend to lay stress on the first aspect or on the second (Trepl 1987). The so-called individualistic concept of the plant community (Gleason 1926) regards plant communities mainly as a product of chance. The availability of resources is seen as the most important factor, and competition, therefore, as the main relation between the members of the community. In an individualistic concept, any plant able to colonize a given area, and to utilize the resources, is regarded as part of the community. In the Clementsian tradition, the community is conceptualized rather holistically, as an organism (Clements 1916). In this concept, all species form part of the community as a whole, and this whole is more than the mere sum of its parts. Hence, the removal or addition of one part necessarily seems to afflict the whole, the community. Bearing an organismic concept in mind, a new species can only be perceived as damaging.

Such an organismic concept of the community is hidden in the concept of 'resistance'. Only an organism has mechanisms to repulse intruders that don't belong to it.

Originally belonging to a medical context, the concept of immunity is related to the concept of disease. It is not only descriptive, but normative. Immunity is understood as an active process of an organism and requires discrimination between 'own' and 'alien', between friend and enemy (Zimmermann 1996). Its weakness or breakdown renders possible the entrance of germs. In the (normative) context of health and disease, the successful intrusion of aliens into an organism means damage. The term "ecosystem-health" (Rapport 1989, Kolasa and Pickett 1992) clearly refers to this medical context, too.

Traditional images of nature

Assessments of species introductions not only depend on the concept of the community, but also on certain images of nature. Elton (1958) regarded natural communities as adapted to the potential of their habitat. This means, after a sufficient time of evolution, every possible niche is occupied. In such a saturated community, there is no room for newcomers. It is, therefore, resistant to invasions. The stability of a natural community is interpreted as a result of species diversity. This concept is clearly related to the idea of a balance of nature. This idea is quite often reflected in ecological theories and even more often in their popular receptions (Jansen 1972, Egerton 1973, Pimm 1993). It resembles the premodern idea of a harmonic order of nature. This kind of Nature has a *telos* and, therefore, is a normative concept (Heiland 1992). According to this concept, human behaviour has to be oriented following the order of the cosmos. The supposition of a natural order necessarily leads to the evaluation of human-made changes, like the introduction of new species, as causing disorder, as destructive.

   The notion of an intrinsically good 'natural' state can also be found in the concept of disturbance. Disturbances are seen as the most important means to overcome the resistance of a natural community.

> "There is no invasion of natural communities without disturbance" (Fox and Fox 1986: 65).

Although there is evidence today that natural habitats are open to invasions, disturbance is still considered one of the most important factors for the success of introduced species:

> "There is abundant (although largely anecdotal) evidence supporting the assertion that plant communities are generally more invasible when they are subject to some form of disturbance" (Burke and Grime 1996).

The concept of disturbance is value-laden, too. The term itself is easy to associate with disruption, destruction, intervention, and the like. Also the common scientific definition of 'disturbance' comprises the term 'disruption':

> "A disturbance is any relatively discrete event in time that *disrupts* ecosystem, community, or population structure and changes resources, substrate availability, or the physical environment" (Pickett and White 1985: 7; italics by UE).

One of the meanings of the term 'disrupt' is "to throw into disorder" (Webster's 1993). This means, the term 'disruption' is related to some kind of order. Explaining successful invasions by disturbance, theories on biological invasions tend to refer to a natural order that is thought to prevent invasions. Fox and Fox (1986) explain the effect of disturbance with the creation of spare resources. These render possible the establishment of new species:

> "New resources may be *utilised* by native species in the community or may be *exploited* by new species, either native or introduced" (Fox and Fox 1986: 57; italics by UE).

In contrasting "utilisation" and "exploitation" native species are conceived as part of the natural balance, whereas new species are regarded as disrupting this order. This theory obviously refers to Elton's concept of saturated natural communities. It can, therefore, be interpreted as a hint at the underlying image of a harmonically ordered nature that humans can only destroy.

Introduced species, nature conservation and ethics

The presented textanalysis revealed some assumptions in the literature on invasions that cannot withstand a critical discussion. Nevertheless, the spread of invasive species might threaten objectives of nature conservation. These aims will be discussed in the following section.

What does nature conservation mean?

The above cited statement by Fox and Fox (1986) suggested that natural communities are free from disturbances. Contrary to this opinion, there seems to be evidence today that natural systems are in permanent change (Pickett and White 1985). Landscapes are heterogeneous both in space and time. What, then, can the idea of the conservation of nature mean (Sprugel 1991)? Especially, what can it mean in the context of species introductions? If a new species invades an area where it hasn't grown before and, due to its competitive ability, replaces the former vegetation this can be seen as a natural process. However, it is a process that is regarded as undesirable in terms of nature conservation. So what are the foundations of this value-judgement?

Most of the arguments against the introduction of new species come down to this one: that introduced species are a threat to global and local species diversity and to endangered species (Elton 1958; Jäger 1977; MacDonald *et al.* 1989; Starfinger 1991; Trepl and Sukopp 1993; Cronk and Fuller 1995; McNeely *et al.* 1995; Kowarik 1996). To regard the replacement of native plants or plant communities by an introduced one as negative, one must consent to the claim that we should preserve as many species as possible. Preservation of species diversity is a reasonable aim for conservationists, if reasoned from a utilitarian or aesthetic or moral basis. If this aim is accepted and if it is afflicted by a certain introduced plant, this can provide a strong argument for the control of the plant's further spread. However, this assessment is only valid for one particular species in one particular region. Every new case requires a new evaluation.

In Central Europe, the extinction of native species by introduced plant species is not the main problem of nature conservation. Some of the most endangered species themselves are not native but owe their existence in the area to human activities. In the European context, the appropriate argument, therefore, is not the preservation of some 'natural' state or of species diversity in general, but of the traditional cultural landscape and of endangered species. The preservation of such a particular state of nature requires arguments that involve human interests and needs.

The idea of preservation of nature is based on two contradictory images of nature. The first is the ideal of a pristine nature that is regarded as intrinsically good, sometimes even in a moral sense of the word. Human activities are supposed to spoil this natural integrity. I've shown above that one can find this image in some ecological concepts, too. Valuable nature, in this sense, is nature that doesn't show human marks. If natural areas are invaded by non-indigenous species, they can no longer be perceived as pristine. Revealing the touch of humans, introduced plants spoil the illusion of natural purity and, thus, diminish the value of the place.

On the other hand, nature is regarded as wild and dangerous. To survive, it is of existential importance for humans to gain control over the forces of nature. From this perspective, only nature under man's dominion, the cultivated landscape, can be valuable. The idea of 'Heimat' is still very helpful to understand this aspect of nature conservation. Historically, in Germany, the idea of nature conservation has been part of the broader concept of the conservation of 'Heimat'. 'Heimat' means the place, where people feel at home. It is not pure nature but a place where humans and nature live together in harmony, dependent on each other. 'Heimat' not only provides a home for her inhabitants, but also grants security and identity.

Possibly the main reason for the negative assessment of the expansion of introduced plants is that they do not fit into any of our favorite images of nature. First, they obviously are not pristine nature. If they are natural at all, they are a symbol of 'bad' nature. Thus, they spoil the idea of a natural harmony. Monospecific stands of an introduced plant do not fit into the ideal of a balanced, diverse and stable nature. In resisting control efforts, they are nature beyond human control, a threat not only to ecosystems but also to humans. In short, 'invasive' species are not too unnatural, but too natural to be an object of conservation. Second, spreading non-indigenous plants are not a part of 'Heimat' in every sense of the word. They are 'aliens', they "don't belong" (Smith 1989), they are unfamiliar to the people. They seem to change the landscape more rapidly than humans are able to adapt to. Thus, they afflict the major function of Heimat: to guarantee stability, safety and identity.

Ethical considerations

The assessment of species introductions is deeply related to much broader questions, such as: What is nature, what are humans, and what kind of relationship should humans and nature have? Such questions are the objective of environmental ethics as well. Answers to these questions are part of everybody's personal belief system. In a pluralistic world, there is no reason to assume that there is one universally right answer. Therefore, questions of right and wrong in environmental ethics and policies still are, and necessarily have to be, a matter of an ongoing discourse.

Scientists play an important role in the making of public opinion. It surely is part of scientists' responsibility to raise the public awareness of risks related to human actions. By providing a better understanding of ecological consequences of species introductions, ecologists and field biologists can, and should, contribute to political decisions concerning these species. But, due to the great influence on public opinion, scientists should be aware of the historic and social context in which they speak and act. This means, they should carefully choose their language and consider the possibility of a biased perception before they draw political conclusions from their theoretical and empirical knowledge.

Some statements in discussions about control efforts or measures of prevention show an unfortunate and unnecessary undertone of 'native is best', that is unappropriate to the question at stake. It is also part of scientific responsibility to object to such statements.

Another unhelpful undertone in the debate about species introductions is the notion of antihumanism. There is a tendency in the environmentally concerned literature, by scientists as well as by philosophers, to regard humans as a pest. Charles Elton interpreted so-called human overpopulation as the real problem behind biological invasions:

> "The reason behind this, the worm in the rose, is quite simply the human population problem. The human race has been increasing like voles or giant snails, and we have been introducing too many of ourselves into the wrong places" (Elton 1958: 144).

In a more recent book about 'biological pollution', Warren Wagner writes:

> "The species *Homo sapiens* itself is without question the super invader of all time" (Wagner 1993: 3).

As I have shown above, such a misanthropic view is the result of opposing humans to a harmonic and balanced nature. From an ethical perspective, it is not very helpful to regard humans as a species that simply has outgrown natural regulatory mechanisms. Obviously, humans are also a biological species and their behaviour might still be influenced by their evolutionary inheritance. But the interpretation of environmental problems like air pollution, habitat degradation and species extinctions, merely in terms of biology represents an inadmissible oversimplification. Population growth, landuse-systems, industrial production and species introductions depend much more on economy and politics than on population biology. If we want to solve these problems, we have to presuppose that humans are more than a peculiar animal species. Responsibility is exclusively human and not natural. To have an addressee for our moral and political demands, we have to accept this.

Conclusions

Which nature do we want?

Assessments are a matter of judgement and not a matter of fact. Therefore, the ecological description of a phenomenon, and its evaluation as desirable or undesirable from a practical perspective, should be kept apart. The evaluation of the expansion of a given species as damaging has to be based on certain values or criteria. To assess a plant as a weed requires a clear articulation of management goals. Managers of parks, preserves etc. should clearly articulate what they are managing for and only then determine which species interfere with management goals (Randall 1997). Instead of generalized condemnations of non-indigenous species, a case-by-case-approach should be taken, that should mainly consider the actual situation:

> "The most important question is the impact of the species on the site in question rather than whether it is unwanted elsewhere or indigenous to the site" (Randall 1997).

The claim for a clear articulation of management goals implies the approval of their arbitrariness. There is not only one nature but many and we have to decide which one we want. In this decision, we tend to rely on traditional images of nature. I have presumed that one reason for the negative assessment of invasive alien plants is that they do not fit into our favourite images of nature. Expanding non-indigenous plants neither fulfil the ideal of unspoiled pristine nature nor that of a harmony within nature or between humans and nature. Introduced plants are not harmful to nature as such, but to the nature we want to preserve.

Values related to the assessment of plant introductions

The ethically required separation of facts and values is hard to accomplish. Science borrows its concepts and terms from everyday language. Thereby, it also imports the related values as connotations. These implicit values can influence value judgements without being noticed. The terminology, as well as some basic concepts, used to describe and explain plant introductions and their consequences are heavily value-laden. Terms like 'alien', 'invasion', 'resistance' or 'immunity' are associated with the idea that the entrance of something extraneous means harm to a community. If borders are supposed to be part of a natural order, their overcoming seems to be undesirable. Both, the non-acceptance of the naturalization of foreigners and the clinging to a rigid order, have parallels in social life that should be critically borne in mind.

Realizing that some of the involved criteria are problematic, because they reflect a xenophobic bias, does not mean to deny that introduced species can threaten objectives of nature conservation. However, to be aware of the political and social context in which this discussion takes place should result in a more precise denomination of the problems at stake. If ecologists do not want to feed common prejudices against foreigners, they should refrain as far as possible from biased terminology and from highly emotionalised presentations of the consequences of species introductions. Whereas local or global species diversity, as well as the uniqueness of a particular ecosystem or landscape, are legitimate arguments, the nativity of a species should play a minor role in the discussion.

Acknowledgements

I thank Thomas Potthast, Tübingen, and two anonymous reviewers for helpful comments on earlier drafts of this paper. This work is part of an interdisciplinary research programme 'Ecology and Conservation Ethics' that is funded by the BMBF (German Federal Ministry of Education and Scientific Research; FKZ 0339561).

References

Binggeli, P. 1994. Misuse of terminology and anthropomorphic concepts in the description of introduced species. Bulletin of the British Ecological Society 25 (1): 10-13.

Brown, J. H. 1989. Patterns, Modes and Extents of Invasions by Vertebrates. In: Drake, J. A., Mooney, H. A., Di Castri, F., Groves, R. H., Kruger, F. J., Rejmanek, M. and Williamson, M. (eds.), Biological Invasions. A Global Perspective. pp. 85-109. John Wiley & Sons, Chichester, New York, Brisbane.

Burke, M. J. W. and Grime, J. P. 1996. An experimental study of plant community invasibility. Ecology 77 (3): 776-790.

Clements, F. E. 1916. Plant Succession. An analysis of the development of vegetation. Carnegie Institution of Washington, No. 242.

Cronk, Q. C. B. and Fuller, J. L. 1995. Plant invaders – The threat to natural ecosystems. Chapman & Hall, London, Glasgow, Weinheim.

De Waal, L. C., Child, L. E., Wade, P.M. and Brock, J. H. (eds.) 1994. Ecology and management of invasive riverside plants. John Wiley & Sons, Chichester, New York, Brisbane.

Di Castri, F., Hansen, A. J. and Debussche, M. (eds.) 1990. Biological Invasions in Europe and the Mediterranean Basin. Kluwer, Dordrecht.

Drake, J. A., Mooney, H. A., Di Castri, F., Groves, R. H., Kruger, F. J., Rejmanek, M. and Williamson, M. (eds.) 1989. Biological invasions. A global perspective. John Wiley & Sons, Chichester, New York, Brisbane.

Eco, U. 1977. Zeichen. Einführung in einen Begriff und seine Geschichte. Suhrkamp, Frankfurt/M.

Egerton, F. N. 1973. Changing Concepts of the Balance of Nature. The Quaterly Review of Biology 48: 322-350.

Elton, C. S. 1958. The ecology of invasions by animals and plants. Methuen, London.

Eser, U. and Potthast, T. 1997. Bewertungsproblem und Normbegriff in Ökologie und Naturschutz aus wissenschaftsethischer Perspektive. Zeitschrift für Ökologie und Naturschutz 6: 181-189.

Finck, H. 1991. Die grünen Besatzer. Ausländer auf Erfolgskurs. Fremde Pflanzen überwuchern deutsche Kräuter. Experten setzen auf die Integrationskraft der heimischen Flora. natur 2/91: 52-54.

Fleck, L. 1993. Entstehung und Entwicklung einer wissenschaftlichen Tatsache. Einführung in die Lehre vom Denkstil und Denkkollektiv/ hg. v. L. Schäfer u. T. Schnelle. Suhrkamp, Frankfurt/M.

Fox, M. D. and Fox, B. J. 1986. The susceptibility of natural communities to invasion. In: Groves, R. H. and Burdon, J. J. (eds.), Ecology of biological invasions. pp. 57-66. Cambridge University Press, Cambridge, London. New York.

Garthwaite, P. E. 1993. End to Term 'native'. Journal of Forestry 87: 59-60.

Ginzburg, C. 1980. Spurensicherung. Der Jäger entziffert die Fährte, Sherlock Holmes nimmt die Lupe, Freud liest Morelli – die Wissenschaft auf der Suche nach sich selbst. Freibeuter 3; 4: 7-17; 11-36.

Gleason, H. A. 1926. The individualistic concept of the plant association. Torrey Bot. Club Bull. 53: 7-26.

Groves, R. H. and Burdon, J. J. 1986. Foreword. In: Groves, R. H. and Burdon, J. J. (eds.), Ecology of biological invasions. pp. vi-viii. Cambridge University Press, Cambridge, London, New York.

Hard, G. 1969. Das Wort "Landschaft" und sein semantischer Hof. Zu Methode und Ergebnis eines linguistischen Tests. Wirkendes Wort 19: 3-14.

Hard, G. 1995. Spuren und Spurenleser. Zur Theorie und Ästhetik des Spurenlesens in der Vegetation und anderswo. Rasch, Osnabrück.

Harper, J. L. 1965. Establishment, aggression, and cohabitation in weedy species. In: Baker, H. G. and Stebbins, G. L. (eds.), The Genetics of Colonizing Species. Proceedings of the First International Union of Biological Sciences Symposia on General Biology. pp. 243-268. Academic Press, New York.

Hartmann, E., Schuldes H., Kübler, R. and Konold, W. 1994. Neophyten. Biologie, Verbreitung und Kontrolle ausgewählter Arten. Ecomed, Landsberg.

Heiland, S. 1992. Naturverständnis. Dimensionen des menschlichen Naturbezugs. Wissenschaftliche Buchgesellschaft, Darmstadt.

Jäger, E. 1977. Veränderungen des Artenbestandes von Floren unter dem Einfluß des Menschen. Biologische Rundschau 15: 287-300.

Jansen, A. J. 1972. An analysis of "balance in nature" as an ecological concept. Acta biotheoretica 21: 86-114.

Joenje, W. 1987. The SCOPE programme on the ecology of biological invasions. An account of the Dutch contribution. Proceedings of the Koninklijke Nederlandse Akademie van Wetenschappen, Series C 90 (1): 3-13.

Kolasa, J. and Pickett, S. T. A. 1992. Ecosystem stress and health. An expansion of the conceptual basis. J. Aquat. Ecosystem Health 1: 7-13.

Kornberg, S. H. and Williamson, M. H. (eds.) 1986. Quantitative aspects of the ecology of biological invasions. Philosophical Transactions of the Royal Society London B 314: 501-742.

Kowarik, I. 1996. Auswirkungen von Neophyten auf Ökosysteme und deren Bewertung. Texte des Umweltbundesamtes 1996/58: 119-155.

Kuhn, T. S. 1976. Die Struktur wissenschaftlicher Revolutionen. Suhrkamp, Frankfurt/ M.

MacDonald, I. A. W., Kruger, F. J. and Ferrar, A. A. (eds.) 1986. The ecology and management of biological invasions on Southern Africa. Oxford University Press, Cape Town.

MacDonald, I. A. W., Loope, L. L., Usher, M. B. and Hamann, O. 1989. Wildlife conservation and the invasion of nature reserves by introduced species. A global perspective. In: Drake, J. A., Mooney, H. A., Di Castri, F., Groves, R. H., Kruger, F. J., Rejmanek, M. and Williamson, M. (eds.), Biological Invasions. A Global Perspective. pp. 215-255. John Wiley & Sons, Chichester, New York, Brisbane.

McNeely, J. A., Gadgil, M., Leveque, C., Padoch, C. and Redford, K. 1995. Human influences on Biodiversity. In: Heywood, V. H. (ed.), Global Biodiversity Assessment. pp. 711-821. Cambridge University Press.

Mooney, H. A. and Drake, J. A. (eds.) 1986. Ecology of biological invasions of North America and Hawaii. Springer. New York, Berlin, Heidelberg.

Mooney, H. A. and Drake, J. A. 1989. Biological invasions. A SCOPE program overview. In: Drake, J. A., Mooney, H. A., Di Castri, F., Groves, R. H., Kruger, F. J., Rejmanek, M. and Williamson, M. (eds.), Biological Invasions. A Global Perspective. pp. 491-506. John Wiley & Sons, Chichester, New York, Brisbane.

Pickett, S. T. A. and White, P. S. (eds.) 1985. The ecology of natural disturbance as patch dynamics. Academic Press, New York.

Pimm, S. L. 1993. Biodiversity and the Balance of Nature. In: Schulze, E.-D. and Mooney, H. A. (eds.), Biodiversity and Ecosystem Function. pp. 347-359. Springer. Berlin, Heidelberg, New York.

Pyšek, P. 1995. Recent trends in studies on plant invasions 1974-93. In: Pyšek, P., Prach, K., Rejmanek, M. and Wade, M. (eds.), Plant Invasions. General aspects and special problems. pp. 223-226. Academic Publishing, Amsterdam.

Pyšek, P., Prach, K., Rejmanek, M. and Wade, M. (eds.) 1995. Plant Invasions. General aspects and special problems. Academic Publishing. Amsterdam.

Randall, J. M. 1997. Defining weeds of natural areas. In: Luken, J. O. and Thieret, J. W. (eds.), Assessment and management of plant invasions. pp. 18-25. Springer, New York.

Rapport, D. J. 1989. What constitutes ecosystem health? Perspect. Biol. Med. 331: 120-132.

Reichholf, J. H. 1996. Wie problematisch sind Neozoen wirklich? In: Neophyten, Neozoen – Gefahr für die heimische Natur? Beiträge der Akademie für Natur- und Umweltschutz Baden-Württemberg 22. pp. 86-90. Stuttgart.

Schwabe, A. and Kratochwil, A. 1991. Gewässerbegleitende Neophyten und ihre Beurteilung aus Naturschutz-Sicht unter besonderer Berücksichtigung Südwestdeutschlands. NNA-Berichte 41: 14-27.

Smith, C. W. 1989. Non-native plants. In: Stone, C. P. and Stone, D. B. (eds.), Conservation Biology in Hawai'i. pp. 60-69. Univ. of Hawaii Press, Honolulu.

Sprugel, D. G. 1991. Disturbance, equlibrium, and environmental variability. What is 'natural' vegetation in a changing environment? Biological Conservation 58: 1-18.

Starfinger, U. 1991. Nicht-einheimische Pflanzenarten als Problem für den Artenschutz. In: Kaule, G. and Henle, K. (eds.), Arten- und Biotopschutzforschung für Deutschland. pp. 225-233. Berichte aus der ökologischen Forschung 4. KVA, Jülich.

Steen, W. J. van der 1995. Facts, values, and methodology. A new approach to ethics. Rodopi, Amsterdam, Atlanta.

Sukopp, H. 1995. Neophytie und Neophytismus. In: Böcker, R., Gebhardt, H., Konold, W. and Schmitt-Fischer, S. (eds.), Gebietsfremde Pflanzenarten. Auswirkungen auf einheimische Arten, Lebensgemeinschaften und Biotope. Kontrollmöglichkeiten und Management. pp. 3-32. Ecomed, Landsberg.

Strohschneider, R. 1991. Einsatz und unkontrollierte Ausbreitung fremdländischer Pflanzen – Florenverfälschung oder ökologisch bedenkenlos? NNA-Berichte 41: 4-5.

Thellung, A. 1918/19. Zur Terminologie der Adventiv- und Ruderalfloristik. Allgemeine Botanische Zeitschrift 24/25: 36-42.

Thompson, D. Q., Stuckey, R.L. and Thompson, E. B. 1987. Spread, Impact, and Control of Purple Loosestrife *Lythrum salicaria* in North American Wetlands. U.S. Fish and Wildlife Service. Washington D.C.

Trepl, L. 1987. Geschichte der Ökologie. Vom 17. Jahrhundert bis zur Gegenwart. Athenäum. Frankfurt/ M.

Trepl, L. and Sukopp, H. 1993. Zur Bedeutung der Introduktion und Naturalisation von Pflanzen und Tieren für die Zukunft der Artenvielfalt. In: Rundgespräche der Kommission für Ökologie Bd. 6. Dynamik von Flora und Fauna – Artenvielfalt und ihre Erhaltung. pp. 127-142. Pfeil, München.

Wagner, W. H. 1993. Problems with biotic invasives. A biologist's viewpoint. In: McKnight, B. N. (ed.), Biological pollution. pp. 1-8. Indiana Academy of Science, Indianapolis.

Weber, M. 1917. Der Sinn der 'Wertfreiheit' der soziologischen und ökonomischen Wissenschaften. In: Max Weber. Gesammelte Aufsätze zur Wissenschaftslehre. pp. 489-540. Mohr, Tübingen.

Webster's New Encyclopedic Dictionary 1993. BD&L, New York

Zimmermann, B. 1996. Wie das Reden vom Immunsystem leibhaftig wird. In: Frauengruppe gegen Bevölkerungspolitik (ed.), Lebensbilder – Lebenslügen. Leben und Sterben im Zeitalter der Biomedizin. pp. 77-87. VLA, Hamburg.

PLANT INVASIONS IN NORTHERN GERMANY: HUMAN PERCEPTION AND RESPONSE

Ingo Kowarik and Hartwig Schepker
Universität Hannover, Institut für Landschaftspflege und Naturschutz, Herrenhäuser Straße 2, D 30419 Hannover, Germany; e-mail: kowari@laum.uni-hannover.de, schepk@laum.uni-hannover.de

Abstract

In Germany, the attitude towards non-native plant species is ambiguous. Some are included in Red Data Books if declining, whereas others are controlled if suspected to cause damage. The public debate is often emotionally biased and there is no consensus on the relevance of plant invasions as threats to biological diversity or as sources of other conflicts. This paper presents the results of an inquiry among local authorities, started in 1995 in Niedersachsen (northern Germany). Firstly, this project aimed to provide a comprehensive picture of the perception of problematic plant invasions by public authorities (nature conservation, forestry, water management, coastal protection, urban green) and NGOs related to nature conservation. Secondly, the human response to invasive species was analysed. The main results are:

- A majority (72%) of 350 replies confirmed invasion conflicts in the area of personal responsibility of the addressee ($\geq$ 80% in forestry, nature conservation and coastal protection compared to 41% in urban green and 60% in water management departments).
- In 55% of 592 specified cases vegetation changes were named as an unwanted invasion effect. Only 7% of these cases involved endangered plant species. A quarter of all conflicts (27%) were related to economic problems: 5% affected the maintenance of watercourses, 22% forest management (mainly the recruitment of native trees).
- Invasion problems were attributed to 31 non-native species, but >80% of all problematic invasion events were related to (in decreasing importance) *Prunus serotina*, *Heracleum mantegazzianum*, *Reynoutria* (incl. *R. japonica*, *R. sachalinensis*, *R. x bohemica*), *Impatiens glandulifera*, *Elodea* (incl. *E. canadensis*, *E. nuttallii*), *Vaccinium corymbosum* x *angustifolium*.
- Half of all invasion events (49%) classified as problematical were subject to control. The majority of control activities were ineffective: In only 23% was control evaluated as successful.
- The majority of all respondents expressed a need for control of plant invasions. In detail, 38% voted for a response in single cases, 29% for a generally high priority for future control. Only 14% expressed no need for control.

The results show that a single case approach is adequate for assessing plant invasions in central Europe.

Introduction

The Trondheim UN Conference on alien species in 1996 resulted in a consensus that biological invasions are one of the most severe threats to biological diversity (Sandlund *et al.* 1996). This is close to a restatement of Elton's insight in his landmark study of 1958. However, in the interim, both the scientific and political handling of the phenomenon have changed. Biological invasions are studied worldwide with increasing intensity (e.g., Drake *et al.* 1989, Pyšek 1995, Williamson 1996) and, in many parts of the world, the response to the introduction and spread of non-native species has be-

Plant Invasions: Ecological Mechanisms and Human Responses, pp. 109–120
edited by U. Starfinger, K. Edwards, I. Kowarik and M. Williamson
© 1998 Backhuys Publishers, Leiden, The Netherlands

come an important part of local nature conservation policies (e.g., U.S. Congress OTA 1993, Pyšek *et al.* 1995, Brock *et al.* 1997, Luken and Thieret 1997).

In Germany, however, there is a lack of a consensus in both official nature conservation policies and in public debate on which strategies should be implemented concerning biological invasions. The federal nature conservation act does provide legal measures against the release of non-native species, but these instruments remain virtually unused in practice (Doyle *et al.* 1998). Simultaneously, the general aim of the law is to preserve the entire biological diversity that exists, at least in part, due to the cultural transformation of central European habitats. As part of this process, biological invasions have contributed for some thousand years to the biological richness of central Europe. Consequently, German Red Data Books include non-native species when they become threatened (Kowarik 1992, Schnittler and Ludwig 1996). Conversely, 20-30 non-native species are controlled because they are suspected to cause damage (Kowarik 1998).

German public opinion, influenced by conspicuously spreading species such as *Heracleum mantegazzianum,* sometimes delivers a general verdict against all non-native species. This is manifested in demands for a general control of alien species (e.g., Disko 1996), as well as in symptoms of a "mania for native plants" (Gröning and Wolschke-Bulmahn 1992) in landscaping and garden design. The resulting controversies are usually biased emotionally and by a mixing of scientific analysis and anthropocentric evaluation of invasion effects (Kowarik 1996, Eser 1998).

Central European studies of plant invasions date back to the early 19th century ("Adventivfloristik", see Trepl 1990), and there are several well documented single case studies. Nevertheless, disagreement on the effects of biological invasions and how they should be treated still prevails in Germany. The relevance of invasions for nature conservation and for other applied disciplines, such as agriculture or forestry, can be examined in two ways. Firstly, case studies on alien species can summarise the impact of a single alien species (e.g, for Germany, Lohmeyer and Sukopp 1992, Hartmann *et al.* 1994, Böcker *et al.* 1995, Kowarik 1996, 1998). Secondly, one can analyze the perception of invasions by those whose interests are possibly affected by alien plants.

This paper employs the second approach. It aims at providing a comprehensive status-quo analysis of plant invasions as a source of 1) conflicts in northern Germany and 2) the subsequent control activities. The study refers to the results of an inquiry started in 1995. This was addressed to all local authorities in Niedersachsen, which are potentially concerned by biological invasions and are responsible for control and other activities in this field. Additionally, members of some NGOs were included. In particular, the study focused on the following questions:

1. To which extent do local authorities and NGO members perceive invasions of non-native plant species as a problem?
2. What conflicts are assumed as results from plant invasions?
3. How many and which species are suspected to be responsible for these conflicts?
4. How frequently are control activities performed and with what success?
5. Is there a need for future control of non-native plant species?

Approach of the inquiry

In October 1995, we started an inquiry on the perception of plant invasions in Niedersachsen, a federal state in northwest Germany covering 47,431 km^2. The recipients of our questions were all authorities that might be concerned by the spread of alien species and the resulting effects. These were the authorities for forestry (private and public), water management, nature conservation, coastal protection, and urban green departments. Additionally, NGO members working in the field of nature conservation were included in the study (see Table 1). The agricultural sector was excluded, because alien plants in arable fields are not usually treated specifically, but only as components of the total weed flora. (*Cyperus esculentus*, which has been controlled specifically in the Netherlands (ter Borg *et al.* 1998), is of only minor importance in Germany, Schroeder and Wolken 1989).

The questionnaire included 15 questions on alien plant species occurring in the area of responsibility of each authority (see Appendix). We stressed clearly the exclusive focus of the inquiry on non-native species that had been evaluated as troublesome by the single addressee. To avoid a bias towards particular species, we did not ask about the effects of specific aliens. Instead, general questions were asked so to allow the respondents to indicate the problems and ecological or economic effects caused by particular alien species. Additionally, we asked about control activities and their success, and invited the addressees to estimate species' rate of spread.

After a pre-test with five authorities, 414 authorities and 214 NGO members received the questionnaire. After 6-8 weeks all authorities which had not yet replied were called by telephone. As a result, 240, or 58%, of the official recipients replied. For the members of the NGOs, the rate of reply was only 19%. Therefore, the results are considered representative for the public authorities, but not for the NGOs. Some addressees returned more than one questionnaire. This was mainly due to large forest authorities ("Forstämter"), which involved their subdivisions ("Forstreviere") in answering. In total, the results of the inquiry are based on 350 questionnaires (see Table 1). χ^2 tests are used to test the differences between the groups.

For data processing, species which may have been misidentified by the addressees were summarised to genus to avoid overinterpretation at the species level (*"Elodea"* = *E. canadensis* and *E. nuttallii;* *"Reynoutria"* [*Fallopia*] = *R. japonica, R. sachalinensis,* and *R. x bohemica;* *"Solidago"* = *S. canadensis* and *S. gigantea*).

Since the study was aimed at the individual perception of invasion effects, the results do not provide any proof of invasion effects, but may contribute to a better understanding of human perception of, and response to, the invasion phenomenon.

Perception of problematic invasions

The opening question of the questionnaire referred to the existence of troublesome invasion events in the personal area of responsibility of each recipient. A large majority (72%) confirmed problems with non-native plant species (Table 1). Invasions were highly problematic for representatives of nature conservation (both the public authorities and NGO sector), forestry, urban green departments and coastal protection. Invasions were significantly less troublesome (p = 0.001) for the water management authorities. However, even here 41% of their answers revealed concerns about invasions.

Table 1. Perception of troublesome invasion events by local authorities and NGOs in Niedersachsen, northern Germany, resulting from an inquiry in 1995. The number of questionnaires does not equate to the number of replies, because some addressees returned more than one questionnaire (see explanation in the text).

addressees (number)	rate of reply	question-naires	perception of problematic invasion events	
			Yes	No
authorities (414)	58%	303	71%	29%
including departments of				
• forestry (135)	79%	163	80%	20%
• water management (190)	42%	76	41%	59%
• nature conservation (55)	71%	49	88%	12%
• urban green (28)	36%	10	60%	40%
• coastal protection (6)	83%	5	80%	20%
NGO members (214)	19%	47	81%	19%
in total	45%	350	72%	28%

Identification of conflicts and problematic invaders

The 350 questionnaires provided data on 457 problematic invasion events by a total of 31 species, and on 592 named conflicts (including multiple answers per species or invasion event). Table 2 gives a summary of these concerns. In 12% of all cases, the invasion phenomenon itself was already regarded as a problem, i.e., without relating it to any distinct consequence. Vegetation changes, due to invasive species, were feared in 55% of the cases. This number included 30 cases of rare native species listed in the regional Red Data Book (Garve 1993), possibly being outcompeted by alien species. In consequence, a high proportion of the respondents estimated vegetation changes as the most relevant invasion problem. Specific problems with the conservation of endangered plant species were seen in only 5% of all named invasion conflicts.

Economic problems in land use accounted for 27% of the perceived conflicts. Specifically, invasions affected the maintenance of watercourses in 5%, and forest management in 22% of the cases. For the rest (6%), health risks, mainly attributed to *Heracleum mantegazzianum*, were mentioned. Foresters mainly feared that alien species may reduce the regeneration of forest trees (29%). Interestingly, the perception of invasion problems was virtually the same by both official and NGO conservationists (p = 0.696). This was also true for the ranking of problematic species (see Table 3). In contrast, there were significant differences (p = 0.001) between the responses of foresters, conservationists (taking both groups together) and water managers.

The problems mentioned were attributed to 31 non-native species. However, more than 80% of them related to *Prunus serotina, Heracleum mantegazzianum, Reynoutria* (incl. *R. japonica, R. sachalinensis, R.* x *bohemica*), *Impatiens glandulifera, Elodea* (incl. *E. canadensis, E. nuttallii*), and *Vaccinium corymbosum* x *angustifolium*. Table 3 shows the most problematic species involved in a total of 457 invasion events. There are conspicuous differences in perception by the different groups (significant differences with p = 0.001 between foresters, water managers and the sum of conservationists, but no differences between official and NGO conservationists). In forestry, *Prunus*

Table 2. Conflicts perceived as resulting from biological invasions in Niedersachsen, Germany. The results are based on 350 questionnaires (see Table 1) including information on 592 individual conflicts. Detailed information is given for the sectors of forestry, nature conservation (including coastal protection), water management and urban green.

perceived conflicts	invasion conflicts perceived by											
	all address.		forestry		nature conservation				water managem.		urban green	
					public		NGOs					
	abs.	%	abs.	%	abs.	%	abs.	%	abs.	%	abs.	%
• invasion phenomenon	74	12	25	8	24	17	16	16	6	11	3	27
• vegetation changes	323	55	127	45	104	72	74	75	15	27	3	27
• maintenance of watercourses	32	5	1	<1	1	<1	2	2	28	50	-	-
• recruitment of native trees	89	15	82	29	5	3	1	1	1	2	-	-
• forest management	40	7	39	14	1	<1	-	-	-	-	-	-
• health risks	34	6	8	3	9	6	6	6	6	11	5	45
in total (= 100%)	592		282		144		99		56		11	

Table 3. Non-native species perceived as troublesome invaders by different groups in Niedersachsen, Germany. The results are based on 350 questionnaires (see Table 1) including information on 457 invasion events by a total of 31 alien species. Detailed information is given for the sectors of forestry, nature conservation (including coastal protection), water management and urban green.

invasions by	invasive species perceived as problematic by											
	all address.		forestry		nature conservation				water managem.		urban green	
					administr.		NGOs					
	abs.	%	abs.	%	abs.	%	abs.	%	abs.	%	abs.	%
Prunus serotina	147	32	106	59	22	16	14	15	5	12	-	-
Heracleum mantegazzianum	82	18	18	10	31	23	19	20	8	20	6	75
Reynoutria spp.	81	18	23	13	32	24	21	23	4	10	1	13
Impatiens glandulifera	29	6	9	5	11	8	3	3	5	12	1	13
Elodea spp.	19	4	-	-	5	4	2	2	12	29	-	-
Vaccinium corymb. x *angustifolium*	16	4	7	4	5	4	4	4	-	-	-	-
Solidago spp.	15	3	3	2	6	4	5	5	1	2	-	-
Impatiens parviflora	11	2	6	3	1	1	4	4	-	-	-	-
Rosa rugosa	10	2	1	1	6	4	3	3	-	-	-	-
Robinia pseudoacacia	7	2	-	-	2	1	2	2	3	7	-	-
Helianthus tuberosus	4	1	-	-	3	2	1	1	-	-	-	-
others	36	8	7	4	1	1	15	16	3	7	-	-
in total (100%)	457		180		135		93		41		8	

serotina was the dominant problem: less in the highlands, but virtually omnipresent in lowland plantations of *Pinus sylvestris*. Within nature conservation, about two thirds of the invasion events concern three taxa: *P. serotina*, the *Reynoutria* species and *H. mantegazzianum*. The latter was the prevailing problematic species in urban areas. In addition to *Elodea*, the maintenance of water courses was influenced by the species named above affecting access.

Perception of invasion dynamics

Very often, the problematic species were believed to spread quickly. Of 422 responses, virtually all estimated a high or medium rate of spread (Table 4, section b). Those species which were perceived as the most troublesome were estimated as spreading most quickly. The ranking echoes that in Table 3: *Prunus serotina* was evaluated as spreading very rapidly by 63%, *Heracleum mantegazzianum* by 58% and the *Reynoutria* species by 37%. In contrast, the spread of both *Impatiens* species was reported as similar, although *I. parviflora* was recognised as being much less troublesome than its congener.

In Table 4 (section a), the personal evaluations of the rate of spread are contrasted with results derived from gridded maps showing species' increase during the last 16 years. There are striking differences between the perception of species' spread and the floristic data. Comparing the real rates of spread with the personal evaluations reveals no significant correlation (Kendall's Tau, p = 0.16). The personal estimates of the spread of *Heracleum* and *Prunus serotina* are almost identical. There is indeed a conspicuous increase in grid squares with *Heracleum* (+281%), but that of *Prunus serotina* is relatively minor (+37%). Spread of the latter species (and of *Solidago* and *Impatiens parviflora*) have been overestimated, while that of *Impatiens glandulifera* was underestimated. These divergences may be a consequence of different categories of species' success. *Prunus serotina*, for example, may already be present in most of the grid squares affording suitable habitats, but may nevertheless increase its abundance within these squares. The clear underestimation of *Impatiens glandulifera*, however, may point to another reason: the similar perception of invasion dynamics and invasion problems.

Control activities and their success

Control activities were carried out in about 50% of all invasion events (222 of 457; Table 5). All these cases were classified as problematic by the local authorities (including NGO members). In only 41% did the mention of control activities coincide with specifying any invasion effects that may have stimulated control. Of these, the reasons were economic in 44 cases (20%; e. g., impairment of sylvicultural practices); in 39 cases (18%) there were specific concerns about nature conservation (e. g., competition with endangered species in the local flora), and other concerns were mentioned in 7 cases. No specific invasion effects were mentioned in 132 cases of control (59%).

We received detailed information on the success of 188 of 222 control activities. Control was described as successful when the population was eliminated or at least declined quickly. Table 5 lists the species for which control measures were undertaken and shows the success of the treatments. The control measures were aimed at 17 spe-

Table 4. The rate of spread of problematic alien species perceived by the addressees of the inquiry (see Table 1) and compared with results of grid mappings. a) grid square records (mappings by Haeupler and Schönfelder 1989, Niedersächsisches Landesamt für Ökologie, Dez. Naturschutz, n. p.; data analysed by Költzsch); b) personal evaluations by the addressees (species not differentiated within the genus *Solidago* and *Reynoutria*)

alien species	a) grid square records			b) evaluations on the rate of increase			
	1980	1996	+/-	high	medium	low	n
Heracleum mantegaz.	47	179	+281%	58%	33%	9%	77
Impatiens glandulifera	87	165	+90%	36%	44%	20%	25
Prunus serotina	187	256	+37%	63%	31%	6%	142
Reynoutria japonica	196	230	+21%	37%	56%	7%	71
R. sachalinensis	58	63	+9%	/	/	/	/
Solidago gigantea	215	221	+3%	33%	67%	-	12
S. canadensis	186	190	+2%	/	/	-	/
Impatiens parviflora	212	204	-5%	36%	45%	18%	11

cies, but the main targets were *Prunus serotina, Heracleum mantegazzianum* and *Reynoutria*. In these taxa, the rate of control was significantly different (p = 0.001), but there were no differences in the success of control (p = 0.595).

In 77% of all cases, control measures were described as unsuccessful. Mechanical treatments were applied to 124 stands of alien species, 10 chemically and 56 by combining both approaches; there was no further information on the rest. Within the different control approaches, the combination of mechanical and chemical treatments was more successful than the others.

Attitude towards future activities

The question on future activities in regard to plant invasions revealed a broad consensus on the necessity of responding to the invasion phenomenon. A large majority (86%) voted for future activities. Only 14% preferred a laissez-faire-strategy. In detail, the attitudes varied widely (Table 6). Future control activities were listed as a hight priority by 29% of the respondents, while 13% considered it as a medium priority. Control actions would be initiated by a case by case basis by 38% of the respondents. Invasion problems seemed to be most urgent within the forest sector. Here, 42% of all replies gave invasive control as a high priority, compared to about 20% in the other groups. Within nature conservation, the spectrum of opinions was more divergent in the NGO group. Compared with the official conservationists, more replies were in favour both of high priority and against control. However, these differences were not significant (p = 0.123). In addition, no differences were found in the attitude of conservationists and of water managers (p = 0.47). In contrast, the responses of foresters differed significantly from those of both conservationists (p = 0.001) and water managers (p = 0.03). Apart from forestry, most respondents voted to control in only single cases. Additionally, about 10% of the replies proposed more rigid legal measures.

Table 5. Control activities on problematic stands of alien plant species and their success. The results are based on 350 questionnaires (see Table 1). The assessment of the success of control activities is based on information on 188 of 222 controlled stands.

alien species	problematic stands			success of control activities					
	all	controlled		n	successful			no success	
		abs.	%		abs.	%		abs.	%
Prunus serotina	147	113	77	103	28	27		75	73
Heracleum mantegaz.	82	52	63	42	9	21		33	79
Reynoutria ssp.	81	24	30	17	3	18		14	82
Impatiens glandulifera	29	3	10	1	1	100		0	0
Elodea ssp.	19	8	42	7	0	0		7	100
Vaccinium cory. x _ang._	16	4	25	3	0	0		3	100
Solidago ssp.	15	3	20	3	0	0		3	100
Impatiens parviflora	11	-	-	-	-	-		-	-
Rosa rugosa	10	2	20	1	0	0		1	100
Robinia pseudoacacia	7	4	57	3	1	33		2	67
Helianthus tuberosus	4	1	25	-	-	-		-	-
other species	36	8	4	8	2	13		6	87
all species (n = 17)	457	222	49	188	44	23		144	77

Table 6. Attitude towards the necessity of future control activities aimed at alien species in Niedersachsen, Germany. The results are based on 350 questionnaires (see Table 1). Detailed information is given for the sectors of forestry, nature conservation and water management (without coastal protection and urban green).

future control ?	need for control expressed by									
	all address.		forestry		nature conservation administr.		NGOs		water managem.	
	abs.	%	abs.	%	abs.	%	abs.	%	abs.	%
◆ no need	59	14	13	8	16	14	21	23	5	14
◆◆ only in single cases	163	38	51	30	56	47	37	41	16	43
◆◆◆ generally low priority	26	6	8	5	8	7	4	4	5	14
◆◆◆◆ gen. medium priority	56	13	25	15	18	15	7	8	3	8
◆◆◆◆◆ gen. high priority	125	29	71	42	20	17	21	23	8	22
in total	429		168		118		90		37	

Discussion and conclusions: a need for single case approaches

1. The inquiry revealed that different groups of local authorities perceive plant invasions as highly troublesome (Table 1). That 49% of the 457 problematic invasion events were subjected to control stresses the virulence of the perceived problems, but does not verify their existence. In a majority of cases (77%), control was unsuccessful (Table 5). The detailed reasons for this failure are unknown. In some cases, the chosen control approaches were obviously inadequate. There is additional evidence for a deficiency in ecological and economic impact analysis preceding control. The conflicts that had stimulated control could generally be identified. In about 60% of cases, however,

reports on control were not combined with a specification of the ecological or economic invasion effects. This may indicate a hasty response to an invasion event without preceding impact analysis (see the concerns of Edwards 1998 on activism in this field).

2. Due to the perceived dimension of invasion problems in northern Germany, responses to plant invasions are urgent. However, the design of appropriate measures should consider additional aspects. We did not ask about ecological, cultural or economic benefits that may outweigh negative invasion effects (e.g., Lasserre 1996, Williams 1996), nor about negative effects of native species that may also be subjected to control (e.g., Williamson 1998). The results of the study stress the perception of unwanted invasions, but do not validate the existence of the assumed problems. All of these aspects should be considered in a comprehensive assessment of the invasion phenomenon. Each decision either for control or a laissez-faire approach should be preceded by a detailed ecological and economic impact analysis and by explicit evaluation.

3. The general perception of plant invasions as a source of conflicts contrasts with the fact that named problems are confined to a rather small group of non-native species. Only about 10 species were believed to be responsible for virtually all of the problems (Table 3). This is about a third of all non-native species that are specifically controlled in Germany, less than 3% of all alien plant species established in Germany and less than 1‰ of the 12,000 plant species which are supposed to be introduced to central Europe (Kowarik 1996). In consequence, a general ban on all groups of cultivated or spreading non-native plant species seems inappropriate for Germany, because the majority of both introduced and established species is not regarded as causing problems. The results of our study support a single case approach as a response to plant invasions.

4. There is additional support for a single case approach when information at the species level is considered. The invasiveness of a given species and subsequent unwanted invasion effects may vary geographically even in a small area such as Germany. Here, only a few species are generally regarded as problematic. *Heracleum mantegazzianum* and the *Reynoutria* species belong to this group, and are considered problematic also in the wider perspective of central, west and north Europe (see papers in Pyšek *et al.* 1995). *Robinia pseudoacacia, Helianthus tuberosus* and the *Solidago* species are widely distributed, but are of only minor importance as troublesome invaders in northern Germany (Table 3, Table 5), although classified as highly problematic in warmer parts of southern Germany (Hartmann *et al.* 1994, Böcker *et al.* 1995). Conversely, invasions by the commonly planted *Rosa rugosa* are only problematic in coastal dunes, and by *Vaccinium corymbosum* x *angustifolium* exclusively in bogs (Schepker and Kowarik 1998). Invasions by *Prunus serotina,* the most troublesome species in Niedersachsen, occur also in other parts of Germany, but are virtually confined to sandy soils (Starfinger 1997).

In summary, a multioptional attitude towards plant invasions seems adequate for landscapes with a cultural tradition as long as in central Europe. Generalisations covering the entire group of non-native species should be replaced by differentiated assessments within a single case approach. We advocate firstly a clear definition of aims of the policy towards invasions, secondly local information on the variety of invasion impacts and the risks and benefits involved, and thirdly explicit evaluation of the social, economic and ecological effects of invasions.

Acknowledgements

Sincere thanks are due to all participants of the inquiry. Wulf Tessin provided constructive comments on the methodological approach of the inquiry, Wiebke Dierks and Leonhard Peters supported the data sampling and processing. Michael Weichert helped with statistics. Florian Költzsch analysed unpublished floristic data which had been kindly provided by Eckhard Garve, Niedersächsisches Landesamt für Ökologie, Dezernat Naturschutz. Special thanks are due to Keith Edwards, Uwe Starfinger and Mark Williamson for stimulating comments on previous versions of the paper. David Reid, London, kindly improved our English. The study had been supported by a grant of the Federal state of Niedersachsen.

References

Böcker, R., Gebhardt, H., Konold, W., Schmidt-Fischer, S. (eds.) 1995. Gebietsfremde Pflanzenarten. Auswirkungen auf einheimische Arten, Lebensgemeinschaften und Biotope. Kontrollmöglichkeiten und Management. 215 pp. Ecomed, Landsberg.

Brock, J. H., Wade, M., Pyšek, P. and Green, D. (eds.) 1997. Plant Invasions: Studies from North America and Europe. 223 pp. Backhuys Publishers, Leiden.

Disko, R. 1996. In dubio contra reum! Mehr Intoleranz gegen fremde Arten. Nationalpark 4/91: 38-42.

Doyle, U., Fisahn, A., Ginsky, H. and Winter, G. 1998. Current legal status regarding release of non-native plants and animals in Germany. In: Starfinger, U., Edwards, K., Kowarik, I. and Williamson, M. (eds.); Plant invasions: ecological mechanisms and human responses, pp. 71-83, Backhuys Publishers, Leiden.

Drake, J.A., Mooney, H.J., Di Castri, F., Groves, R.H., Kruger, F.J., Rejmanek, M. and Williamson, M. (eds.) 1989. Biological invasions: a global perspective. Chichester etc.

Edwards, K. 1998. A critique of the general approach to invasive plant species. In: Starfinger, U., Edwards, K., Kowarik, I. and Williamson, M. (eds.); Plant invasions: ecological mechanisms and human responses, pp. 85-94, Backhuys Publishers, Leiden.

Elton, C.S. 1958. The ecology of invasions by animals and plants. 181 pp. Methuen, London.

Eser, U. 1998. Assessment of plant invasions: theoretical and philosophical fundamentals. In: Starfinger, U., Edwards, K., Kowarik, I. and Williamson, M. (eds.); Plant invasions: ecological mechanisms and human responses, pp. 95-107, Backhuys Publishers, Leiden.

Garve, E. 1993. Rote Liste der gefährdeten Farn- und Blütenpflanzen in Niedersachsen und Bremen, 4. Fassung vom 1. 1. 1993. Inform. d. Natursch. Nieders. 1: 1-37.

Groening, G. and Wolschke-Bulmahn, J. 1992. Some notes on the mania for native plants in Germany. Landscape Journal 11 (2): 116-126.

Haeupler, H. and Schönfelder, P. (eds.) 1989. Atlas der Farn- und Blütenpflanzen der Bundesrepublik Deutschland. - 768 pp. Ulmer, Stuttgart.

Hartmann, E., Schuldes, H., Kübler, R. and Konold, W. 1995. Neophyten. Biologie, Verbreitung und Kontrolle ausgewählter Arten. 301 pp. Ecomed, Landsberg.

Kowarik, I. 1992. Berücksichtigung von nichteinheimischen Pflanzenarten, von "Kulturflüchtlingen" sowie von Vorkommen auf Sekundärstandorten bei der Aufstellung "Roter Listen". Schr.Reihe Vegetationskde. 23: 175-190.

Kowarik, I. 1996. Auswirkungen von Neophyten auf Ökosysteme und deren Bewertung. Texte des Umweltbundesamtes 58/96: 119-155.

Kowarik, I. 1998. Neophyten in Deutschland. Quantitativer Überblick, Einführungs- und Ausbreitungswege, ökologische Folgen und offene Fragen. Texte des Umweltbundesamtes (in press).

Lasserre, P. 1996. Ethics, education and information: necessary elements in assessing risks and benefits of alien species. In: Sandlund, O. T., Schei, P. J., Viken, A. (eds.): Proceedings of the Norway/UN Conference on alien species, pp. 10-12. Trondheim.

Lohmeyer, W. and Sukopp, H. 1992. Agriophyten in der Vegetation Mitteleuropas. Schr.Reihe Vegetationskde. 25: 1-185.

Luken, J.O. and Thieret, J.W. (eds.) 1997. Assessment and Management of Plant Invasions. 324 pp. Springer, New York, Berlin, Heidelberg.

Pyšek, P. 1995. Recent trends in studies on plant invasions (1974-1993). In: Pyšek, P., Prach, K., Rejmanek, M. and Wade, M. (eds.): Plant invasions. General aspects and special problems, pp. 223-236. SPB Academic Publishing, Amsterdam.

Pyšek, P., Prach, K., Rejmanek, M. and Wade, M. (eds.) 1995. Plant invasions. General aspects and special problems. 263 pp. SPB Academic Publishing, Amsterdam.

Sandlund, O.T., Schei, P.J. and Viken, A. (eds.) 1996. Proceedings of the Norway/UN conference on alien species. 233 pp. Directorate for Nature Management and Norwegian Insitute for Nature Research, Trondheim.

Schepker, H. and Kowarik, I. 1998. Invasive North American Blueberry Hybrids (*Vaccinium corymbosum* x *angustifolium*) in northern Germany. In: Starfinger, U., Edwards, K., Kowarik, I. and Williamson, M. (eds.): Plant invasions: ecological mechanisms and human responses, pp. 253-260, Backhuys Publishers, Leiden.

Schroeder, C. and Wolken, M., 1989. Die Erdmandel (*Cyperus esculentus* L.). Ein neues Unkraut im Mais. Osnabrücker Naturwissenschaftliche Mitteilungen 15: 83-104.

Schnittler, M. and Ludwig, G. 1996. Zur Methodik der Erstellung Roter Listen. Schr.Reihe Vegetationskde. 28: 709-739.

Starfinger, U. 1997. Introduction and naturalization of *Prunus serotina* in Central Europe. In: Brock, J. H., Wade, M., Pyšek, P. and Green, D. (eds.): Plant Invasions: Studies from North America and Europe, pp. 161-172. Backhuys Publishers, Leiden.

ter Borg, S. 1998. (Im)migration of Cyperus esculentus in N.W. Europe: invasion on a local, regional and global scale. In: Starfinger, U., Edwards, K., Kowarik, I. and Williamson, M. (eds.); Plant invasions: ecological mechanisms and human responses, pp. 261-273, Backhuys Publishers, Leiden.

Trepl, L. 1990. Research on the anthropogenic migration of plants and naturalization. Its history and current state of development. In: Sukopp. H., Hejny, S. and Kowarik, I. (eds.): Urban ecology. Plants and plant communities in urban environments, pp. 75-97. SPB Academic Publ., The Hague.

U.S. Congress OTA (Office of Technology Assessment) 1993. Harmful non-indigeneous species in the United States. OTA-F-565. U.S. Government Printing Office, Washington D.C.

Williams, C.E. 1996. Potential ecological functions of nonindigenous plants. In: Luken, J.O. and Thieret, J.W. (eds.): Assessment and Management of Plant Invasions, pp. 26-34. Springer, New York, Berlin, Heidelberg.

Williamson, M. 1996. Biological Invasions. 244 pp. Chapman and Hall, London etc.

Williamson, M. 1998. Measuring the impact of plant invaders in Britain. In: Starfinger, U., Edwards, K., Kowarik, I. and Williamson, M. (eds.); Plant invasions: ecological mechanisms and human responses, pp. 57-68, Backhuys Publishers, Leiden.

Appendix:

Questions of the Inquiry [* = open response possible]

1. Which non-native species are causing problems in your area of responsibility? [*]
2. What kind of conflicts do you relate to these species? [*]
3. Can you qualify and quantify the impact of the problematic species (e. g. name of outcompeted plants, yield loss, erosion problems...)? [*]
4. What is the size of the problematic stands? [three size classes]
5. How much did the problematic alien species increase in the last 5 years? [very high, high, medium, low, no increase, decrease]
6. Are there any problematic stands of these non-native species in nature reserves? [yes/no; name of the reserve]
7. Which biotope types are containing problematic stands of alien species? [List of 21 biotope types] Please give an example for an outstanding example of a problematic stand (including the number of grid square etc.). [*]
8. Which starting-point do the problematic stands have (e.g., plantation, sown by bee-keepers, garden escape ...) ? [*]
9. Are there any control activities in your area of responsibility?

10. Are there control activities only in nature reserves or also outside of protected areas?
11. With which treatment, at what time, how often, and by whom have these control activities been carried out? [*]
12. For how many years have these control activities been carried out? [*]
13. How was the success of the control activities? [population eliminated, strong decline, few decline, no decline, increase of population]
14. Please estimate the costs of the control activities. [*]
15. How is your attitude towards future activities concerning the invasion of problematic non-native species? [control with generally high, medium, low priority, only in single cases, no need for control; need for more rigid legal measures; multiple responses possible]

CASE STUDIES I: SPECIES

INVASION, ECOLOGY AND MANAGEMENT OF *ELAEAGNUS ANGUSTIFOLIA* (RUSSIAN OLIVE) IN THE SOUTHWESTERN UNITED STATES OF AMERICA

John H. Brock
Environmental Resources, School of Planning and Landscape Architecture, Arizona State University, Tempe, AZ 85287-2005 USA; e-mail: john.brock@asu.edu

Abstract

Elaeagnus angustifolia (Russian olive) was introduced to the plains of North America in the 19th century. Since that time, it has spread throughout the western United States because of its ornamental value in urban landscapes, into natural areas by direct planting as a conservation plant, and then by natural processes. *E. angustifolia* is now naturalized and in the southwest provides the greatest negative ecosystem impacts to riparian habitats. At five sites in northeastern Arizona, *E. angustifolia* averaged 726 plants/ha and 58.2 % canopy cover. This invasive species is difficult to control with fire or mechanical techniques, however the herbicides triclopyr, glyphosate and imazapyr provide much potential. No biological control treatments are currently prescribed for *E. angustifolia*, although a canker disease may hold some promise, while insect agents are apparently unexplored. *E. angustifolia* is still planted as an ornamental and conservation plant, despite its negative impacts. Native plants that could provide the same attributes in the landscape are recommended as a way to curb its continued invasion of natural landscapes in the western United States.

Introduction

People have been moving plants around the globe for thousands of years. Unfortunately, some of these plants that were valuable in their native environments become weedy as they are moved to similar environments. Alien plants that become invasive are considered by Vitousek *et al.* (1996) to be another symptom of global environmental change. When the invasive plant species enters stands of native vegetation, the presence of the alien species can have very negative effects. In many cases the invasion can threaten or eliminate native species, contribute to degraded environmental conditions (such as a decrease in surface cover for soil protection), and degrade environmental processes (Vitousek 1986; Brock and Farkas 1997). The objective of this paper is to describe the invasion, ecology and management of *E. angustifolia* (Russian olive), an alien woody plant that invades riparian and other habitats in the southwestern United States of America. This species has become naturalized in much of the western United States (Olson and Knopf 1986a), in areas disturbed by anthropogenic actions, such as: dewatering of streams for irrigation, regulation of stream flow with building of reservoirs, poor grazing management, cutting of stream side vegetation that was considered to be water-wasting, recreational activities that caused bare soils on streamsides, and planting of exotic vegetation as wildlife food sources or for soil conservation purposes.

Plant Invasions: Ecological Mechanisms and Human Responses, pp. 123–136
edited by U. Starfinger, K. Edwards, I. Kowarik and M. Williamson
© *1998 Backhuys Publishers, Leiden, The Netherlands*

Elaeagnus angustifolia

Fig. 1. Morphology of *E. angustifolia* stems, leaves, and flowers (Cronquist *et al.* 1997).

Taxonomic classification

E. angustifolia (*L*). is a member of the Elaeagnaceae (Oleaster) family (Kearney and Peebles 1968). The genus *Elaeagnus* has about 40 species of shrubs and trees (Olson 1974), many of which are grown as ornamentals and wildlife plants (Bailey 1949). *E. angustifolia* is a small tree up to 10 m in height. This plant is often thorny, has lanceolate or oblong-lanceolate leaves (Fig. 1) that are bright green above and silvery below (Kearney and Peebles 1968; Young and Young 1992). The effect of the leaf coloration is that the canopy provides a silvery-green component in the landscape. Leaves on older tissue have the lanceolate shape, while shade leaves and leaves on juvenile stems commonly have the oblong-lanceolate shape. In both cases the leaf margins are entire. The scurfy-pubescent with stellate haired leaves are arranged alternately along branches that are often reddish brown in color. The fragrant pale yellow flowers are without petals, are regular, perfect, pollinated by insects in the spring, and borne in

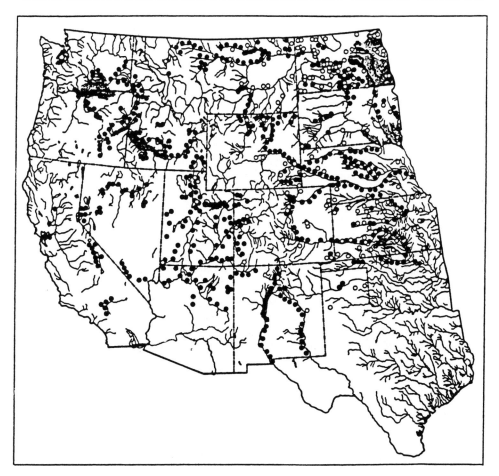

Fig. 2. Distribution of *E. angustifolia* in the western United States of America. Modified from Olson and Knopf (1986a).

small clusters. Plants of this species begin producing fruits at three to five years of age (Young and Young 1992; Borell 1962). The fruit is a dry achene (simulating a drupe) enveloped in a fleshy pericarp (Fig. 1) (Kearney and Peebles 1968; Young and Young 1992). The hard seeds have dark brown stripes and are distributed mainly by birds, although Ireland (1997) reported channel catfish had *E. angustifolia* seeds as a component of their stomach contents. *E. angustifolia* spreads primarily by seeds, but is reported to spread by root stalks (Bovey 1965) and by layering of branches, as observed by the author of this paper.

Invasion history

E. angustifolia has been widely planted in the western part of North America, where it escaped cultivation and is now naturalized in several areas (Turner and Karpiscak 1980; Olson and Knopf 1986a; Young and Young 1992) (Fig. 2). *E. angustifolia* most likely arrived in the United States of America with Prussian Mennonites, who had migrated to southern Russia in the late seventeenth century seeking religious freedom, and then entered North America in the nineteenth century (Tellman 1997). In their change of homelands, they brought with them many crops, of most importance was hard winter wheat and several invasive alien plants, including *E. angustifolia. E. angustifolia* was recommended in South Dakota by the Agricultural Extension Service in 1901 as a drought, animal and frost tolerant plant which was available in plant nurseries at this time (Tellman 1997).

The plant arrived in the southwestern USA, in the Utah and Arizona area, among communities established by the Mormons about 1900, although Christensen (1963) states that *E. angustifolia* was not planted in Utah prior to that time. Apparently people passed cuttings among themselves with encouragement of W. H. Crawford of St. George, UT, who introduced many plants to the Virgin River area of southern Utah and northern Arizona (Welsh *et al.* 1971; Tellman 1997). The tree had become established in nature in central Utah by 1924 (Christensen 1963).

Agents of the extension service recommended *E. angustifolia* for Arizona in 1909, where it continues to be available as a cultivated landscape plant. *E. angustifolia* was present at the Tucson, Arizona Plant Materials Center operated by the Soil Conservation Service agency in 1935 for evaluation and dispersal (Mark Pater, pers. comm. 1998). In Oak Creek Canyon of north-central Arizona, *E. angustifolia* was recognized as an escaped plant in 1941. In 1964, the National Park Service planted *E. angustifolia* in Canyon de Chelly in northeastern Arizona (Fig. 3) for soil stabilization. Ten years later it was one of the dominant trees of the canyon bottom (Tellman 1997).

E. angustifolia has been widely available from nurseries and some state and federal agencies as a landscape and conservation plant. Olson and Knopf (1986b) reported a 1984 survey that showed *E. angustifolia* available in 16 western states with the sole exception being Utah. It remains available on the open market.

E. angustifolia began showing its invasive nature in a very short time after its introduction to North America. The average time lag for alien trees in Germany to show their invasive nature was 184 years with only 6 % of the introductions becoming invasive in a time span less than 50 years (Kowarik 1995). It is fairly obvious that a long time lag did not apply to this species. However, unlike *Tamarix chinensis* (synonyms *T. gallica* and most recently *T. ramosissima*) which was introduced to North America (Brock 1997) about the same time as *E. angustifolia, E. angustifolia's* range does not extend into the hot deserts of central and southern Arizona, particularly not in the lower watersheds of the Salt and Gila rivers (Fig. 2). Presently, *E. angustifolia* is well established in riparian habitats of the four corners region (Utah, Colorado, Arizona and New Mexico) and continues to spread, not only along riparian habitats, but also into abandoned cropland and other waste places.

Ecology of *Elaeagnus angustifolia*

Germination and vegetative propagation

Prompt germination of *E. angustifolia* seeds required nine to twelve weeks of after-ripening at a temperature of 5° C (Hogue and LaCroix 1970). Fresh seeds have a dormancy induced by coumarin-like chemical inhibitors. Nearly complete germination was

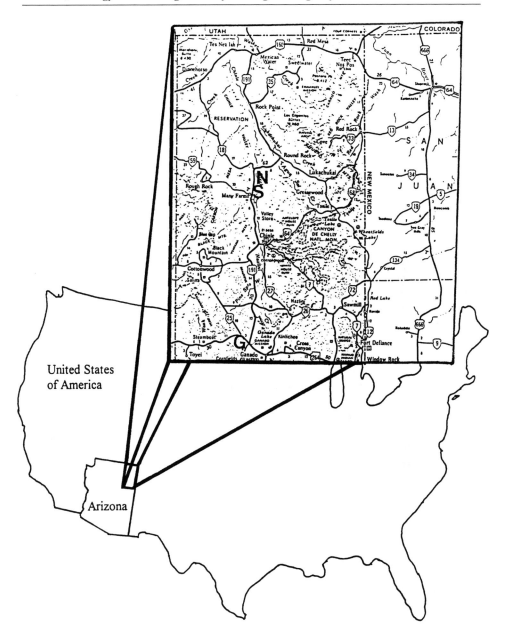

Fig. 3. Location of study sites for *E. angustifolia* stands along Chinle Wash (sites N and S) and the Pueblo Colorado Wash (site G) in northeastern Arizona. Canyon de Chelly National Monument is located, to the east, between the Chinle Wash and Pueblo Colorado Wash sites.

obtained by removing the endocarp and seed coats of *E. angustifolia* seeds (Hogue and LaCroix 1970). Acid treatment is common for hard seeds and improves germination of *E. angustifolia* seeds. Soaking the seeds in sulfuric acid for one hour provided germination percentages ranging from 66 to 98 percent (Heit 1967). Shafroth *et al.* (1995) found that *E. angustifolia* germinated at various times of the year under different combinations of elevation and light and, after germinating, suffered little mortality of seedlings. In nature, the most common method of seed dispersal is ingestion by birds. With passage through the bird's digestive system the germination viability of the seeds should be increased. Young and Young (1992) indicate that seed viability is from one to three years. No literature concerning seed bank viability of *E. angustifolia* was found. With other riparian species, it is common that seed viability in nature is of much shorter duration than under laboratory conditions (Seigel and Brock 1990). Borell (1962) stated that nursery transplants may grow 1.25 to 2 m in height the first year and be 4 m in height after the second year. Plants three to five years after transplanting can be expected to produce fruit (Borell 1962).

Literature specifically describing vegetative propagation of *E. angustifolia* was not encountered in research for this paper. Tellman (1997) reported information that cuttings were commonly used to transfer plant material of *E. angustifolia,* especially as it spread among people in the southwestern United States in the nineteenth century. Bovey (1968) states that *E. angustifolia* spreads from root stalks and was observed to do so by this author, however published material indicating the extent and viability of root sprouting as a regenerative process was not found. Brock (1997 field notes) observed *E. angustifolia* to have the potential to propagate new plants by vegetative means through the process of layering. Branches that were covered with sediments were noted to have numerous adventitious roots on buried stem portions.

Adaptive characteristics

E. angustifolia has a suite of characteristics that enhances its adaptability to the environment of the western United States. It will grow in either full light or shade, but seems to prefer full sunlight. In the greenhouse, *E. angustifolia* had light saturation at 1400 μmol m^{-2}s^{-1} PAR, net photosynthesis measured at 25° C ranging from 14.3 to 17.3 μmol CO_2 m^{-2}s^{-1} , and rates of dark respiration ranging from 0.79 to 1.11 μmol CO_2 m^{-2}s^{-1} (Cote *et al.* 1988).

While *E. angustifolia* is often observed along streams, it is a facultative riparian species. The plant is drought tolerant, as is evidenced by its wide use as a shelterbelt plant, where it exists on ambient precipitation (Carmean 1976). Shelterbelts are rows of trees/shrubs used to reduce soil erosion by wind in the plains states and provinces of North America.

E. angustifolia grows well in a wide range of soils (Carmean 1976; Knopf and Olson 1984), but does not do well in acid soils (Hayes 1976). It tolerates a wide range of soil nutrient levels and has symbiotic actinomycetes species of the genera *Frankia* associated with its root system for atmospheric nitrogen fixation (Miller and Baker 1985; Cote *et al.* 1988). The mode of infection is intercellular penetration of the epidermis and apoplastic colonization of the root cortex (Miller and Baker 1985). *E. angustifolia* tolerates salty soils which provides another adaptive advantage in the semi-arid and arid regions of the western United States. In comparison with twenty other woody perennial trees or shrubs, *E. angustifolia* rated the highest in all categories for field survival, tissue survival under plasmolysis tests, and with tetrazolium salts (Monk and Wiebe 1961). Other plants with similar salt tolerance included: *Rhus trilobata* (squaw bush), *Gleditisa triancanthos* (honey locust), *Robinia psuedoacacia* (black locust), and *Tamarix gallica* (salt cedar). In the southwestern USA, where *Tamarix* and *Elaeagnus* are the most invasive of alien woody species along water courses, the exudation of salts from soils and geological materials in the landscape is a common feature. The reasons *E. angustifolia* easily adapts to sites and becomes competitive to native plants include: (a) withstands drought conditions, (b) tolerates saline soils, (c) has fairly rapid germination of weathered or scarified seeds, (d) germination during moist periods of a growing season, (e) symbiotic nitrogen fixation, (f) rapid growth rates, and (g) an early age for flowering.

Beneficial ecological interactions

Many alien plants make positive contributions to their environment, with these beneficial attributes being the reason they were introduced to new sites. This is quite true for *E. angustifolia,* as it provides edible fruits for wildlife, is a nectar source, provides cover and nesting sites, and was often planted for soil conservation purposes. The fruits of *E. angustifolia* have high value for wildlife, including over 50 game and non-game birds and mammals (Borell 1976; Olson and Knopf 1986a). It has been identified in stomach contents of fishes (Ireland 1997), and serves as a spring time nectar source for insects especially bees (Hayes 1976). *E. angustifolia* has been widely planted in shelterbelts to help control wind aided soil erosion (Borell 1976), where it provides multiple uses as cover and nesting sites for wildlife. This plant was also planted directly into floodplains as an attempt to stabilize streambanks, as reported by Tellman (1997) for Canyon de Chelly in northeastern Arizona. While the short term goal of streambank stabilization was achieved, the introduced population provided a seed source for down stream invasion.

Freehling (1982) reviewed the literature related to *E. angustifolia* and found it to be an important cover and nesting plant, particularly for avifauna species in the middle Rio Grande valley in New Mexico. It is the belief of many biologists that some bird populations have increased with the presence of *E. angustifolia,* especially mourning doves (Freehling 1982).

Negative ecological interactions

Like some alien plants, *E. angustifolia* has become a weedy species that threatens the habitats it has invaded. In the southwestern USA, it is becoming more frequent in the vegetative composition of woody species along streams. Since the 1930's, it has become a codominant with *Populus fremontii* (Fremont cottonwood) in the Rio Grande valley of New Mexico (Freehling 1982; Hink and Ohmart 1984). The increase in the number of *E. angustifolia* was quite apparent to Mount *et al.* (1995), who studied vegetation changes over an eleven year period along another reach of the Rio Grande river in New Mexico. On some sites, it forms nearly monotypic stands (to be described later in this paper) and crowds out native species (Waring and Tremble 1993).

The presence of *E. angustifolia* can result in loss of biodiversity as it tends to support fewer invertebrates than native species (Waring and Tremble 1993) and, as a result, has fewer resources for higher trophic levels. In some studies, authors have found fewer birds and less species richness in stands dominated with *E. angustifolia,* compared to stands of native trees (Brown 1990). When *E. angustifolia* is present, it can be a deterrent to livestock grazing of forage materials. Borell (1976) states that it makes an excellent living fence and sometimes was planted as a hedge type species to confine livestock or exclude them from an area. On many rangeland areas, the presence of

E. angustifolia is considered to decrease forage availability by its physical presence and in thinning herbaceous vegetation cover.

In addition to its negative ecosystem impacts, *E. angustifolia* can affect public health. During the spring pollination period, pollen from this species has been reported to be an allergen to many people (Kernerman *et al.* 1992).

Management and control

The desire to control *E. angustifolia* comes primarily from rangeland managers who see this species as competitive with forage species. More recently, the need to explore management actions for *E. angustifolia* is being realized by a wider range of persons interested in natural resources management. This was manifested by a two day Woody Plant Wetland Workshop in September 1997 that dealt with the ecology and management of it *E. angustifolia* and *T. ramossisima*.

Control of *E. angustifolia* is viewed by many as being difficult because of its resprouting capability and seed bank. The earliest documented efforts to control this plant was in Nebraska in the 1960's, where it had invaded native grasslands. This research showed that nearly 100 percent top kill was achieved with aerially-applied herbicide sprays of 2,4,5-T, 2,4,5-TP (silvex) and 2,4-D aplied alone and in a series of 1:1 herbicide ratios (example 2,4,5-T and 2,4-D) in a solution of diesel and water (Bovey 1965). Repeat applications of the herbicide rates of about 2.2 and 4.4 kg/ha were necessary to attain best results on large trees because of their regrowth potentials. The average level of control for all treatments was 96 percent. Research in Kansas with similar herbicides applied of near saturation to foliage, plus dicamba and picloram provided total root kill (Ohlenbusch and Ritty 1978). In the Kansas research, basal sprays of 2,4,5-T, 2,4,5-TP, dicamba, and triclopyr provided excellent control as did saturation of the plant base with diesel oil. Soil applied herbicide formulations of dicamba, picloram, and tebuthiuron were less effective. These soil treatments averaged 30 percent plant kill with tebuthiuron having the highest level of success at 86 percent (Ohlenbusch and Ritty 1978). Glyphosate was included in the Kansas work and killed *E. angustifolia*, but the authors reported damage of understory vegetation and did not recommend its use in similar conditions. More recently, Parker and Williamson (1996) recommend the use of directed basal spray with 25 percent triclopyr herbicide mixed with diesel or other crop oils for effective and selective control of *E. angustifolia*. They report that this treatment worked best on young stems with smooth bark and that on older plants, part of the treatment should include stem portions with smooth bark tissue. Recently, the New Mexico Agricultural Extension Service found that using imazapyr and glyphosate may effectively control *E. angustifolia* (Keith Duncan, pers. comm. 1997). Many of the sites invaded by of *E. angustifolia* in the southwestern USA do not have a rich herbaceous understory. Thus, the use of herbicides such as glyphosate and imazapyr, which could damage desirable herbaceous plants is less of a concern in this region compared to other areas of North America. Along with the foliage and basal spray applications, applying triclopyr herbicide as a cut-stump treatment should also provide effective control, but is more labor intensive.

Fire and mechanical treatment for *E. angustifolia* control typically results in dense regrowth of the plant. The sole exception for mechanical control was hand pulling (Virginia Native Plant Society 1997). Any treatment that removed the top of the plant, such as direct cutting, resulted in denser regrowth. Fire during the dormant season provides vigorous resprouting of *E. angustifolia* and is not often recommended for vegetation management in riparian areas. No literature concerning the effects of mechanical treatments, such as bull dozing/tree grubbing or root plowing, on *E. angustifolia* was found: however, these treatments are usually highly effective in controlling other small trees with regrowth potential.

A fungal disease that attacks *E. angustifolia* provides some promise for biological control. Morehart *et al.* (1980) reported *Phomosis* canker being effective in controlling *E. angustifolia* trees two and three years of age. However, in instances where this plant is still considered valuable, such as eastern Kansas, the disease is considered very damaging (Tisserat 1995). The disease agent *Phomosis* causes a canker to develop on stem tissues of *E. angustifolia*. Injury to the stem was necessary to promote canker formation by *Phomosis elaeagni* (Morehart *et al.* 1980). Of eight *Phomosis* species tested, only *P. elaeagni* produced disease. Tree death resulted from damage to secondary phloem and xylem tissues and the roots. In more mature trees, lower branches are infected first and after a couple of years, the upper canopy becomes infected and the tree dies.

In eastern Kansas, the causative fungus has been identified as *P. arnoldii* (Tisserat 1995). The disease produces cankers on the young stems that restricts water movement and leads to rapid foliage wilting (Tisserat 1995). This disease is restricting planting of *E. angustifolia* in eastern Kansas and is most devastating in closely spaced plantings with poor air movement. In addition to *Phomosis,* Tisserat (1995) states that *E. angustifolia* is also susceptible to verticillium wilt. The potential for biological control with this fungal disease is present; however no literature reporting insect damage to *E. angustifolia* was found.

Northeastern Arizona populations

Methods

Five stands of *E. angustifolia* growing in northeastern Arizona were sampled for woody plants in May 1997. One stand, designated G, (Fig. 3) was near the community of Ganado along Pueblo Colorado Wash, and the other four stands, along Chinle Wash near the community of Many Farms, are designated N and S (Fig. 3). At the N and S sites, two stands were sampled from each side of Chinle Wash. Site N was approximately three km north of site S.

The sites along Chinle Wash were chosen because natural resource managers of the Bureau of Indian Affairs (BIA) provided evidence that these sites were virtually free of *E. angustifolia* in 1982. That year, the BIA placed gabions and wood baffles, secured to steel posts, perpendicular to the channel of Chinle Wash on its first flood terrace to arrest streambank erosion. Along with these physical structures, cuttings of *Salix exigua* (coyote willow) were planted to provide biological remediation. Only the west floodplain of Chinle Wash contained the soil conservation structures. Following the soil conservation treatment, plants of *E. angustifolia* were observed to be colonizing the site. The gabions and baffles have allowed deposition of about one meter of sediment and have essentially arrested streambank erosion. The sites now support dense stands of woody riparian vegetation dominated by *E. angustifolia* with *T. ramosissi-*

ma, some *S. exigua,* and infrequent plants of *P. fremontii.* Understory vegetation was sparse at the Chinle and Pueblo Colorado Wash sites. The study site near Ganado was chosen because of its accessibility and because it supported a dense stand of *E. angustifolia* representative of many of the streamsides in the southwestern portion of the United States.

Ten macroplots, 25 by 4 m in dimension, were sampled for presence of woody riparian plants in each stand. Each macroplot was randomly located on the floodplain, perpendicular to the stream beginning at the active stream channel. The number of woody plants per plot and their canopy diameter (cm) was recorded by size class. Four size classes were utilized in this study. The smallest plants, seedlings - saplings were designated as size class 1(newly emerging seedlings of very recent germination were not included) and were less than 1 m in height. Size class 2, 3 and 4 increased in stature within the stand by height to 2 m, 3 m and > 4 m respectively. The reason for placing the plants into size classes was to provide a way to assess the degree of recruitment within the *E. angustifolia* populations.

Results

Plant numbers at each site ranged from 1150 plants/ha for the west stand at the S site to 430 plants/ha for the stand of *E. angustifolia* plants on the east side of Chinle Wash at the N site (Table 1). The west stand at the S site had been treated for soil stabilization approximately 15 years ago and contained 470 plants/ha on average of *S. exigua* (Table 1.) The average number of *E. angustifolia* in all stands was 726 plants/ha. The stands also support numerous *T. ramosissima,* growing as a shrub in the understory to the larger *E. angustifolia.* In areas where the winter season is colder, *T. ramosissima* often does not attain tree size and, when closely spaced, it is difficult to identify individual plants. Along the Chinle and Pueblo Colorado Washes, shrubs of *T. ramosissima* were very common, except at the N site; the average for the five study sites was 1830 *T. ramosissima* plants/ha. The range in plant numbers for *T. ramosissima* was from 2910 to 240 for the NE and N sites respectively. While *P. fremontii* is a common tree along many desert streams of the American southwest, only a few saplings of this species were found in the fifty macroplots examined.

Table 1. Average number of alien and native woody riparian plants (number/ha) estimated in May 1997 for five study sites along the Chinle and Pueblo Colorado Washes in northeastern Arizona.

Stand / Site	Species (number/ha)			
	Elan	*Tara*	*Saex*	*Pofr*
North	630	240	0	0
South	1150	2090	470	0
Southeast	680	2500	0	0
Northeast	430	2910	80	10
Ganado	740	1410	0	0
Average per hectare	726	1830	110	2

Elan = Elaeagnus angustifolia, Tara = Tamarix ramosissima, Saex = Salix exigua, and *Pofr = Populus fremontii*

Tree number per size class

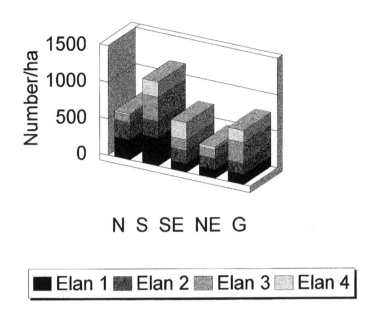

Fig. 4. Total density (number/hectare) and density by size class of *E. angustifolia* plants in 1997 from five stands in northeastern Arizona.

The age distribution of *E. angustifolia* plants, as indicated by size class, was fairly consistent. Averaged for the five stands, class 1 contained 28 % of the population, class 2, 3 and 4 contained 27.6, 26.7 and 17.6 % of the population respectively. These data indicate populations that are still undergoing recruitment with fairly even numbers of juvenile and more mature plants. The west stand at the S site contained the most class 1 plants. All sites contained large and maturing *E. angustifolia* plants but the stands on the west and east side of Chinle Wash at the N site contained the fewest class 4 plants, 20 and 40 plants/ha (Fig. 4).

The average canopy diameters for *E. angustifolia* plants in class 1 through 4 were 1.05, 2.31, 3.29, and 4.25 m. Greatest percent canopy cover by *E. angustifolia* was 77.9 % in the east stand at the S site of Chinle Wash followed closely by 77.6 % for the Pueblo Colorado Wash site (G) (Fig. 5). As would be expected, the greatest contribution to canopy cover at those sites was from the largest (class 4) trees. The stand with the least canopy cover provided by *E. angustifolia* was the east side of Chinle Wash, at the N site, with only 27.4 percent. It is interesting to note that plant numbers, and in this case canopy cover, was lowest for the north site. The working hypothesis is that the *E. angustifolia* invading Chinle Wash could be from seed sources in Canyon de Chelly, which is part of the headwaters of the wash, from plantings made in 1964 by the National Park Service. However, mature *E. angustifolia* are present in the agricultural landscape of the Chinle valley, which could also serve as point type seed sources,

Canopy cover by size class

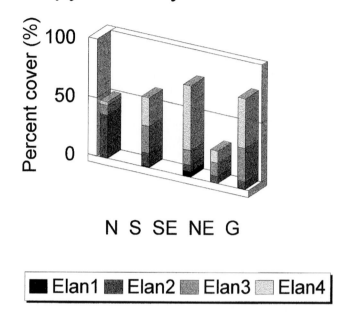

Fig. 5. Percent canopy cover by stand and canopy cover by size class of *E. angustifolia* in 1997 from five stands in northeastern Arizona.

of which the seed would need to be transported to the wash corridor. It is my opinion that the seeds of *E. angustifolia* are being moved along Chinle Wash by fluvial processes more than by transport from adjacent uplands.

Conclusion

E. angustifolia has naturalized to streams and abandoned cropland in the western United States. The greatest impact from this invasive alien has been in riparian areas. Ecosystem perturbations, such as removal of native plant cover, regulation of stream flows, interruption of natural fire regimes, change in the composition of atmospheric gases, global warming, and direct human intervention by planting in wildland areas, have all contributed to its distribution. Ecosystem processes, such as sediment transport during flood flows and seed dispersal by wildlife species, also contributes to its spread once it has been introduced on the landscape.

  Cessation of direct planting of *E. angustifolia* as a conservation plant will help decrease the spread of this species. The desirable attributes of *E. angustifolia,* including drought tolerance, dense canopy or spiny canopy to repel livestock, and a fleshy fruit utilized by a variety of wildlife, could, with careful selection be replaced by native species with similar attributes. Such tree species might include; *Celtis reticulata* (netleaf hackberry) in warmer habitats and *C. occidentalis* (hackberry) in more temperate areas, or species of *Crataegus* (hawthorn). Native shrubs for planting could include *Prunus virginiana* (chokecherry), *P. angustifolia* (chickasaw plum) and *Rhus trilobata* (skunk brush) these plants provide good cover for small mammals and birds, along with fleshy fruits. Along streams, a shrub, *Ribes cereum* (wax currant) or other *Ribes* species with the common name of gooseberry, produce fleshy fruits. As the focus for conservation plants turns from exotic to native species, the need for more plant materials research becomes apparent.

Future research concerning *E. angustifolia* will deal with estimates of seed rain, and seed banks. The area of vegetation management of *E. angustifolia* should also prove fruitful for natural resource scientists. Detailed research with herbicides needs to be conducted, and the study of biological control with insects and pathogens for the control of *E. angustifolia* seems to be open. The ecosystem impacts of *E. angustifolia* are becoming well understood. The next decision is if management of this alien invasive plant is to be undertaken. Then land managers will need to decide what species of plant will replace it in the landscape.

References

Bailey, L. H. 1949. Manual of Cultivated Plants. MacMillan Co., New York, NY.

Borell, A. E. 1976. Russian-olive for wildlife and other conservation uses. Soil Conservation Service. USDA Leaflet No. 517, 8 p.

Bovey. R. W. 1965. Control of Russian olive by aerial application of herbicides. Journal of Range Management 18:194-195.

Brock, J. H. 1997. The ecological role of tamarisk and Russian olive; Interaction with native plant species and other exotic plants. Paper presented at: Woody Plant Wetland Workshop, Colorado State University, Cooperative Extension Service, Grand Junction, CO, USA

Brock, J. H. and Farkas, M. C. 1997. Alien woody plants in a Sonoran Desert urban riparian corridor: An early warning system about invasiveness? In: Brock, J. H., Wade, M., Pyšek, P. and Green, D. M. (eds.), Plant Invasions: Studies from North American and Europe. pp. 19-35. Backhuys Publishers, Leiden, The Netherlands.

Brown, C. R. 1990. Avian use of native and exotic riparian habitats on the Snake River, Idaho. MS Thesis. Department of Fishery and Wildlife Biology, Colorado State University, Fort Collins, CO. 60 p.

Carmean, W. H. 1976. Soil conditions affect growth of hardwoods in shelterbelts. North Central Forest Experiment Station, Forest Service, USDA. Research Note NC-204.

Christensen, E. M. 1963. Naturalization of Russian olive (*Eleagnus angustifolia* L.) in Utah. American Midland Naturalist 10:133-137.

Cote, B., Carlson, R. W. and Dawson, J. O. 1988. Leaf photosynthetic characteristics of seedlings of actinorhizal *Alnus* spp and *Elaeagnus* spp. Photosynthesis Research 16:211-218.

Freehling, M. D. 1982. Riparian woodlands of the middle Rio Grande Valley, New Mexico: A study of bird populations and vegetation with special reference to Russian-olive (*Elaeagnus angustifolia).* Report to: US Fish and Wildlife Service, Albuquerque, NM. 35 p.

Hayes, B. 1976. Planting the *Elaeagnus* Russian and autumn olive for nectar. American Bee Journal 116:74 and 82.

Heit, C. E. 1967. Propagation from seed. Part 6: Hardseededness - a critical factor.   American Nurseryman 125:10-12, 88-96.

Hink, V. C. and Ohmart, R. D. 1984. Middle Rio Grande biological survey. A report to: Army Corps of Engineers. 193 p.

Hogue, E. J. and LaCroix, L. J. 1970. Seed dormancy of Russian olive (*Elaeagnus angustifolia* L.). Journal of American Horticultural Society 95:449-452.

Ireland, T. 1997. The ecological role of tamarisk and Russian olive on threatened & endangered species. Paper presented at: Woody Plant Wetland Workshop, Colorado State University, Cooperative Extension Service, Grand Junction, CO, USA.

Kearney, T. H. and Peebles, R. H. 1968. Arizona Flora. University of California Press, Berkeley, CA.

Kernerman, S. M., McCoullough, J., Green, J. and Ownby, D. R. 1992. Evidence of cross-reactivity between olive, has, privet, and Russian olive tree pollen allergens. Annals of Allergy 69:493-496.

Knopf, F. L. and Olson, T. E. 1984. Naturalization of Russian-olive: Implications to Rocky Mountain wildlife. Wildlife Society Bulletin 12:289-298.

Kowarik, I. 1995. Time-lags in biological invasions with regard to success and failure of alien species. In: Pyšek, P., Prach, K. , Rejmanek, M. and Wade, M. (eds.), Plant Invasions: General Aspects and Special Problems, pp. 15-38. SPB Academic Publ., Amsterdam.

Miller, I. M. and Baker, D. D. 1985. The initiation, development and structure of root nodules in *Elaeagnus angustifolia* L. (*Elaeagnaceae*). Protoplasma 128:107-119.

Monk, R. W. and Wiebe, H. H. 1961. Salt tolerance and protoplasmic salt hardiness of various woody and herbaceous ornamental plants. Plant Physiology 36:478-482.

Morehart, A. L., Carroll, R. B. and Stewart, M. 1980. *Phomopsis* canker and dieback of *Elaeagnus angustifolia*. Plant Disease. 64:66-69.

Mount, J., Krausman, W., and Finch, D. M. 1995. Riparian habitat change along the Isleta-Belen reach of the Rio Grande. In: Shaw, D. W. and Finch, D. M. (eds.) Desired future conditions for Southwestern riparian ecosystems: Bring interests and concerns together. Forest Service, USDA. General Technical Report RM-GTR-272. p. 58-61.

Ohlenbusch, P. D. and Ritty, P. M. 1978. Russian olive control - a preliminary look. Proceedings: North Central Weed Control Conference. Annual Meeting. 33:132.

Olson, D. F. 1974. *Elaeagnus*. In: Seeds of the Woody Plants of the United States. pp. 224-226. Forest Service, USDA, Washington, DC.

Olson, T. E. and Knopf, F. L. 1986a. Naturalization of Russian-olive in the western United States. Western Journal of Applied Forestry 1:65-69.

Olson T. E. and Knopf, F. L. 1986b. Agency subsidization of a rapidly spreading exotic. Wildlife Society Bulletin 14:492-493.

Parker, D. and Williamson, M. 1996. Low-impact, selective herbicide application for control of salt cedar and Russian-olive. Southwest Region, Forest Service, USDA, Albuquerque, NM. 2 p.

Shafroth, P. B., Auble, G. T. and Scott, M. L. 1995. Germination and estabishment of native plains cottonwood (*Populus deltoides* Marshall subsp. *monilifera*) and the exotic Russian-olive (*Elaeagnus angustifolia* L.). Conservation Biology 9:1169-1175.

Seigel, R. S. and Brock, J. H. 1990. Germination requirements of key southwestern woody riparian species. Desert Plants 10:3-8,34.

Tellman, B. 1997. Exotic pest plant introductions in the American southwest. Desert Plants 13:3-10.

Tissert. 1995. *Phomopsis* canker of Russian olive. Cooperative Extension Service. Kansas State University.

Turner, R. M. and Karpiscak, M. M. 1980. Recent vegetation changes along the Colorado River between Glen Canyon Dam and Lake Mead, Arizona. Geological Survey Professional Paper, 1132, USDI.

Virginia Native Plant Society. 1997. Autumn olive (*Elaeagnus umbellata* Thurnberg) and Russian olive (*Elaeagnus angustifolia* L.) Internet Fact Sheet. 2 p.

Vitousek, P. M. 1986. Biological invasions and ecosystem properties, Can species make a difference? In: Mooney, H. A. and Drake, J. A. (eds.), Ecology of Biological Invasions of North American and Hawaii. Ecological Studies 58, pp.163-176. Springer-Verlag, New York.

Vitousek, P. M., D'Antonio, C. M., Loope, L. L. and Westbrooks, R. 1996. Biological invasions as global environmental change. American Scientist. 84:468-478.

Waring, G. L. and Tremble, M. 1993. The impact of exotic plants on faunal diversity along a southwestern river. Unpublished Report to: Nature Conservancy. 33 p.

Welsh, S. L., Nelson, C. R. and Thorne, K. H. 1971. Naturalization of plant species in Utah. In: James, L. F., Evans, J. O., Ralphs, M. H. and Childs, R. D. (eds.), Noxious Range Weeds, pp. 17-29. Westview Press, Denver, CO, USA.

Young, J. A. and Young, C. G. 1992. Seeds of woody plants in North America. Dioscorides Press. Portland, OR.

ECLIPTA PROSTRATA (L.) L. AS A NEW WEED OF RICE FIELDS IN SARDINIA (ITALY)

Giuseppe Brundu, Vincenzo Satta and Tullio Venditti
*Dipartimento di Botanica ed Ecologia vegetale, Università di Sassari. Via Muroni,
25-07100 Sassari, Italy; e-mail: gbrundu@tin.it, sattav@tin.it,
tulliovend@hotmail.com*

Abstract

The spread of *E. prostrata* in Sardinia can be dated to very recent times and it is related to the expansion of rice cultivation. The weed finds refuge in the raised banks of artificial channels and in the wet areas surrounding the rice fields. Its establishment is enhanced by inadequate and/or inefficient cleaning of the channels. The weed produces a great amount of seeds and, thus, control techniques should not be limited only to cultivation areas. The spread of *E. prostrata* in Sardinia has been encouraged by neglect.

Since the first records, in 1985, the plant has actually invaded almost all Sardinian rice fields. The colonisation process starts from the raised banks of channels. It invades the rice field during the dry phase of cultivation, shortly before harvesting, especially when other weeds have been eliminated by herbicides. Apart from competing directly with the rice plants, the alien is very harmful because its sprouts clog the operating mechanisms of the harvesting machines.

Introduction

After first being recorded in Sardinia in 1985, *Eclipta prostrata* is now a undesirable invader of rice fields. As well as competing directly with the rice plants, its main harm is, during harvesting, while still green, clogging harvesting machines.

Changes in agricultural practices (*i.e.* planting pattern density, length and frequency of dry periods, fertilisations, chemical control of weeds) have promoted the invasiveness of this alien.

This new weed has to be controlled. In the last 50 years many exotic species have been recorded in the rice fields of Sardinia (Viegi 1993): *Cyperus difformis* L., *Cyperus eragrostis* Lam., *Echinochloa crus-galli* (L.) Beauv., *Echinochloa oryzoides* (Ard.) Fritsch, *Elatine triandra* Schkuhr, *Heteranthera limosa* (Sw.) Willd., *Heteranthera rotundifolia* (Kunth) Grisebach, *Paspalum dilatatum* Poiret, *Paspalum paspaloides* (Michk.) Scribner, *Schoenoplectus mucronatus* (L.) Palla, *Ammania auricolata* Willd. Most of these aliens may have been introduced by seed imported from South-East Asia (Chiappini 1962; Marchioni Ortu and De Martis 1982; Marchioni Ortu *et al.* 1988).

General aspects of the species

Eclipta prostrata (L.) L. (=*E. alba* (L.) Hasskarl, incl. var. *neapolitana* N. Terr.), (*Compositae*) is an annual aquatic emergent plant, with standing or twining stems,

Plant Invasions: Ecological Mechanisms and Human Responses, pp. 137–141
edited by U. Starfinger, K. Edwards, I. Kowarik and M. Williamson

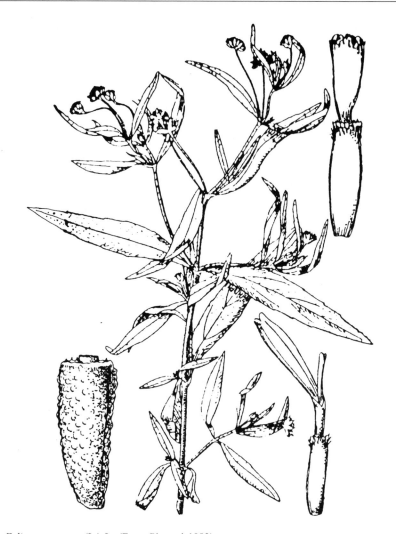

Fig. 1. Eclipta prostrata (L.) L. (From Pignatti 1982).

depending on the water depth (Fig. 1). It can produce secondary roots from lateral branches when they reach the soil, forming new plants and occupying the surrounding space. A significative polymorphism, possibly adaptive to different environments, is commonly present.

It is thought to originate in America, but nowadays it has a very wide geographical range, being present in almost all the tropical and sub-tropical regions of the world (Terracciano 1902; Kupicha 1975; Tutin *et al.* 1976; Feinbrun-Dothan 1978; Lisowski 1991).

Growing only in wet sites like channels, it has become an aggressive weed in many rice fields and in other crops like peanut, soybean and cotton. As a result of rice cultivation, this plant has extended its range in south Europe.

E. prostrata is used as food in tropical Africa, after boiling, to avoid the effect of toxic sulphur compounds derived from tyophene (Feinbrun-Dothan 1978). It also contains alfa-terthienyl, a compound toxic to the nematode *Pratylenchus penetrans* (Heywood and Harborna 1977), a harmful pest to many cultivated plants. There are also poli-acethylens, which might have fungicidal properties and alkaloids such as nicotine (Pal and Narasimhan 1943; Swain and Williams 1977).

E. prostrata in Italy

The first record in Italy, in 1900, (Terracciano 1902) was near Naples. The alien was called *Eclipta alba* (L.) Hasskarl. Terracciano suggested that the seeds might have been carried by migrating birds, from Egypt or Asia Minor, the only places in the Mediterranean where the species was formerly present. He noticed some differences between the *typus* and the samples, and so described a new variety *neapolitana*.

In spite of further field research, there were no further finds at these stations. In 1958, Anzalone found a site in central Italy, near the coast, under a road bridge. In his opinion the alien was quite rare in Italy, being present mostly in small sites along the Tyrrhenian coast.

The first record for Sicily (Giardina 1992) was near the sandy shores of the stream *Buttaceto*, inside the Nature Reserve *Oasi del Simeto*, at 1 m a.s.l.

Viegi *et al.* (1974) describe the species as randomly adventive in Italy, Tutin *et al.* (1976) regard it as locally naturalized, Pignatti (1982) as naturalized in central Italy, even though it has disappeared from the former sites.

E. prostrata in Sardinia

E. prostrata was found for the first time in Sardinia in 1985, by Marchioni *et al.* (1988), but no data were collected at that time regarding the extent of its invasion. Some samples were later collected by Satta (1994) who noticed its invasivity in the rice fields of *Oristano*, in central-western Sardinia, where the introduction was from seed stocks coming from the Philippines.

These rice fields with the weed are on soils locally called *Gregori*, which are Typic and Lithic Palexeralfs, according to USDA Soil Taxonomy (Aru and Baldaccini 1961; Aru *et al.* 1991). They are rich in pebbles, have a high clay content, are slightly saline and markedly hydromorphic because of poor drainage and are not well-suited for agriculture, with the exception of rice cultivation.

In 1994 the damage to production was 100% because the harvesting was not possible due to the presence of the weed. *E. prostrata* was distributed with a mosaic-like pattern in the rice fields, with 1-2 plant m^{-2}, and an average covering of 20-30%. In the years since then the population has been controlled by chemicals and soil tillage. But inadequate cleaning of channels provide a shelter for the weed that can re-invade the fields from year to year, to different extents.

E. prostrata it is well adapted to the cycle of rice cultivation in Sardinia. Plants that have survived the winter (these survivors are shrubby chamaephytes, with resting buds below 50 cm and partially lignified branches) start to grow in late spring, because of rising temperature. At the same time, the seeds start to germinate in wet habitats such as channels and abandoned fields. Germination is normally over by the first half of June, so that after then, the weed is resistant to anti-germinating agents.

It can tolerate treatments against *Echinochloa* sp. pl. (e. g. Molinate, Molinate with Thiobencarb) and treatment in the crop tillering phase against *Cyperaceae* and *Alismataceae* (e.g. Bentazon: 1.6 kg ha^{-1} a.p.). *E. prostrata* also takes advantage of fertilisation, mostly by phosphate but also nitrogen. The former is spread before planting rice, at about 100-120 kg ha^{-1} of P_2O_5, the latter at 90-100 kg ha^{-1} of N (normally as urea).

The weed grows more vigorously outside rather than within the rice fields. This may be due to the competition of the rice plants and the weed is normally more upright and weaker. During the flooding of the crop, the weed is easily controlled and harmless. But dry periods of cultivation promote the prostrate growth form and rooting by those branches that reach the soil. *E. prostrata* covers the soil to a great extent and also grows vigorously within the crop.

In the second half of July the rice fields are flooded for the last time. The water depth in the field ranges between 8-10 cm, the air temperature is about 24°C (average daily air temperature for the month of July).

These conditions, create a sort of tropical microclimate and promote vegetative growth and flowering, mostly between August and September. In the milk-phase of rice, in the dry phase of cultivation, the weed invades the fields and propagates vegetatively (branch rooting) and by seeds. The achenes germinate readily; the dormant phase is very short or nearly absent (Marchioni *et al.* 1988).

When the crop is ready for harvest, there is a set of *E. prostrata* plants in different phases of phenology.

Sardinian rice production represents about 1% of the total Italian yield. The rice fields occupy an area of 35 km². Cultivation is mostly concentrated in the south-west of the island, and recently, some trial plots have been planted in alluvial plains of northern Sardinia. The yields range between 6.5 and 7 tons per hectare and the production has a high economic value, because most of it is sold and used as certified seed.

Conclusions

We analysed some aspects of the ecology and spread of *E. prostrata* in the rice fields of Sardinia. The weed was introduced in 1985 to rice fields by seeds imported from the Philippines. *E. prostrata* it is well adapted to the cycle of rice cultivation in Sardinia and its spread is related to the expansion of rice cultivation. The weed finds refuge in the channels and in other wet areas, thus the spread in Sardinia is encouraged by neglect.

References

Anzalone, B. 1958. *Eclipta alba* Hassk. var. *neapolitana* N. Terr. N. Giorn. Bot. It. 65: 878-879.
Aru, A. and Baldaccini, P. 1961. Contributo alla pedologia dell'Oristanese. I suoli sulle alluvioni del Tirso e sui detriti di falda del Monte Arci. C.R.A.S. Soc. Poligr. Sarda, Cagliari.
Aru, A., Baldaccini, P. and Vacca, A. 1991. Nota illustrativa alla carta dei suoli della Sardegna. Regione Autonoma della Sardegna, Cagliari.
Chiappini, M. 1962. Diffusione del *Paspalum distichum* L. ssp. *paspalodes* (Michx.) Thell. in Sardegna. Annali di Botanica 27 (2): 331-336.

Feinbrun-Dothan, N. 1978. *Eclipta* L. In: Zohary, M. and Feinbrun-Dothan, N. (eds.), Flora Palaestina. pp. 324-325. The Israel Academy of Sciences and Humanities, Jerusalem.

Giardina, G. 1992. *Eclipta prostrata* (L.) L. Avventizia nuova per la Sicilia. Inf. Bot. It. 24: 200-201.

Heywood, V.H. and Harborne, J.B. 1977. An overture to the *Compositae*. In: Heywood, V.H., Harborne, J.B. and Turner, B.L. (eds), The Biology and Chemistry of the *Compositae*. pp. 15-16. Academic Press.

Kupicha, F.K. 1975. *Eclipta* L. In: Davis, P.H. (eds.), Flora of Turkey and the East Aegean Islands. pp. 45-46. University Press, Edinburgh.

Lisowski, S. 1991. Les *Asteraceae* dans la flore d'Afrique centrale. vol 1-2. Fragmenta floristica et Geobotanica. 36 (1/1): 223-225.

Luppi, G. and Finassi, A. 1989. Riso (*Oryza* sp.). In: Baldoni, R. and Giardini, L. (eds.), Coltivazioni erbacee. pp. 221-270. Patròn Editore, Bologna.

Marchioni, A. and De Martis, B. 1982. Su alcune avventizie nuove per la flora di Sardegna. Atti Soc. Tosc. Sci. Nat., Mem., Serie B 98: 61-66.

Marchioni Ortu, A., De Martis, B., Ortu, M. and Scintu, G. 1988. Ecologia della germinazione in *Eclipta prostrata* (L.) L. (*Compositae*). Thalassia Salentina 18: 335-342.

Pal, S.N. and Narasimhan, N.J. 1943. The alkaloid in *Eclipta alba* (L.) Hassk. J. Indian Chem. Soc. 20, 181.

Pignatti, S. 1982. Flora d'Italia. Edagricole, Bologna.

Satta, V. 1994. *Eclipta prostrata* (L.) L. Nuova avventizia delle risaie in Sardegna. Inf. Bot. It. 26: 215-216.

Spanu, A. and Milia, M. 1980. Possibilità produttive e prospettive di sviluppo della risicoltura in Sardegna. L'Informatore Agrario 36 (26): 11221-11224.

Spanu, A., Pruneddu, G. and Deidda, M. 1989. Primi risultati sulla coltivazione del riso (*Oryza sativa* L.) irrigato per aspersione. Rivista di Agronomia 4: 378-384.

Swain, T. and Williams, C.A. 1977. *Heliantheae*. Chemical review. In: Heywood, V.H., Harborne, J.B. and Turner, B.L. (eds.), The Biology and Chemistry of the *Compositae*. pp. 688-696. Academic Press.

Terracciano, N. 1902. Il genere *Eclipta* nella flora italiana. Bull. Soc. Bot. It. 15: 65-69.

Tutin, T.G. *et al.* 1976. Flora Europea. Cambridge University Press.

Viegi, L. 1993. Contributo alla conoscenza della biologia delle infestanti delle colture della Sardegna nord-occidentale. I. Censimento delle specie esotiche della Sardegna. Boll. Soc. Sarda Sci. Nat. 29: 131-234.

Viegi, L., Garbari, F. and Cela Renzoni, G. 1974. Le esotiche avventizie della Flora italiana. Inf. Bot. It. 6: 274-280.

COST EFFECTIVE CONTROL OF *FALLOPIA JAPONICA* USING COMBINATION TREATMENTS

Lois Child[1], Max Wade[2] and Martin Wagner[3]

[1]*International Centre of Landscape Ecology, Department of Geography, Loughborough University, Loughborough, Leicestershire LE11 3TU. United Kingdom;* [2]*Department of Environmental Sciences, University of Hertfordshire, Hatfield AL10 9AB. United Kingdom;* [3]*Thames Water Utilities, Environment and Science, Nugent House (RBH2), Vastern Road, Reading, Berkshire RG1 8DB. United Kingdom, e-mail: L.E.Child@lboro.ac.uk*

Abstract

An area of monoculture *Fallopia japonica* was selected in north east London, UK in order to explore the application of combined treatment methods to reduce both the length of time taken and the high costs involved in eradicating the plant. Six plots were assigned treatments in a fully crossed two factor field experiment to assess the effects on the growth of the plant of combined mechanical digging and spraying treatments using the herbicide glyphosate. The trial was conducted over two growing seasons and was extended for a further season to assess the effect of an additional spray application. Significantly greater reduction in plant density, plant height, stem diameter and number of leaves was recorded in plots subjected to a combination of treatments. The combination of disturbance due to digging followed by a single application of herbicide achieved a high level of control in 18 months. Following a second herbicide application, complete control was achieved. The implications of this combination of treatments on infested sites awaiting redevelopment in the urban environment are discussed.

Introduction

Fallopia japonica (Houtt.) Ronse Decraene (Japanese knotweed) (syn. *Reynoutria japonica* (Houtt.), *Polygonum cuspidatum* (Sieb. and Zucc.)) is a vigorous invasive plant native to Japan and northern China which was introduced to Europe in the mid-nineteenth century as an ornamental and fodder plant (Conolly 1977). Since its naturalisation in 1886, it has spread rapidly throughout the British Isles causing serious management problems in a variety of habitats, including river and stream corridors, road verges and along railway embankments and has become a particular problem in the urban environment, especially on waste industrial sites (Child *et al.* 1992; Child and de Waal 1997; Child and Wade 1997).

F. japonica is a rhizomatous perennial plant which exhibits vigorous growth in spring, producing tall stems up to 2.5 m in height by early summer. The plant flowers in August to September and dies back late in the season, leaving erect stems which persist over the winter. In the British Isles, *F. japonica* does not produce viable seed. Although the plant is functionally dioecious, all *F. japonica* plants recorded to date have been found to be male sterile (Bailey 1990). Spread of *F. japonica* in the British Isles is purely by vegetative means, primarily by rhizome fragments but also by cut stems. As

Plant Invasions: Ecological Mechanisms and Human Responses, pp. 143–154
edited by U. Starfinger, K. Edwards, I. Kowarik and M. Williamson

little as 0.7 g of rhizome material has been shown to give rise to a new plant (Brock and Wade 1992; Brock *et al.* 1995).

The disposal of *F. japonica* stems and rhizomes in the UK is subject to Environmental Protection (Duty of Care) Regulations (Department of the Environment 1991) and as such must be disposed of at a licensed landfill site (Ministry of Agriculture, Fisheries and Food 1995). The use of herbicides is regulated in the UK by the Control of Pesticides Regulations 1986 (Ministry of Agriculture, Fisheries and Food 1995). Only two herbicides, glyphosate and 2,4-D amine, are approved for use in or near water and are effective against *F. japonica* (Roblin 1988; Beerling 1990; de Waal 1995). De Waal (1995) explored the combined effect of cutting and spraying with glyphosate at different times throughout a growing season and achieved good control with three glyphosate applications over three growing seasons. Other translocated herbicides which have been shown to be effective against the plant are picloram, triclopyr (Harper and Stott 1966; Scott and Marrs 1984) and imazapyr (Welsh Development Agency 1991). These herbicides are suitable for use on paving, hard surfaces, non-amenity areas and land without wildlife value and may be suitable for use on redevelopment sites. They are not approved for use near water and are persistent in the soil, which may delay replanting with substitute species.

On redevelopment sites, construction works can be adversely affected by *F. japonica* causing damage to paving and hard surfaces and treatment of the plant is necessary prior to work commencing (Welsh Development Agency 1994). An alternative method of decontamination of sites has been employed in cases where a site has not been available for pre-treatment of *F. japonica*. This involves removal from the site of top soil containing plant fragments to a depth of 2 m and disposing of the spoil in a landfill site. The cost of this extreme measure is very high and is wasteful in terms of land resources. A recent estimate of the cost of removing 2,730 m^3 of soil contaminated with *F. japonica* rhizomes from a redevelopment site in Cardiff, South Wales and subsequent disposal to landfill, was in the region of £ 0.5 million (pers. com. Kajima Engineering, Cardiff).

Redevelopment of waste industrial sites is often delayed when *F. japonica* is growing on site. Financial appraisal of redevelopment sites includes an element for the length of time during which the land remains unused before development, known as the waiting time, where investment costs, in terms of land purchase cost and interest payments (the Total Finance Costs), accumulate. If this waiting time could be cut by rapid treatment of *F. japonica*, the costs associated with the pre-treatment of sites would fall significantly. An estimation of associated costs are described in Table 1. In cases where land does not become available to a potential redeveloper until building work is due to commence, suitable areas may be found on site for transfer of infested soils for later treatment. Where no space is available for transfer of infested soils, e.g. on small sites and those where compulsory purchase orders necessitate the acquisition of a minimum land area such as highway improvements, rapid treatment on site is required to avoid the excessive costs of removal of infested material to landfill. It is against this background that the present paper seeks to provide a cost effective and rapid treatment for the control of *F. japonica*.

The rationale behind the trial was the hypothesis that a combination of digging to a depth of 0.5 m, thoroughly fragmenting the rhizome system and spraying the resulting regrowth with a translocated herbicide would significantly accelerate the rate at which

Table 1. Summary of comparative treatment costs for development sites adjusted for finance costs. Assumptions: Land value costed at £75 m^{-2}; interest rate of 10% year $^{-1}$ rolled up quarterly. Treatment costs from Spon's Architects' and Builders' price book (Davis *et al.* 1992).

TREATMENT	DURATION OF TREATMENT (months)	TOTAL TREATMENT COST (£ m^{-2})	TOTAL FINANCE COST (£ m^{-2})	TOTAL COST OF TREATMENT ADJUSTED FOR FINANCE COST (£ m^{-2})
DIG AND SPRAY (DIG + 2 SPRAYS)	18	1.93	11.98	13.91
CONVENTIONAL SPRAY (2 TREATMENTS ANNUALLY FOR AT LEAST 3 YEARS)	36	1.32	25.87	27.19
(i) EXCAVATE (to 2m depth), CART AWAY AND LANDFILL	3	20.90*	1.88	22.78*
(ii) As (i) + IMPORT SOIL AND COMPACT	3	49.00*	1.88	50.88*

* excluding Landfill Tax currently at £7.00 m^{-3}

F. japonica was controlled. The trial site consisted of a monoculture stand of *F. japonica* adjacent to water and therefore the herbicide glyphosate was selected.

Experimental design

Mill Meads, an area of waste industrial land at Ordnance Survey grid reference TQ 385830, is infested with *F. japonica* and was identified as a suitable site for the trial. The site is adjacent to Abbey Creek at its confluence with the Channelsea River at Stratford, north east London and therefore, only those herbicides approved for use near water by Ministry of Agriculture Fisheries and Food (1995) can be used.

An area of approximately 420 m^2 of monoculture *F. japonica* was divided into six treatment plots. The plots were allocated to various treatments to be carried out initially over a two year period. One of the plots received no treatment and was used as a control. Two treatments DIGGING and SPRAYING were applied at two levels in a fully crossed factorial design. Two replicate plots were assigned to dug plots so that these could be further separated in year three for a second spray treatment if necessary. All plots, including the control, were strimmed (cut with a metal bladed strimmer) to remove dead stems at the end of the growing season in October 1993 and late September 1994. The plots and treatments for years one and two (1993, 1994) are indicated in Table 2. Throughout the trial the plots were monitored according to the procedure described below.

Table 2. Timing of treatments and monitoring for initial trial and follow up showing allocation of plots to two treatments (DIGGING and SPRAYING) at two levels (DIGGING and NO DIGGING, SPRAYING and NO SPRAYING).

Date	Treatment	Plot number
INITIAL TRIAL		
September 1993	site preparation – strimming	all plots
October 1993	treatment 1 – digging	
	no digging	plot 1, plot 4
	digging	plot 2, plot 3,
		plot 5, plot 6
May 1994	monitoring	all plots
May 1994	treatment 2 – spraying	
	no spraying	plot 4, plot 5, plot 6
	spraying	plot 1, plot 2, plot 3
September 1994	strimming	all plots
may 1995	monitoring	all plots
FOLLOW UP TREATMENT		
June 1996	monitoring	plot 2, plot 3
	(access difficult to Plots 1, 4, 5 and 6 due to plant height and density)	
June 1996	strimming	all plots
July 1996	monitoring	all plots
July 1996	follow up treatment:	
	no spraying	plot 4, plot 5
	spraying	Plot 1, Plot 6
	spot treatment	plot 2, plot 3
November 1996	monitoring	all plots

Methods

Site preparation, digging and spraying were carried out by contractors on the assigned plots according to the procedures described below. Timings of treatments and post-treatment monitoring are shown in Table 2. A follow up treatment was applied after the initial trial to investigate the effect of strimming and a late spray on one of the dug and unsprayed plots and of spot treatment of surviving plants in previously sprayed plots. Strimming was undertaken to remove old stems at the beginning of the trial and in 1994 at the end of the season. Strimming carried out in the final year (June 1996) was undertaken on actively growing plants in order to remove excessively tall plant material, as access to plots for monitoring was impeded. Strimming was carried out over all plots in order to exclude additional treatment effects.

Site preparation

The site was prepared in late September 1993, when dead stems of *F. japonica* over all plots were cut to a height of 200 mm using a metal bladed strimmer. The cut material was left *in situ*. On 12 October 1993, plots 2, 3, 5 and 6 were dug using a JCB mechanical excavator. An area of approximately 370 m^2 was completed in 2.5 hours. Digging was carried out according to the digging procedure described below. The site was left in this condition over the winter 1993-4.

Procedure for digging

Digging, carried out using a JCB excavator with a 0.24 m^3 bucket, involved:
(i)   scraping of surface crowns and rhizomes into a pile
(ii) cultivation (excavation to a depth of 500 mm and replacement of soil)
(iii) spreading piled crowns and rhizomes back over the cultivated area

Procedure for spraying

Spraying was carried out by an approved contractor using a standard knapsack sprayer fitted with a very low volume nozzle. Glyphosate was applied at a rate of 5 l ha^{-1} (2,154 g active ingredient ha^{-1}), using a low water volume (80 l ha^{-1}), when the plant had reached a height of 0.75-1.00 m and sufficient leaf area was available to take up the herbicide. It was important that the whole of each plot was sprayed and that no stems were missed. It was equally important to the success of the trials that spray did not drift onto plots which were not allocated to receive spray treatment

Procedure for monitoring

In order to assess the continued effectiveness of treatment, a non-destructive sampling technique was adopted and no cropping was undertaken. The following variables were measured within each treatment plot throughout the trial, using three 1 m^2 quadrats in each plot:

- shoot density, i.e. number of shoots per m^2
- shoot height
- stem diameter
- number of leaves per stem
- assessment of approximate leaf area

Care was taken to minimise any edge effects from adjoining treatments by placing each of the three quadrats at measured distances along the centre line of each plot.

The total number of stems in each quadrat was counted and recorded and only those stems which lay completely inside the quadrat frame were counted. Nine plants were selected for measurement in each 1 m^2 quadrat. Plant height was measured from ground level to the base of the apical shoot and stem diameter was measured using callipers at approximately 10 mm above the second node from the ground. The number of leaves per plant was counted including only those leaves which were fully open. The length x width of a leaf (at the widest part) was measured on a plant selected as representative of the plants measured within each quadrat to approximate average leaf size per quadrat. This was used purely as a visual observation during the trial.

Procedure for calculation of costs

Treatment costs (Table 1) were taken from Spon's Architect's and Builder's price book (Davis *et al.* 1992). Prices are as quoted for "measured work – minor work" (i.e. prices apply to a small project costing in the region of £70,000 in the outer London area) and are as follows:

(a)

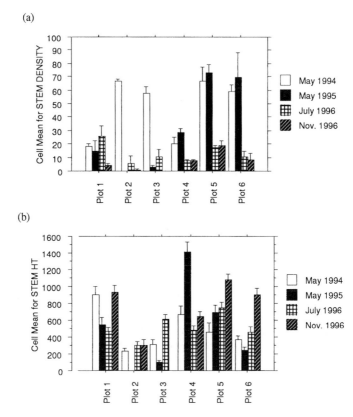

(b)

Fig. 1. Results of (a) mean stem density, (b) mean stem height for plots 1-6 throughout the trial period (Error bars show ±1 Standard Error).

- Digging at £1.49 m^{-3}. Treatment defined as: Excavation and filling – Excavation using a wheeled hydraulic excavator with a 0.24 m^{-3} bucket; excavating to reduce levels; not exceeding 1 m deep.

- Spraying at £0.22 m^{-2}. Treatment defined as: Clearing site vegetation.

- (i) Dig, cart away and landfill at £20.90 m^{-3}. Treatment defined as: Excavation to reduce levels up to 2 m depth at £3.98; Mechanical disposal of excavated materials to landfill not exceeding 13 km distance using lorries at £14.10 m^{-3}. Surcharge for extra distance at £2.82 m^{-3}.

- (ii) Dig, cart away and landfill, import soil and compact at £49.00 m^{-3}. Treatment defined as (i) plus: Mechanical filling with imported soil, deposition; compacting in layers at £28.10 m^{-3}.

Data analysis

Results were entered onto an EXCEL database for further analysis. Data were transformed to log$_{10}$ to equalise variances according to Day and Quinn (1989). Analysis of

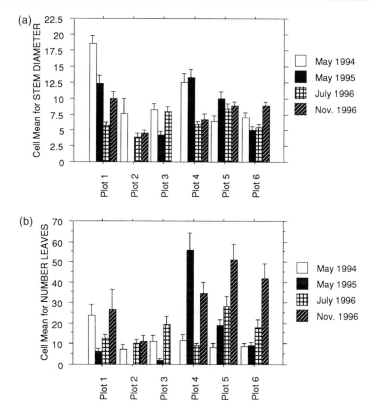

Fig. 2. Results of (a) mean stem diameter, (b) mean number of leaves for plots 1-6 throughout the trial period. (Error bars show ± 1 Standard Error).

results from the 1994 monitoring were carried out using single factor Analysis of Variance (ANOVA). Remaining data were analysed using two factor ANOVA in StatView (Abacus Concepts Inc. 1992-95). Multiple comparisons of means were made *a posteriori* using Tukey's test for highest significant difference (hsd). From these analyses an assessment of the efficacy of treatments was made and compared against the control.

Results

Monitoring 1994

Plots were monitored in May 1994 between treatments 1 and 2 in order to assess the effect of digging on plant growth. Results are presented in Figures 1 and 2. The plots subjected to the digging treatment (plots 2, 3, 5 and 6) showed a significantly higher plant density than undug plots (plots 1 and 4) ($p<0.01$). The plants in these plots also achieved significantly less height than those in undug plots ($p<0.05$), had significantly smaller stem diameter ($p<0.01$) and significantly lower number of leaves ($p<0.05$).

Visual inspection of the plots identified differences in the vegetation composition of

Table 3. Observations of species appearing in plots subjected to digging, by plot number

Species	1994	Monitoring date 1995	1996 June	1996 November
Alliaria petiolata				2,3
Anthriscus sylvestris	2,3,5,6	2,3,5,6	2,3,5,6	2,3,5,6
Ballota nigra			2,3	2,3
Buddleja davidii				2,3
Cirsium sp.	2,3,5,6	2,3,5,6	2,3,5,6	2,3,5,6
Galium aparine			2,3	2,3
Heracleum sphondyllium	2,3,5,6	2,3,5,6	2,3,5,6	2,3,5,6
Impatiens glandulifera		5,6		
Rubus fructicosus agg.				2,3
Rumex sp.	2,3,5,6	2,3,5,6	2,3,5,6	2,3,5,6
Silene vulgaris			2,3	2,3
Tussilago farfara				2,3
Urtica dioica	2,3,5,6	2,3,5,6	2,3,5,6	2,3,5,6

the plots. In dug plots, species other than *F. japonica* were present (Table 3), whereas in plots not subjected to digging, only *F. japonica* was recorded. Growth of *F. japonica* was more even over dug plots, i.e. the undug plots had more clumps of the plant growing from established crowns. Regrowth on dug plots was from rhizomes which had been dug and re-spread over the plot.

The plots were visually inspected in August 1994 following the May spraying treatment. Little damage was recorded to the plants and it appeared that the spray had had no effect.

Monitoring 1995

All plots were monitored in May 1995 one year after spraying. The results are presented in Figures 1 and 2. Visual inspection of the plots identified almost total control of the plant in plots subjected to digging and spraying (plots 2 and 3). In plot 2, no *F. japonica* plants were recorded in any of the three quadrats and in plot 3, no plants were recorded in one of the quadrats with only two and five plants in the other two quadrats respectively. Undug plots were still monoculture *F. japonica*, whereas plots which had been dug had an increasing number of native species (Table 3). These native species had replaced the *F. japonica* in plots 2 and 3 with 100% cover. In plots 5 and 6 (dug but not sprayed) a reduced cover of native species was recorded as above with the addition of *Impatiens glandulifera* which had been and remained abundant outside the test plots particularly along the river banks but had not been previously recorded within the plots.

There were significant differences ($p<0.05$) between the mean stem densities of plots 2 and 3 and those of 5 and 6. There was a significantly lower stem density ($p<0.05$) in plots 2 and 3 (dug and sprayed), when compared against the control plot and a significantly greater stem density ($p<0.05$) in plots 5 and 6 (dug not sprayed), when compared against the control plot. The mean stem density of plants in plot 1 (not dug and sprayed) was not significantly different to that recorded in the control plot.

Monitoring 1996

The monitoring in early June 1996 was impossible to complete due to very dense growth of *F. japonica* on plots 1, 4, 5 and 6 with plant heights in excess of 2 m. Plots 2 and 3 however, were relatively clear of *F. japonica*, with two of the three quadrats on each plot recording *F. japonica* stem densities of 0. Measurements of the few remaining plants on these plots were taken. A mean of three plants was recorded for plots 1, 4, 5 and 6 for plant height, stem diameter and leaf number for comparison purposes.

Other species were becoming well established on plots 2 and 3 (Table 3). Following the June 1996 visit, all plots were strimmed and were re-monitored in July 1996 (Figures 1 and 2). No significant difference was detected in plant density between plots 1, 4, 5 and 6. Many of the new shoots in the sprayed plots were showing signs of herbicide treatment, (small bushy plants with narrow pointed leaves and red coloured stems). Much of the regrowth since strimming was from cut stems. Mean plant height was significantly greater in plot 5 than all other plots ($p<0.01$). Mean stem diameter and mean leaf number followed the same trend.

Following this monitoring visit, plots 1, 2, 3 and 6 were subjected to a second follow up spraying treatment in July 1996, to spot treat any remaining plants.

In November 1996, the plots were monitored (Figures 1 and 2). Visual inspection of Plots 2 and 3 recorded a good level of control with only two *F. japonica* plants recorded in total for both plots. Twelve species of native plants had established a complete cover of replacement vegetation in plots 2 and 3 (Table 3) whereas in plots 5 and 6, only five additional species were recorded and *F. japonica* remained dominant.

Plants which had been sprayed (plots 1 and 6) were showing extensive yellow/ brown spotting on leaves and no new growth. Plants in plots 4 and 5 (unsprayed) were green and growth luscious.

A significantly higher plant density ($p<0.05$) was recorded for plot 5 than for all other plots. No significant differences in plant height, stem diameter or leaf number were recorded between plots 1 and 6 when compared with the control, plot 4.

Discussion

One of the main problems associated with the control of *F. japonica* is the extent of the underground rhizome system, which may extend up to 7 m away from a parent plant and reach a depth of up to 2 m (Richards, Moorehead and Laing 1990). An estimate of the underground biomass in the upper 250 mm of soil is given by Brock (1994) as 14,000 kg ha^{-1} dry weight. Consequently, when deciding on a control method, the treatment of the rhizome system must be of prime consideration. Up till now treatment has been either with a translocated herbicide or by mechanical methods such as mowing or strimming, although the latter does not stop the plant regenerating from underground rhizomes despite reducing above ground vigour. This study has investigated the effect of integrating the use of a translocated herbicide and digging in an attempt to reduce the time needed to achieve control.

Digging had a significant impact on the growth of *F. japonica* by increasing stem density, but resulted in a decreased plant height, stem diameter and number of leaves. This would be expected due to the breaking up of the rhizome system, which provides resources to the plants early in the season. Where the rhizome system was intact, plants had greater height, greater stem diameter and more leaves but lower stem density. Adachi *et al.* (1996) describe in detail the morphology of *F. japonica* rhizomes. The dormancy of buds appears to be regulated by apical dominance. When the integrity of the rhizome system is broken, it would appear that dormant lateral buds will be stimulated to produce shoots whereas, with an intact rhizome system, an hierarchical system of bud development is seen with only apical rhizome buds and subterranean winter buds at the basal part of the shoot producing aerial shoots. This is confirmed in the present study, where fragmented rhizomes gave rise to a significantly greater number of shoots than those which were intact.

F. japonica is reported to be sensitive to frost (Conolly 1977). Dug rhizomes were left on the soil surface over winter with the intention that frost would cause damage. The results from the 1994 monitoring suggest that exposure of rhizomes to frost over winter did not damage the rhizomes sufficiently to significantly affect the regeneration of the plant. The winter of 1993-4 was mild although mean minimum temperatures in November 1993 and February 1994 were below average. Although some desiccation may have occurred in surface rhizomes, a significant number of rhizome fragments regenerated successfully. These appeared to be below the soil surface rather than within the litter layer.

Visual assessment of the plots in August 1994, following the first spraying treatment, suggested that the treatment had not been effective. However, on monitoring the following May, good control had been achieved in the plots subjected to digging and spraying (Plots 2 and 3) and significantly reduced plant height and number of leaves were recorded on plants in Plot 1 (sprayed only) when compared to the control plot (Plot 4). These results confirm the observations of de Waal (1995), who reported green and flowering plants 73 days after treatment with glyphosate. When monitoring ten months later, she reported that the plants showed a marked reduction in plant height and leaf size when compared to the control plot. The effectiveness of glyphosate is often not realised until the following growing season. This is due to its mode of action, being translocated to the rhizome system and disrupting new growth.

The results of these trials suggest that a single application of herbicide without digging is insufficient to control *F. japonica*. The findings agree with those of de Waal (1995), Roblin (1988) and Beerling (1990) where several applications of herbicide were required before effective control was achieved. Beerling (1990) and Roblin (1988) suggest that two herbicide applications annually over a number of years may be necessary to completely eliminate the plant. The Welsh Development Agency (WDA) guidelines (1991) recommend that the plant is treated for a minimum of three years with follow up monitoring to ensure complete control.

As described in the methods, the impact of strimming is only applicable to the 1996 data. Mean stem density ranging from $7.3 - 28$ stems m^{-2} was recorded in this study for the control plot over the period of the trial. This figure is comparable with that reported by de Waal (1995) of $10 - 20$ stems m^{-2} on strimmed plots.

The implications of the results of these trials are that a combination of digging followed by a single spray treatment gives a good level of control of *F. japonica*. A second application of herbicide achieved complete control in plots which had been dug and sprayed. These results indicate that a shorter treatment period could be achieved by combination treatments. This would be cost effective both in terms of treatment costs but more importantly, in terms of 'waiting time' on redevelopment sites. Such land in many urban areas in the UK is infested with *F. japonica* and in some cases the land is adjacent to water courses which restricts the range of herbicides which can be used. The application of glyphosate would be relevant in these situations. Where sites are not adjacent to water, other translocated herbicides such as picloram, triclopyr or imazapyr could be used. The length of waiting time significantly increases the cost of pre-treatment of sites by increasing the 'Total Finance Costs' (Table 1). The combination of mechanical digging and spraying treatments has the potential for greatly reducing the time required for treatment and hence total finance costs.

As a winter exposure did not encourage frost damage, a digging treatment carried out in early spring, followed by a spraying treatment on the regrowth, could result in a good level of control achievable in several months. With a second spray later in the year, total control may be possible in one year. In order to refine this technique, further research focusing on timing of treatments, critical depth of digging and the effects of intermittent/ continuous disturbance would be of interest.

Acknowledgements

The support of Thames Water Utilities is acknowledged in providing the trial site and assisting with site visits.

References

Abacus Concepts Inc. 1992-95. StatView. Abacus Concepts Inc.

Adachi, N., Terashima, I. and Takahashi, M. 1996. Central die-back of monoclonal stands of *Reynoutria japonica* in an early stage of primary succession on Mount Fuji. Annals of Botany 77: 477-486.

Bailey, J.P. 1990. Breeding behaviour and seed production in alien Giant knotweed in the British Isles. Conference Industrial Ecology Group of the British Ecological Society: 121-129.

Brock, J.H. 1994. Technical note: Standing crop of *Fallopia japonica* in the autumn of 1991 in the United Kingdom. Preslia 66: 337-343.

Brock, J.H., Child, L.E, Waal, L.C. de and Wade, P.M. 1995. The invasive nature of *Fallopia japonica* is enhanced by vegetative regeneration from stem tissues. In: Pyšek, P., Prach, K., Rejmánek, M. and Wade, M. (eds.), Plant invasions: general aspects and special problems. pp. 131-139. SPB Academic Publishing, Amsterdam.

Brock, J.H. and Wade, P.M. 1992. Regeneration of *Fallopia japonica*, Japanese knotweed, from rhizome and stems: Observations from greenhouse trials. IXe Colloque International sur la Biologie des Mauvaises Herbes. Dijon (France): 85-94.

Beerling, D.J. 1990. The ecology and control of Japanese knotweed and Himalayan balsam on river banks in South Wales. PhD. Thesis, University of Wales, Cardiff.

Child, L.E. and Waal, L.C. de 1997. Management of *Fallopia japonica* in the urban environment. In: J.H. Brock, M. Wade, P. Pyšek and D. Green. (eds.), Plant Invasions: Studies from North America and Europe. pp. 207-220. Backhuys, Leiden.

Child, L.E. and Wade P.M. 1997. Reasons for the successful invasion of *Fallopia japonica* in the British Isles and implications for management. Proceedings International Workshop on Biological Invasions of Ecosystem by Pests and Beneficial Organisms, February 1997, Tsukuba, Japan: 253-267.

Child, L.E., Waal, L.C. de, Wade, P.M. and Palmer, J.P. 1992. Control and management of *Reynoutria* species (Knotweed). Aspects of Applied Biology 29: 295-307.

Conolly, A.P. 1977. The distribution and history in the British Isles of some alien species of *Polygonum* and *Reynoutria*. Watsonia 11: 291-311.

Davis, Langdon and Everest 1992. Spon's Architects' and Builders' price book. E. and F.N. Spon, London.

Day, R.W. and Quinn, G.P. 1989. Comparisons of treatments after an analysis of variance in ecology. Ecological Monographs 59: 433-463.

Department of the Environment 1991. Waste Management – The Duty of Care, a code of practice. Environmental Protection Act 1990, HMSO, London.

Ministry of Agriculture, Fisheries and Food. 1995. Guidelines for the use of herbicides on weeds in or near watercourses and lakes. MAFF Publications, London.

Harper, C.W. and Stott, K.G. 1966. Chemical control of Japanese knotweed. Proceedings 8th British Weed Control Conference: 511-515.

Richards, Moorehead and Laing. 1990. Japanese knotweed (*Reynoutria japonica*) in Wales. 1, Main Text. Director of Land Reclamation, Welsh Development Agency, Cardiff.

Roblin, E. 1988. Chemical control of Japanese knotweed (*Reynoutria japonica*) on river banks in South Wales. Aspects of Applied Biology 16: 201-206.

Scott, R. and Marrs, R.H. 1984. Impact of Japanese knotweed and methods of control. Aspects of Applied Biology 5: 291-296.

Waal, L.C. de 1995. Treatment of *Fallopia japonica* near water – a case study. In: Pyšek, P., Prach, K., Rejmánek, M. and Wade, M. (eds.), Plant invasions general aspects and special problems. pp. 203-212. SPB Academic Publishing, Amsterdam.

Welsh Development Agency. 1991. Guidelines for the control of Japanese knotweed (*Reynoutria japonica*). Welsh Development Agency, Cardiff.

Welsh Development Agency. 1994. Model Tender Specifications for the eradication of Japanese knotweed. Welsh Development Agency, Cardiff.

FRUIT AND SEED PRODUCTION IN *GLEDITSIA TRIACANTHOS*

Federico Colombo Speroni and Marta. L. de Viana
Universidad Nacional de Salta – Facultad de Ciencias Naturales
Cátedra de Ecología – Buenos Aires 177, 4400, Salta, Argentina.
e-mail: deviana@ciunsa.edu.ar.

Abstract

Gleditsia triacanthos is a legume that invades mountain forest gaps in northwestern Argentina. Invasion success depends on both: the features of the environment and the biological characteristics of the alien plants (seed production, dispersal, and establishment). We studied *G. triacanthos* fruit and seed production in four invaded gaps, sampling ten adult trees in each one. Fruit production starts at 10cm DBH and there were no further increments in production with DBH. Similar results were obtained with relation to fruit number and seed weight. These results agree with the characteristics described by Rejmánek and Richardson (1996) for invasive species, such as early fruit production that remains constant allover the life cycle. The results are compared with other invasive and native species.

Introduction

Biological invasions may be defined as the movement of an organism from beyond its previous range of distribution into new geographic areas. As such, invasive species impact the ecology as well as the economy of a colonized area (Kareiva 1996). These invasions which affect directly biodiversity at global, regional and local scales, are a consequence of human-caused disruption of the biogeographical barriers which prevent species spread (D'Antonio and Vitousek 1992). Invasion success depends on both: the biological characteristics of the alien plants such as short juvenile periods, short intervals between large seed crops, early and consistent reproduction, and rapid population growth (Baker and Stebbins 1965; Rejmánek and Richardson 1996), and the features of the environment (Baker and Stebbins 1965; Drake *et al.* 1989). However, there are disagreements about the role of disturbances as well as anthropogenic habitats in invasion success (Kornberg and Williamson 1987; Lodge 1993; Williamson 1996).

   Gleditsia triacanthos, is a common colonizer of disturbed areas in the eastern deciduous forest of the USA (Burton and Bazzaz 1995). It was first introduced in the Province of Salta in the early fifties as ornamental. This legume rapidly invaded the mountain forests specially in areas with a high level of anthropogenic disturbances (periodic fires, logging, cattle grazing and tourism), which are not colonized by native trees. Because the invasion occurs through gap colonization, the only way to do it is by seed dispersal. The objective of this work was the study of *G. triacanthos* fruit and seed production.

Plant Invasions: Ecological Mechanisms and Human Responses, pp. 155–160
edited by U. Starfinger, K. Edwards, I. Kowarik and M. Williamson
© *1998 Backhuys Publishers, Leiden, The Netherlands*

Table 1. Tree species of the San Lorenzo's mountain forest (*alien species)

Species	Family	Leaf habit	Height (m)	DBH (m)	Fruit	Fruit size (mm)	Seeds/ fruit	Seed size (mm)	Dispersal
Podocarpus parlatorei	Podocarpaceae	evergreen	20	1.5	drupaceous	5-6	1	4	?
Juglans australis	Juglandaceae	deciduous	20	0.5	drupe	30-40	1	25	?
Alnus acuminata	Betulaceae	deciduous	15	0.5	preudostrobile	20-25	some	2	wind
Celtis spinosa	Ulmaceae	deciduous	12	0.4	drupe	6-9	1	4-6	bird
Phoebe porphyria	Lauraceae	evergreen	30	1.5	berry	13-18	1	9-13	bird
Prunus tucumanensis	Rosaceae	deciduous	10	0.5	drupe	13	1	10	bird
Polylepis australis	Rosaceae	deciduous	8	0.4	aquenio	4-5	many	8-10	wind
Enterolobion contortisiliquum	Leguminosae	deciduous	20	1	pod	40-70	many	10	?
Erythrina falcata	Leguminosae	deciduous	20	1	coreaceous pod	70-200	3-6	15	?
Parapiptadenia excelsa	Leguminosae	deciduous	20	0.6	coreaceous pod	60-180	6-10	7	wind
Tipuana tipu	Leguminosae	evergreen	30	0.5	samara	40-70	3	10-15	wind
Fagara coco	Rutaceae	evergreen	10	0.5	follicle	50-70	1	5-7	?
Cedrella lilloi	Meliaceae	deciduous	15	1.2	capsule	30-40	1	15-23	wind
Allophylus edulis	Sapindaceae	evergreen	8	0.3	drupe	8	1	4	bird
Blepharocalyx gigantea	Mirtaceae	evergreen	30	1	drupe	8	1-2-4	5	bird
Myrcianthes mato	Mirtaceae	evergreen	8	0.4	drupe	15-20	1	10	bird
Myrcianthes pseudo-mato	Mirtaceae	evergreen	12	0.8	drupe	15	1	6	bird
Duranta serratifolia	Verbenaceae	deciduous	7	0.2	drupe	8-10	4-5	2-3	bird
Jacaranda mimosifolia	Bignoniaceae	deciduous	15	0.7	capsule	50 -70	many	7-9	wind
Tecoma stans	Bignoniaceae	deciduous	6	0.3	silique	100-150	many	7	wind
Sambucus peruviana	Caprifoliaceae	deciduous	8	0.4	berry	8	3-6	3	bird
Gleditsia triacanthos*	Leguminosae	deciduous	20	0.5	pod	200	20	7-10	cattle
Ligusrtrum lucidum*	Oleaceae	evergreen	12	0.7	drupe	5	1	2	bird
Morus sp*	Moraceae	deciduous	15	1	berry	3	many	1	bird

Study area

San Lorenzo's mountain forest is a Yungas community, that spreads on the oriental slopes of the western mountains of Lerma Valley, between 1300 and 1800 masl. Weather is subtropical with a wet summer period; mean annual rainfall is 1390 mm. More than 80% of the rains fall from november through march. The mean maximum and minimum temperetures in January (hottest month) and in July (coldest month) are 27°C – 16.3°C and 20°C – 2.9°C, respectively (Arias and Bianchi 1996). The forest is a mixture of evergreen and deciduous species most of them have a low growth rate and produce fleshy fruits, dispersed mostly by birds and wind. The most conspicuous species in the forest are *Phoebe porphyria, Podocarpus parlatorei, Blepharocalyx gigantea* which easily reach 30 m in height and 1.5 m in DBH (Table 1).

Methods

G. triacanthos abundance and DBH were recorded in four invaded patches. The patches location was determined using *Geographic Positioning System* (GPS). Density of *G. triacanthos* was assessed by calculating the area of each patch and then counting the number of trees. Using random sampling point in circular field plots (Skalski 1987), ten adult trees (DBH >10 cm) were selected in each patch, except for patch four, where the total population was used (N=12).

Fruit production was assessed by counting them in a randomly selected sample of five branches per tree, that were measured in lenght. 40 fruits were collected from each tree and in lab, fruit an seed weight and seed number were recorded in order to assess fruit and seed biomass allocation. The number of dead seeds was recorded for each fruit. The total number of branches per tree was determined in order to have an estimation of fruit and seed production per tree. Differences in DBH, fruit number and weight and seed number per fruit and seed weight between patches were analized with Kruskal Wallis, while the relationship between DBH and fruit and seed number and weight, with Spearman rank correlation. All the analysis were done using Systat (1992).

Results

There were significant between patch differences in *G. triacanthos* density, and mean DBH. Fruit production per meter branch as well as fruit production per tree were similar between patches. Similar results were obtained with seed weight, seed number per fruit and the proportion of dead seeds per fruit (Table 2).

Fruit production begins early in the life cycle (10 cm DBH) (Colombo Speroni 1996) and there are no further increments in production with DBH. The same was obtained for seed weight. However, fruit weight decreases with DBH while seed number per fruit increases (Fig. 1).

Table 2. Patch characterization. Location (GPS), area, density, number of trees per patch, DBH, fruit number/ branch and tree, fruit weight, seed number/ fruit, seed weight, and percentage of dead seeds per fruit. (The values are mean ± standard error, * KW, P < 0.05)

	Patch			
	1	2	3	4
Location	24°43.331 S	24°43.209' S	24°43.322' S	24°43'17.3" S
	65°30.483' W	65°30.850' W	65°30.752' W	65°30'15.6" W
Area (m²)	3000	2400	8700	2000
Patch Density (/m²) *	0.085	0.007	0.006	0.0066
Trees per patch	255	16	52	12
DBH (cm) *	23.5 ± 6.95	27.5 ±10.97	18.6 ± 6.26	19.1 ± 7.77
Fruit / branch	5.59 ± 3.78	4.31 ± 3.35	3.92 ± 2.82	2.12 ± 4.01
Fruit/tree	487±294	445±334	303±221	288±617
Fruit weight (gr)	7.06 ± 1.69	7.22 ± 1.55	7.23 ± 2.69	6.86 ± 1.45
Seeds/ fruit	20.48 ± 2.49	20.50 ± 3.95	18.31 ± 2.73	21.27 ± 3.04
Seed weight (gr)	0.11 ± 0.02	0.12 ± 0.02	0.09 ± 0.02	0.11 ± 0.01
Dead seeds/fruit (%)	1	0.8	2.1	1.2

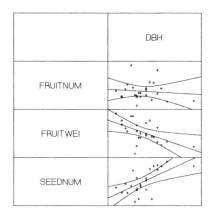

Fig. 1. Regression between DBH versus fruit number (r = -0.052; P>0.05) weight (r = -0.55; P< 0.05) seed number per fruit and weight (r = 0.65; P< 0.05 and 0.28, P> 0.05, respectively).

Discussion

The fact that fruit production (fruit number per meter branch) is constant with DBH, shows a basic alien species characteristic: short juvenile period, early reproduction and a large fruit production that remains constant throughout the species life cycle (Baker and Stebbins 1965). Vegetative reproduction is another important feature related to the invasion success (Green and Johnson 1994; Williamson 1996), is also present in *G. triacanthos* and may be related to the decrease in fruit weight with DBH.

Harper (1977) argued that there must be a compromise between seed number and seed weight , and that reproductive effort does not greatly differ among species, so we may expect an inverse relation between seed production and seed weight. Colonizer species tend to produce a higher number of small seeds (Whitmore 1989). Seed production per tree in *G. triacanthos* is higher than the recorded for species with a similar

Table 3. Seed production per tree and seed weight for seventeen species reported in Greene and Johnson (1994) and three species of San Lorenzo mountain forest.

Species	Seed Number / Tree	Seed weight (mg)
Betula papyrifera	27239	0.6
Betula alleghaniensis	12158	1.6
Picea glauca	7202	2
Picea engelmanii	3671	3.3
Picea rubens	5051	3.4
Liriodendron tulipifera	13509	9.1
Pinus ponderosa	2667	37.7
Abies concolor	5196	40
Acer saccharum	1751	48
Pinus lambertiana	1159	216
Fagus grandifolia	463	247
Quercus velutina	93	1851
Quercus coccinea	525	1930
Quercus rubra	253	2475
Carya glabra	40	3423
Quercus alba	184	3779
Quercus prinus	107	4535
Celtis spinosa	157	120
Duranta serratifolia	250	116
G. triacanthos	7653	120

seed weight (such as *Fagus grandifolia* and *Pinus lambertiana)* and seed weight is higher than the reported for other species with a similar seed production per tree (such as *Picea glauca, P. rubens* and *Abies concolor)* (Greene and Johnson 1994)(Table 3).

G. triacanthos produce a lot of fruits and seeds early in the life cycle, the seeds are big and dispersed by cattle. Rejmánek (1996) suggested that this dispersal by vertebrates gives invasive ability. In this context comes the question: Why the native species do not colonize the forest gaps? In the natural forest there are species that have a high production of small seeds dispersed by wind as *Cedrella lilloi, Parapiptadenia excelsa* and *Tipuana tipu.* However, they show a slow growth rate in comparison to other species such as *Duranta serratifolia, Celtis spinosa* and *Allophylus edulis,* that produce fleshy fruits dispersed by birds (Table 1, 3).

Because the fast growing native species are dispersed by birds, their seeds are found under perches inside the forest or in its edges. The same pattern occurs with other alien species as *Ligustrum lucidum* and *Morus sp.* that do not colonizes open areas, and are also bird dispersed.

G. triacanthos fruits and seeds that pass through the digestive tracts of cattle, remain with the dung in the gaps and subsequent colonization can occur when *G. triacanthos* acts as perch, facilitating community succession.

We assume that *G. triacanthos* success depends on a high fruit and seed production throughout its life cycle, cattle dispersal, and vegetative reproduction. The colonizing failure of the native species can be related to a change in the dispersal pattern in the community as a result of the extensive cattle grazing as well as to the fact that the fast growing native species are dispersed by birds. Most of the wind dispersed species have a low growth rate except *T. stans* and *P. excelsa*, that are not found in the patches. We don't know the reasons for these species failure in colonization, a subject that will be addressed in future works.

References

Arias, M. and Bianchi, A.R. 1996. Estadísticas Climatológicas de la Provincia de Salta. Instituto Nacional de Tecnología Agropecuaria. Estación experimental Agropecuaria Salta.

Baker, H. G. and Stebbins, G.L. (eds). 1965. The genetics of colonizing species. Academic Press, New York, New York, USA.

Burton, P. and Bazzaz, F. 1995. Ecophysiological responses of tree seedlings invading different patches of old-field vegetation. Journal of Ecology 83: 99-112.

Colombo Speroni, F. 1996. Características estructurales de parches con y sin invasión de *Gleditsia triacanthos*. Informe, Cátedra de Ecología. Facultad de Ciencias Naturales. Universidad Nacional de Salta. 7 pp.

D'Antonio, C. and Vitousek, P. 1992. Biologival invasions by exotic grasses, the grass/ fire cycle, and the Global Change. Ann. Rev. Ecol Syst. 23:63-87.

Drake, J.A., Mooney, H.M. di Castri, F. Groves, R.H. Kruger, F.J, Rejmánek, M. and Williamson, M. (eds.) 1989. Biological Invasions. A Global perspective, John Wiley & Sons, Chichester, England.

Greene, D.F and Johnson, E.A. 1994. Estimating the mean annual seed production of trees. Ecology 75:642-647.

Harper, J.L., 1977. The population biology of plants. Academic Press, London.

Kareiva, P. 1996. Contributions of ecology to biological control. Ecology 77: 1963-1964.

Kornberg, F.R. and Williamson, M. (eds) 1987. Quantitative aspects of the ecology of biological invasions. The Royal Society, London, England.

Lodge, D.M. 1993. Biological invasions: Lessons for ecology, Trends in Ecology and Evolution 8: 133-137.

Rejmánek, M. 1996. A theory of seed plant invasiveness: the first sketch. Biological Conservation 78:171-181.

Rejmánek, M. and Richardson, D.M. 1996. What attributes make some plant species more invasive?. Ecology 77: 1655-1661.

Skalski, J.R. 1987. Selecting a random sample of points in circular field plots. Ecology 68: 749.

Whitmore, T.C. 1989. Canopy Gaps and the two major groups of the forest trees. Ecology 70: 536-538.

Williamson, M. 1996. Biological Invasions. Chapman & Hall, London.

IMPATIENS GLANDULIFERA ROYLE IN THE FLOODPLAIN VEGETATION OF THE ODRA RIVER (WEST POLAND)

Zygmunt Dajdok, Jadwiga Anioł-Kwiatkowska, Zygmunt Kącki
Department of Systematics and Phytosociology, Institute of Botany, University of Wrocław, ul. Kanonia 6/8, 50-328 Wrocław, Poland;
e-mail: dajdokz@biol.uni.wroc.pl

Abstract

In many European countries Himalayan Balsam *Impatiens glandulifera* has become increasingly common, both in ruderal and in seminatural or even natural habitats. The rate of expansion of this neophyte in Poland seems to be less than for example in the Czech Republic, but in fact there are no data on the actual number of its localities in our country. Also very poor are data on the types of plant communities with this species.

    I. glandulifera was recorded from both the sources and the mouth of the Odra river, but until the early 70s there were no records from the mid section of the Odra valley. In 3-years we found almost 20 new localities along a 30 km section of the Odra valley (between the town of Oława and Wrocław) and many more in ruderal habitats near villages. *I. glandulifera* competes there with both native and other alien plant species like *I. parviflora* or *Solidago gigantea*. Some individuals of *I. glandulifera* were noted in such annual and perennial plant communities (according to Braun-Blanquet school) as: *Polygono-Bidentetum*, *Calystegio-Archangelicetum litoralis*, *Cuscuto-Convolvuletum*, *Phalaridetum arundinaceae* and *Urtico-Calystegietum*. It was found also in riverside, periodically flooded forests of *Alno-Padion*, but the biggest and thickest stands of *I. glandulifera* were classified as *Impatienti-Calystegietum* (Moor 1958) Soó 1971, which prefers semi-shaded forest edges, not far from the river.

Introduction

The spread of *Impatiens glandulifera* in Europe started in the 1830s after introducing its seeds to botanical gardens in England. In Poland the first report dates from 1961 (Jasnowski 1961), though in Silesia it was seen at the beginning of this century (Schube 1903). The species has reached the stage of holoagriophyte in riverine habitats and is still expanding to new territories.

    In the 1960s *I. glandulifera* was recorded from the mouth of the Odra river-region of Szczecin (Jasnowski 1961) and from the sources of this river, in the Czech Republic (Lhotska and Kopecký 1966; Kopecký 1967). Both localities could be reached by *I. glandulifera* with quite different ways, but considering its high expansion rate – it follows from the calculations of Perrins *et al.* (1993) that the maximum rate of expansion of the species, also by anthropogenic means, is ca. 38 km/year, and thus it could be expected to occur along the whole Odra river valley. However, the distribution map of this species in Poland, (Zając and Zając 1973) shows only few localities on the Odra river valley, especially in its upper and mid section, so we started to study the distribution of this species on the banks of the second largest river in Poland. The aim

Plant Invasions: Ecological Mechanisms and Human Responses, pp. 161–168
edited by U. Starfinger, K. Edwards, I. Kowarik and M. Williamson
© *1998 Backhuys Publishers, Leiden, The Netherlands*

of this paper is to present plant communities with *I. glandulifera*, their habitat conditions and occurrence of other alien plants.

Occurrence of *Impatiens glandulifera* on the Odra river valley

I. glandulifera is one of the species that use stream and river valleys as specific ecological corridors along which they disperse. The process was described by numerous authors, among others from the Czech Republic (Pyšek and Prach 1993, 1995) and Great Britain (Perrins *et al.* 1993). At the same time, besides typical shore habitats, typically ruderal localities are known, also from Poland (Anioł-Kwiatkowska 1974; Fijałkowski 1978). In Moravia the species was observed to enter even railway embankments (Grüll and Vaněčkova 1982).

On the banks of the Odra river *I. glandulifera* was found along a section ca. 30 km long, between Oława and Wrocław, in the mesoregion of the Odra Outwash Valley (Fig. 1). The highest concentration of patches with *I. glandulifera* were in three places: 1. within the administrative borders of Wrocław (Fig. 1a); 2. near Siedlce and Czernica (Fig. 1b); 3. to the S of Jelcz (Fig. 1c). In this section of the Odra river *I. glandulifera* occurs both in the belt of herbs which grow on habitats just next to the river, and at a distance of several dozen or even several hundred meters from the river bank. Phytocenoses where the species was found were subject to phytosociological analysis by taking phytosociological releves of the patches where it grew, using Braun-Blanquet's method. Patches of annual and perennial herbaceous vegetation, near the river, where only single specimens of *I. glandulifera* were found, were classified as: *Polygono-Bidentetum* (Koch 1926) Lohm 1950, *Calystegio-Archangelicetum litoralis* Pass. (1957) 1959, *Cuscuto-Convolvuletum* R. Tx. 1947 et Lohm 1953, *Phalaridetum arundinaceae* (Koch 1926 n.n) Libb. 1931 and *Urtico-Calystegietum* Görs et Th. Müller 1969.

Close to the Odra river bed, Himalayan Balsam occurred also in disturbed, periodically flooded, riverside willow-poplar carrs *Salici albae-Populetum* (R. Tx. 1931) Meijer Drees 1936. Somewhat further from the Odra river bank, stands of *I. glandulifera* were noted also in small fragments of a dry-ground forest where it developed best in sunny places (Table 1, releve 4), competing with other expansive species – *I. parviflora* and native species of herb layer.

In areas neighbouring the Odra river banks *I. glandulifera* appeared not abundantly in clear-felled communities (Table 1, releve 1-3), where the most significant species in the herb layer were those of the class *Artemisietea* Lohm., Prsg. et Tx. 1950, among them *Artemisia vulgaris*, *Carduus crispus*, *Urtica dioica* and *Solidago gigantea* – which also forms dense stands near the river bed.

Most often, however, within the river valley between Oława and Wrocław, apart from ruderal habitats next to buildings, *I. glandulifera* occurred abundantly forming its own community: *Impatienti-Calystegietum* (Moor 1958) Soó 1971 (Table 2), formerly described as *Impatienti-Solidaginetum* Moor 1958 p.p by Kopecký (1967). Phytocenoses *Impatienti-Calystegietum* clearly prefer semi-shaded places on forest edges or near small copses, usually not far from the Odra river. The characteristic species is clear dominant in the patches of *Impatienti-Calystegietum*. *Urtica dioica* often codominates and occurs in all the releves with mentioned, another neophyte, invading riverine habitats – *Solidago gigantea*. Also frequent were: *Calystegia sepium*, *Galium aparine*, *Glechoma hederacea*, *Lamium maculatum*, *Carduus crispus*, *Chaerophyllum bulbosum* and *Phalaris arundinacea*. Average number of species in one releve is 20.8, in 9 releves

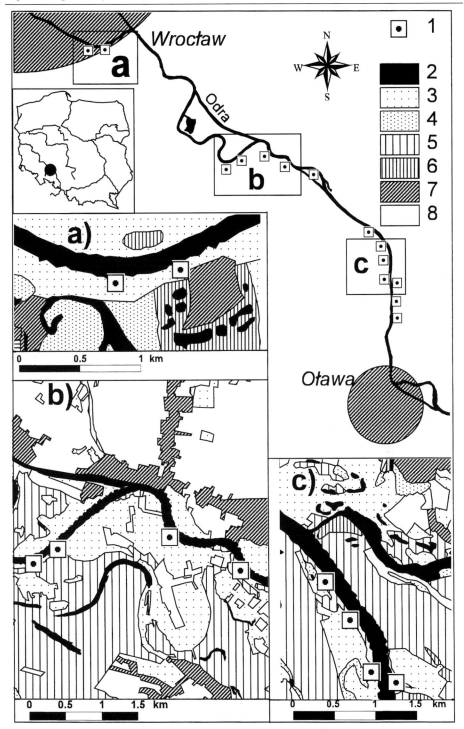

Fig. 1. Localities of *Impatiens glandulifera* on the banks of Odra river between Oława and Wrocław: a – Nowy Dom within the boundary of Wrocław; b – near Jeszkowice and Czernica; c – near Jelcz; 1 – localities of the species; 2 – Odra river, oxbows, ponds; 3 – meadows; 4 – wet meadows; 5 – forests; 6 – wet forests; 7 – urban areas; 8 – fields, barrens.

Table 1.

Community of *Alno-Padion*
Community with *Artemisia vulgaris*

Successive number of releve		1	2	3	4
No of releve		8	9	10	11
Date – year:		95	95	95	95
– month:		07	07	07	07
– day:		19	19	19	19
Area of releve (m^2)		50	50	10	200
Cover tree layer (%)		0	0	70	70
Cover shrub layer (%)		0	0	0	90
Cover herb layer (%)		100	100	100	80
Locality:		JEL	JEL	JEL	SIE
Number of species		25	36	21	29

Ch. All. *Alno-Padion*:

Festuca gigantea		1	+	.	+
Stachys sylvatica		.	.	.	+
Circaea lutetiana		.	.	.	+
Prunus padus	b	.	.	.	+
	c	.	.	.	+

Ch. O. *Fagetalia* et Cl. *Querco-Fagetea**:

Milium effusum		2	1	+	+
Scrophularia nodosa		+	1	+	+
Brachypodium sylvaticum*		.	.	.	2
Euonymus europaeus*	c	.	.	.	+
Viola reichenbachiana		.	.	.	+
Tilia cordata	a	.	.	.	3
	b	.	.	.	2
	c	+	.	.	+
Corylus avellana	b	.	.	.	3
	c	.	.	.	+
Carpinus betulus	a	.	.	.	3
	b	.	.	.	1
	c	.	.	.	1

Ch. Cl. *Artemisietea*:

Artemisia vulgaris		3	3	1	.
Impatiens parviflora		+	1	2	4
Solidago gigantea		+	2	1	+
Impatiens glandulifera		1	+	+	1
Aegopodium podagraria		2	2	.	1
Geum urbanum		1	+	.	+
Urtica dioica		3	.	3	.
Carduus crispus		1	1	.	.
Lamium maculatum		+	1	.	.
Galium aparine		+	+	.	.
Myosoton aquaticum		+	+	.	.

Sporadic species: Alliaria petiolata – 4 (+), Arctium lappa – 2 (+), Cirsium arvense – 2 (+), Cucubalus baccifer – 3 (+), Erysimum cheiranthoides – 3 (+), Geranium robertianum – 4 (+), Glechoma hederacea – 4 (+), Humulus lupulus – 3 (2), Rubus caesius – 3 (+), Rudbeckia laciniata – 2 (+), Silene alba – 1 (+),

Table 1. Continued.

Community of *Alno-Padion*
Community with *Artemisia vulgaris*

Successive number of releve		1	2	3	4
No of releve		8	9	10	11
Date – year:		95	95	95	95
– month:		07	07	07	07
– day:		19	19	19	19
Area of releve (m^2)		50	50	10	200
Cover tree layer (%)		0	0	70	70
Cover shrub layer (%)		0	0	0	90
Cover herb layer (%)		100	100	100	80
Locality:		JEL	JEL	JEL	SIE
Number of species		25	36	21	29

Accompanying species:

Dactylis glomerata		2	1	+	+
Phalaris arundinacea		1	1	3	.
Moehringia trinervia		1	.	+	+
Stellaria media		.	+	+	1
Hypericum perforatum		+	+	+	.
Matricaria perforata		1	1	.	.
Lactuca serriola		1	+	.	.
Agrostis capillaris		+	+	.	.
Sambucus nigra	c	+	+	.	.
Poa palustris		.	+	+	.
Sonchus asper		+	+	.	.

Sporadic species: Ajuga reptans – 4 (+), Capsella bursa-pastoris-2 (r), Convallaria majalis – 4 (+), Cornus sanguinea – 2c (+), Deschampsia cespitosa – 4 (+), Echinochloa crus-galli – 3 (+), Geranium pusillum – 2 (+), Leucanthemum vulgare – 2 (1), Oxalis europaea - 2 (r), Plantago major – 2 (+), Poa annua – 2 (+), Poa trivialis – 4 (+), Polygonum persicaria – 3 (+), Quercus robur – 4a (1), c (+), Ranunculus repens – 2 (+), Rorippa sylvestris – 2 (+), Rumex acetosella – 3 (2), Rumex crispus – 2 (+), Verbascum chaixii – 4 (+), Vicia cracca – 3 (+).

Explanations:
plant layers: a – tree layer, **b** – shrub layer, **c** – herb layer; **Localities: JEL** – Jelcz; **SIE** – Siechnice;

Table 2.

Impatienti-Calystegietum (Moor 1958) Soó 1971

Successive number of releve	1	2	3	4	5	6	7	8	9	
Number of releve	6	5	4	1	12	2	3	7	13	C
Date – year:	95	95	95	95	96	95	95	95	96	o
– month:	07	07	07	07	07	07	07	07	08	n
– day:	19	19	19	19	26	19	19	19	15	s
Area of releve (m^2)	200	200	400	100	100	300	200	200	20	t
Cover tree layer (%)	50	30	40	20	0	10	0	20	0	a
Cover shrub layer (%)	0	0	30	10	0	10	10	40	0	n
Cover herb layer (%)	100	100	100	100	100	100	100	100	100	c
Locality:	SIE	SIE	SIE	JEL	KOT	JEL	SIE	SIE	WRO	y
Number of species	23	23	23	22	16	22	14	28	16	

Ch. Ass. *Impatienti-Calystegietum*:
Impatiens glandulifera	3	2	3	3	4	3	3	3	4	V

Ch. All. *Convolvulion sepium*:
Calystegia sepium	+	1	+	2	1	+	+	+	+	V
Humulus lupulus	+	+	1	+	+	+	1	+	1	V
Solidago gigantea	1	1	2	2	1	2	+	2	.	V
Myosoton aquaticum	+	.	.	+	+	+	.	+	+	IV
Cucubalus baccifer	+	+	+	.	+	.	.	+	.	III
Fallopia dumetorum	.	.	+	.	.	+	.	+	1	III
Senecio fluviatilis	.	.	.	.	.	.	1	1	.	II
Cuscuta europaea	.	.	.	+	.	.	+	.	.	II

Ch. O. *Galio-Calystegietalia sepium*:
Galium aparine	1	+	1	1	+	2	.	1	+	V
Glechoma hederacea	+	+	.	+	1	+	1	+	2	V
Lamium maculatum	2	+	2	+	.	2	3	+	.	IV
Impatiens parviflora	+	+	+	+	+	+	.	+	.	IV
Aegopodium podagraria	1	+	1	+	.	.	.	+	.	III
Rubus caesius	.	+	+	.	.	+	+	.	1	III

Ch. Cl. *Artemisietea*:
Urtica dioica	2	3	1	3	2	2	2	3	2	V
Carduus crispus	1	+	1	+	1	3	2	+	.	V
Chaerophyllum bulbosum	+	1	+	1	+	1	.	1	+	V
Artemisia vulgaris	+	+	+	+	.	+	.	+	.	IV
Elymus repens	.	.	.	+	+	.	.	.	1	II

Sporadic species: Galeopsis tetrahit – 9 (1), Tanacetum vulgare – 8 (+), Torilis japonica – 9 (+),

Accompanying species:
Ch. Cl. *Phragmitetea*:
Phalaris arundinacea	+	3	3	+	+	1	.	2	.	IV
Poa palustris	+	+	+	+	.	+	+	+	.	IV

Sporadic species: Iris pseudacorus – 6 (+), Phragmites australis – 3 (+),

Table 2. Continued.

Impatienti-Calystegietum (Moor 1958) Soó 1971

Successive number of releve		1	2	3	4	5	6	7	8	9	
Number of releve		6	5	4	1	12	2	3	7	13	C
Date - year:		95	95	95	95	96	95	95	95	96	o
- month:		07	07	07	07	07	07	07	07	08	n
- day:		19	19	19	19	26	19	19	19	15	s
Area of releve (m^2)		200	200	400	100	100	300	200	200	20	t
Cover tree layer (%)		50	30	40	20	0	10	0	20	0	a
Cover shrub layer (%)		0	0	30	10	0	10	10	40	0	n
Cover herb layer (%)		100	100	100	100	100	100	100	100	100	c
Locality:		SIE	SIE	SIE	JEL	KOT	JEL	SIE	SIE	WRO	y
Number of species		23	23	23	22	16	22	14	28	16	

Other:

Angelica archangelica												
ssp. litoralis		+	1	+	+	.	.	.	+	.	III	
Salix fragilis	a	4	3	3	2	.	.	.	.	.	III	
	b	.	.	3	1	.	.	.	.	.		
Elymus caninus		.	.	+	1	.	.	+	+	.	III	
Prunus padus	a	1	2	.	.	.	.	.	+	.	II	
Symphytum officinale		.	.	+	.	+	+	.	.	.	II	
Prunus spinosa	b	.	.	2	.	.	.	1	.	.	I	
	c	.	.	.	.	.	+	.	.	.		
Salix alba	a	.	.	.	.	.	2	.	3	.	II	
Salix viminalis	b	.	.	.	.	.	.	.	3	.	II	
	c	.	.	.	.	.	.	.	+	.		
Cornus sanguinea	b	.	.	.	2	.	.	.	.	.	II	
	c	.	.	.	.	.	+	.	.	.		

Sporadic species: Arrhenatherum elatius – 8 (+), Circaea lutetiana – 1 (+), Deschampsia cespitosa – 1 (+), Festuca gigantea – 2 (+), Lactuca serriola – 8 (+), Poa pratensis – 9 (1), Poa trivialis – 5 (+), Populus nigra – 1c (+), Populus tremula – 2a (1), b (2), Salix viminalis – 9 (+), Sambucus nigra – 2b (+), c (+), Sisymbrium loeselii – 8 (+), Symphytum tuberosum – 9 (+).

Explanations:
plant layers: a – tree layer, **b** – shrub layer, **c** – herb layer; **Localities:**
SIE – Siechnice; **JEL** – Jelcz; **KOT** – Kotowice; **WRO** – Wrocław;

49 species were found (Table 2). In those relatively rich in species patches of *Impatienti-Calystegietum* specimens of *Impatiens glandulifera* showed the highest vigour, manifest as size attained, abundance of flowers and formed fruits.

Compared to patches of alliance *Alno-Padion* and class *Artemisietea*, the main differences are in light conditions, which in the forest community are decidedly poorer than in phytocenoses of *Impatienti-Calystegietum*, where *I. glandulifera* reaches the greatest size and blooms most abundantly, and in humidity conditions, which are worse in clear fellings than in the edge patches.

On particular localities specimens of red-pink and light pink flowers were found, plants blooming wine red or white being much less common. The polymorphism results probably from the presence of a few subspecies of *I. glandulifera* in Europe, the subspecies having different flower colour and forming hybrids (Valentine 1971).

The situation after the flood of 1997

In July 1997 in south-western Poland, during the flood, the water in the Odra river valley in places reached more than 2-3 m above ground level during a period of 1-2 weeks. All the localities of *I. glandulifera* known to us were destroyed and no seedlings of this species appeared there after the flood. Single blooming and fruiting individuums – remnants of the old colonies, were observed only on the top of nearby flood banks. Seeds of those plants will probably initiate the process of renewal or re-invasion of *I. glandulifera* in the Odra river valley in future years.

References

Anioł-Kwiatkowska, J. 1974. Flora i zbiorowiska synantropijne Legnicy, Lubina i Polkowic. Acta Univ. Wrat. no 229, Prace Bot. XIX: 1-152. Wrocław.

Fijałkowski, D. 1978. Synantropy roślinne Lubelszczyzny. PAN Warszawa – Łódź.

Grüll, F. and Vaněčková, L. 1982. Přispevek k charakteristice společenstwa s *Impatiens glandulifera* na březich Svitavy u Blanska. Zpr. Čs. Bot. Společ. 17: 135-138. Praha.

Jasnowski, M. 1961. *Impatiens Roylei* Walpers – nowy składnik lasów łegowych w Polsce – *Impatiens Roylei* Walpers – eine neue Auenwaldpflanze in Polen. Fragm. Flor. et Geobot. Ann. VI, Pars 1: 77-80. Kraków.

Kopecký, K. 1967. Die flussbegleitende Neophytengesellschaft *Impatienti-Solidaginetum* in Mittelmähren. Preslia 39: 151-166. Praha.

Lhotská, M. and Kopecký, K. 1966. Zur Verbreitungsbiologie und Phytozönologie von *Impatiens glandulifera* Royle an den Flusssystem der Svitva, Svratka und oberen Odra. Preslia 38: 376-385, Praha.

Perrins, J., Fitter, A. and Williamson, M. 1993. Population biology and rates of invasion of three introduced *Impatiens* species in the British Isles. Journ. of Biogeog. 20: 33-44.

Pyšek, P. and Prach, K. 1993. Plant invasions and the role of riparian habitats: a comparison of four species alien to Central Europe. Journ. of Biogeog. 20: 423-420.

Pyšek, P. and Prach, K. 1995. Historický přehled lokalit *Impatiens glandulifera* na území České republiky a poznámky k dynamice její invaze. Zpr. Čes. Bot. Společ. 29 (1994): 11-31. Praha.

Schube, T. 1903. Die Verbreitung der Gefäßpflanzen in Schlesien preußischen und österreichischen Anteils. Breslau. R. Nischkowsky. pp. 362.

Valentine, D. H. 1971. Flower-colour polymorphism in *Impatiens glandulifera* Royle. Boissiera 19: 339-343.

Zając, E. U. and Zając, A. 1973. Badania nad zasięgami roślin synantropijnych. 3. *Corydalis lutea* DC. 4. *Linaria cymbalaria* (L.) Mill. 5. *Impatiens roylei* Walp. Zesz. Nauk. UJ, Prace Bot. 1: 41-55. Kraków.

THE IMPACT OF ANTHROPOGENIC DISTURBANCE ON LIFE STAGE TRANSITIONS AND STAND REGENERATION OF THE INVASIVE ALIEN PLANT *BUNIAS ORIENTALIS* L.

Hansjörg Dietz[1] and Thomas Steinlein[2]

[1]*Julius-von-Sachs-Institut für Biowissenschaften, Lehrstuhl für Botanik II – Ökophysiologie und Vegetationsökologie, Universität Würzburg, Julius-von-Sachs-Platz 3, D-97082 Würzburg, Germany; e-mail: hjdietz@botanik.uni-wuerzburg.de; [2]Lehrstuhl für Experimentelle Ökologie und Ökosystembiologie, Universität Bielefeld, Universitätsstraße 25, D-33615 Bielefeld, Germany*

Abstract

Seed bank dynamics, stand regeneration and life stage transitions of the perennial alien forb *Bunias orientalis,* which is invasive in lowland limestone regions of Central Europe, were investigated at various habitats differentiated by type and frequency of anthropogenic disturbance. At every phase of its life-cycle, the species profited from anthropogenic disturbance, such as mowing or soil disruption, based on species-specific traits such as rapid exploitation of resources provided by disturbance, high regeneration capacity and high growth rates. These opportunistic traits, combined with a permanent seed bank and polycarpic life-form, apparently contribute to the persistence and propagation of the stands of *B. orientalis* despite the rather low competitive ability of the species. Disturbance regimes of intermediate frequency, which repeatedly provide gaps for regeneration of adults, recruitment of seedlings and establishment of juvenile plants simultaneously seem to be most favorable and, therefore, promote further spread of the species.

Introduction

Up to now, only limited generalisations are possible regarding the factors underlying the expansive spread of introduced species into new areas. This may be due to a lack of quantitative data (Rejmánek 1995, Rejmánek and Richardson 1996) and also the complexity of the processes involved (Orians 1986). However, anthropogenic disturbance, resulting from common land management practice, is widely accepted to be crucial for the establishment of many alien plant species (e.g. Mooney and Drake 1986; Drake *et al.* 1989; Hobbs and Huenneke 1992). Anthropogenic disturbance includes a great variety of factors, among them cutting gaps, aboveground shearing (by mowing), root fragmentation (by digging), spread (by soil translocation) and soil nutrient enrichment (by deposition of nutrient-rich material), all of which may, separately or in combination, favour performance and spreading of alien plants (e.g. cf. Hobbs and Atkins 1988; Hobbs 1989; Brock *et al.* 1995; Burke and Grime 1996; Young *et al.* 1997). Life-cycle traits of many invasive alien species may be adapted to these conditions by a long history of association with human activities (di Castri 1989). Thus, for understanding the processes associated with the spread of invasive plants apparently furthered by anthropogenic disturbance, comparative plant life-history analy-

Plant Invasions: Ecological Mechanisms and Human Responses, pp. 169–184
edited by U. Starfinger, K. Edwards, I. Kowarik and M. Williamson
© 1998 Backhuys Publishers, Leiden, The Netherlands

ses within a range of habitats differing in type and intensity of anthropogenic disturbance are indispensable.

Among the alien plants currently expansive (invasive *sensu* Pyšek 1995) in Central Europe, *Bunias orientalis* L. (Brassicaceae), a nitrophilous polycarpic perennial
forb probably originating in South-West Russia (Tutin *et al.* 1993), belongs to this
group of species profiting from anthropogenic disturbance. *B. orientalis*, which has
spread conspicuously along roads, into meadows, vineyards, and other disturbed ground
in lowland limestone regions during the last decades (e.g. Ullmann *et al.* 1988; Steinlein
et al. 1996), has been demonstrated to benefit from mowing regimes, as the species
retains high reproductive effort and shows higher rosette size growth under these conditions. Further, even small root fragments revealed strong regeneration capacities in
a greenhouse study (Steinlein *et al.* 1996). As a semi-rosette species, *B. orientalis* also
shows a comparatively low competitive ability for light (Dietz and Ullmann 1997;
Dietz *et al.* 1998). Therefore, site conditions may become critical for this species once
tall matrix vegetation forms a closed canopy due to lack of disturbance.

The non-clonal plant may flower in the second year, showing high seed set in favorable habitats. Seed fall, beginning in July and extending to spring of the next year,
is generally local, with the oval pods containing the seeds lacking any dispersal equipment.

This paper presents the results of a comparative study on stand regeneration and
life stage fates of *B. orientalis* at habitats differing in disturbance regime. We focused
on three disturbance regimes commonly experienced by stands of the species: (i) mowing
(mown grassland), (ii) soil deposition or soil disruption (ruderal sites with soil management) and (iii) infrequent disturbance by man (fallow herbaceous vegetation, which
is termed 'undisturbed' below). In particular, we examined seed bank dynamics, seedling recruitment and root regeneration, and life stage transitions in general. Our objective is to determine the reaction of life cycle stages of *B. orientalis* to disturbance and
to evaluate the dependence of stand establishment and development on specific disturbance regimes.

Material and methods

General site description

Except for the root regeneration experiment, investigations were performed within stands
of *B. orientalis* at several field locations in the vicinity of Würzburg (Unterfranken,
Germany, 49°47' N, 9°56' E) and in the Botanical Garden of the University of Würzburg.
The sites were characterized by herbaceous vegetation with subdominant cover of *B.
orientalis* (cover range ca. 5% to 50%) and by relatively moist, loamy, calcareous
soils. The habitats differed mainly in disturbance regime but also with respect to soil
moisture and nutrient availability, as indicated by their species composition (Ellenberg
1988) and productivity (see Table 1).

Seed bank persistence

We monitored the seed bank development at four different locations by analysis of soil
cores collected at different times of the year (see Table 1 and below). The sites

Table 1. Vegetation parameters of the six main different habitat types used for the investigations on seed bank density, seedling recruitment and life stage transitions. The cover and height values shown were analyzed or estimated in June 1995 (Dietz 1996, Dietz and Ullmann 1997). Error ranges denote ± SE. Ae, *Arrhenatherum elatius* (L.) P.B. ex J. et C. Presl; As, *Anthriscus sylvestris* (L.) Hoffm.; Av, *Artemisia vulgaris* L.; Er, *Elymus repens* (L.) Gould; Ga, *Galium album* Mill.; Ph, *Picris hieracioides* L.; Rc, *Rubus caesius* L.; Tv, *Tanacetum vulgare* (L.) Bernh.; Ud, *Urtica dioica* L.

Type of habitat	Vegetation layer			
	Mean total cover [%]	Mean height [cm]	Mean *B. orientalis* cover [%]	Dominant species
Old fallow (OF)	87 ±5	38 ±8	6 ±1	Er, Rc
Ruderal, nutrient rich (RNR)	92 ±4	60 ±32	35 ±7	Ud, Rc, Ae
Fertile Slope 1 (FS1)	80-100	80-120	20-30	Ud, Rc
East- or southward exp. slopes (SES)	82 ±6	39 ±7	16 ±5	Grasses, varying
Wayside vegetation (WV)	> 90	70-90	40-50	Ud, Rc, partly As
Mown roadside vegetation (RVM1)	ca. 80	40-60	5-40	Ph, Av, Ae, Ga
Mown roadside vegetation (RVM2)	ca. 90	50-80	50-60	Ae, Ud
Grassland vegetation mown once a year (GV)	89 ±4	52 ±15	32 ±5	Ae

chosen varied mainly according to (i) frequency of disturbance (mowing) and (ii) density of the *B. orientalis* stands. Two roadside vegetation stands (RVM1 and RVM2, Table 1) were chosen, each mown irregularly, i.e. one or two times a year at different months. At RVM1, the infrutescences of *B. orientalis* plants growing within and up to 2 m adjacent to the sampling area were clipped and removed prior to fruit ripening, in order to prevent direct replenishment of the seed bank during the time of the study. RVM2 was not treated this way. Soil cores were sampled between March 1994 and October 1996 (RVM1), and between March 1994 and October 1995 (RVM2). At each date at RVM1, 33 soil cores were uniformly taken from an area of 20 m x 4 m (sampling positions were spaced 2 m). At RVM2, 10 soil cores were systematically collected from an area of 8 m x 2 m (sampling positions again spaced 2 m).

The other two locations were unmown (fertile slope (FS1) and wayside vegetation (WV), see Table 1) and infrutescences were not removed in these cases. The seed bank was sampled between April 1995 and May 1996 (FS1) or between January 1995 and May 1996 (WV), respectively. Twelve soil cores each were taken from an area of 3 x 2 m (sampling positions were spaced 1 m in these cases). At FS1, the first three sampling positions were located at the upper edge of the slope and the last three sampling positions at the lower edge of the slope. The remaining six sampling positions were located intermediate on the slope.

All soil cores were sampled with a digging cylinder of 10 cm depth (volume ca. 230 cm^3). As 70 to 90% of the seed bank of *B. orientalis* can be found in the upper 5 cm of the soil layer at various habitats (Steinlein, unpublished), a sampling depth of 10 cm should be sufficient for an almost complete inclusion of the seed bank. The soil cores were rinsed and the fruits of *B. orientalis* counted. 'Free' seeds (without pericarp) were not found in the soil cores. Pericarp fragments of broken fruits were disregarded. The number of seeds was calculated by multiplication with 1.4, which is the mean number of seeds per fruit (Steinlein *et al.* 1996). Seed viability tests (Tetrazolium staining, Mac Kay 1972) were performed for a subset of the samplings. 50 seeds

Table 2. Parameters of soil disturbance or soil deposition plots set up at eight different locations. Plot size (normally 1 m^2) was reduced and number of plots increased at very heterogeneous sites. Abbreviations as in Table 1. See text for further information.

Type of habitat	Date of disturbance	Number of plots	Total area [m^2]	Dominant species
Old fallow (OF)	Aug 1995	2	2	Er, Rc
Fertile Slope 1 (FS1)	Aug 1995	2	2	Ud, Rc
Fertile Slope 2 (FS2)	Jul 1995	2	1	Ae, Ud
Wayside vegetation (WV)	Jun 1995	2	2	Ud, Rc, partly As
Moist soil spoil (MSS)	Jun 1995	5	1.5	Ae, Av, Tv
Mown roadside vegetation (RVM)	Jun 1995	7	2	Ph, Av, Ae, Ga
Soil deposition 1 (SD1)	Aug 1994	2	2	-
Soil deposition 2 (SD2)	Aug 1995	1	2	-

were used for each test. The seed bank density was also determined at three other sites which were investigated for seedling recruitment (see Tables 1, 2 and below).

Seedling recruitment

Seedling recruitment was investigated within permanent plots of 1 m^2 area at four different habitats: old fallow (OF, 5 plots), ruderal, nutrient rich habitat (RNR, 8 plots), east- or southward exposed slopes (SES, 5 plots) and grassland vegetation (GV, 8 plots). The permanent plots were placed randomly within sections of *B. orientalis* stands of average population density. Within each of the plots, *B. orientalis* seedlings were censused at the beginning of each month in the period from March to October 1995. Newly emerged seedlings were marked with colored toothpicks to distinguish them from those which were already present in earlier censuses. In October mean leaf length of all surviving seedlings was determined. During the whole growth period, there were no apparent disturbances in the plots other than the monthly censuses.

Additionally, at six locations the soil was locally disturbed once between June and August 1995 (Table 2, upper part), by turning over the uppermost soil layer with a spade to a depth of 5 – 10 cm. In August/September, four to eight weeks after the respective soil disturbance treatments, the plots within the disturbed patches were re-visited and the newly recruited seedlings of *B. orientalis* censused. A second census was performed at the end of October, including the determination of mean leaf lengths of the surviving seedlings. On two occasions, recruitment of *B. orientalis* seedlings from a seed bank contained within soil depositions of unknown origin were followed (see Table 2, lower part). In both cases, soil material was deposited in August to cover approximately 10 m^2 of wayside vegetation. Both events were unrelated, as the first one occurred in August 1994 and the second one in August 1995 at a location ca. 20 km apart. Recruited *B. orientalis* seedlings were counted within permanent plots in September, and mean leaf lengths of the plants were determined in a second census in October.

Root regeneration

From May to August 1994, juveniles of *B. orientalis* and of five nitrophilous peren-nial ruderal plants co-occurring with *B. orientalis* in the field, (*Cirsium vulgare* (Savi) Ten. (Asteraceae), *Arctium tomentosum* Mill.(Asteraceae), *Rumex obtusifolius* L. (Polygo-naceae), *Urtica dioica* L. (Urticaceae) and *Artemisia vulgaris* L. (Asteraceae)) were grown in two-species mixtures in plastic pots (volume 60 l, height 70 cm, diameter 36 cm) filled with sandy loam (diallel competition experiment with nine individuals of species A grown as 'border' plants and three individuals of species B grown as 'cen-ter' plants; further details are given in Dietz *et al.* 1998). The soil and the root systems of the 162 'border' plants per species contained within the pots were poured onto a heap in an unshaded, even area of the Botanical Garden of Würzburg. Care was taken to apply the soil randomly. By using a shovel dredger, the heap was levelled to cover an area of 5 x 5 m (thickness of the soil layer was 20 – 30 cm). The soil dump was left unmanaged until August 1995. For each species, the number of regenerated shoots or shoot clumps (i.e. shoots regenerating from the same point or intermingled shoots forming entities) was counted in October 1994 and, for a second time, in August 1995. For *B. orientalis*, leaf lengths of the regenerated rosettes were measured at both times.

Stage transitions

In May and September 1994, a total of about 400 *B. orientalis* individuals were randomly selected and marked at OF, RNR (including several vegetation patches with regener-ating vegetation resulting from recent local soil disturbance) and GV (cf. Table 1). The plants were marked with labelled tent pegs and their developmental stage (seed-ling/juvenile, vegetative or reproductive) and size (number of shoots, mean rosette leaf length [for plants with < 10 rosette leaves] or rosette diameter [for plants with > 10 rosette leaves]) were recorded. The plants were revisited in May and September 1995 and, if they had survived, their stage and size was recorded again. However, not all labelled tent pegs could be recovered, thus reducing total sample size to 308 indi-viduals.

Results

Seed bank persistence

Both seed bank density and dynamics showed pronounced differences among the lo-cations. The maximum seed bank density was relatively low (< 2000 seeds * m^{-2}) at RVM1, with patchy distribution of *B. orientalis* (cf. Table 1, Fig. 1). High maximum seed bank density (ca. 10000 seeds * m^{-2}) was found at the sites with more homoge-neously dense *B. orientalis* stands regardless of whether they were mown (RVM2) or not (FS1, WV).

    Seasonal dynamics of the seedbank were apparent at all locations studied. Gener-ally, following a depletion of the seed bank in spring (mainly due to germination), summer and/or in winter (due to granivory or mortality), the seed bank was replen-ished after the main seed fall in July to October. These fluctuations were considerably dampened, although not completely eliminated, by infrutescence removal at RVM1. This treatment resulted in a decrease in seed bank density of 62% after 25 months

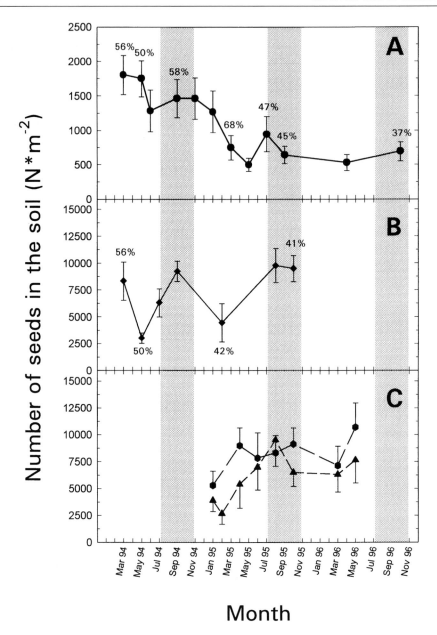

Fig. 1. Seed bank dynamics of *B. orientalis* at four different locations. A: mown roadside vegetation site with infrutescences of *B. orientalis* removed (RVM1); B: mown roadside vegetation site without removal treatment (RVM2); C: fertile slope (FS1, filled triangles), wayside vegetation (WV, filled hexagons). The period of main seed fall is indicated by the grey areas in each panel. Numbers above symbols denote percent viability (Tetrazolium test) of the sampled seeds. Error bars indicate ±SE.

(March 1994 to April 1996, Fig. 1 A). At RVM2, in spring, seed bank densities were decreased to about half or one third of the values observed the year before following seed fall (Fig. 2 B). Apart from seasonal effects, seed bank density remained fairly constant at this site.

At the two unmown stands, FS1 and WV, seed bank depletion was comparatively low between the periods of seed fall; seed bank densities increased in spring and early summer before main seed fall started (Fig. 1 C). Both effects probably are due to delayed seed movement through the relatively thick litter layer present at these sites, prior to its decomposition in the growth period. At the mown sites RVM1 and RVM2, the formation of a litter layer is prevented by litter removal subsequent to mowing. Apart from the seasonal effects at FS1 and WV, seed bank density showed an increasing trend over time.

The viability of the seeds varied between 37% and 68% (Fig. 1 A, B). There was no significant loss in seed viability with decreasing seed bank density at RVM1 (i.e. roughly with time, $P > 0.25$, Spearman's rank correlation).

Seedling recruitment

In all four habitats, seedling recruitment of *B. orientalis* peaked in March in the undisturbed plots (Fig. 2). The seedlings of the March cohort had recruited within four weeks prior to the March census, as no seedlings were observed yet at the beginning of February. Seedling recruitment decreased thereafter and was almost absent from

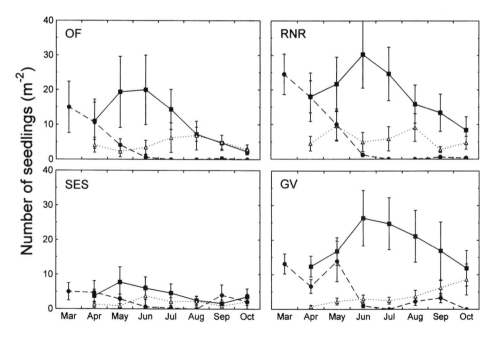

Fig 2. Seedling population dynamics of *B. orientalis* within undisturbed permanent plots at four different habitats, old fallow (OF), nutrient rich habitat (RNR), slopes (SES) and the grassland vegetation (GV). Solid lines/squares denote seedlings which were already present in the census of the preceding month. Dashed lines/circles denote newly recruited seedlings and dotted lines/triangles indicate seedlings which died between the preceding and the current census. Error bars indicate ±SE.

the plots in June and July. However, at GV, there was a second peak of seedling recruitment in the May census that was probably due to the higher gap frequency there in April as compared to the other habitats. Whereas, at OF and RNR, seedling recruitment was negligible in the rest of the growth period from August to October, a weak late-summer peak of seedling recruitment could be observed at SES and GV. Except for RNR, seedling mortality was comparatively low early in the growth period (Fig. 2). At OF, RNR and SES, there was increased seedling mortality in (late) summer and, at GV, the highest seedling mortality was delayed to September and October. The resulting seedling population development showed a similar pattern in all four habitats (Fig. 2, solid lines). Following spring recruitment, seedling population density reached its maximum in May (SES) or June. Subsequently, the seedling populations constantly declined due to reduced recruitment rates and increased mortality, and reached their minimum density in September (SES) or October. The mean total number of seedlings recruited at the different habitats (ranging from 20 to 55 m^{-2}) corresponded to 2% or less of the number of seeds present in the seed bank in the same area (Table 3). In October, the fraction of seedlings recruited in 1995 which survived varied between 6% (OF) and 24% (GV, Table 3). At that time, the seedlings had a mean leaf length between 2.1 and 3.5 cm. At GV, however, mean leaf length was considerably greater (5.8 cm, Table 3).

In the disturbed plots, seedling recruitment following disturbance greatly varied between habitats. Four to eight weeks after disturbance, 32 to approximately 1000 seedlings were counted per square meter (Table 3). Except for FS1 (upper edge) and WV, seedling recruitment following disturbance was several times higher than the total seedling recruitment in the undisturbed plots in 1995. Accordingly, the disturbance-elicited seedling recruitments amounted to a relatively high proportion of the seed number in the seed bank (1.5 to 11%, Table 3). Mean seedling leaf lengths in October varied between 3.3 and 6.0 cm. Overall mean leaf length (4.3 cm) did not significantly differ from that of the undisturbed sites (3.4 cm; P = 0.30, Fisher's two-sample randomization test [Manly 1991]). However, at the disturbed sites, there was a considerable proportion of large seedlings with mean leaf lengths > 7.5 cm (data not shown) whereas, except for GV, the seedlings were homogeneously small under undisturbed conditions. At SD1 and SD2, the seedlings were particularly well developed, with mean leaf lengths of 30 and 15 cm, respectively.

Root regeneration

Regeneration success differed markedly between the six species studied. Regeneration (almost) completely failed for *C. vulgare, U. dioica* and *A. vulgaris* (Table 4). Twenty-two regenerated shoots were observed for *A. tomentosum* in October 1994 (corresponding to regeneration of 14% of the rootstocks), of which approximately 50% survived to August 1995. Regeneration success was higher for *R. obtusifolius*, with 54 regenerated shoot clumps in 1994 (33% of all rootstocks), of which 38 survived until August 1995. *B. orientalis* showed lateral shoot regeneration at multiple points along the main root(s) (cf. Hahn 1993), as opposed to the other species, which were restricted to terminal regeneration from the rootstock. This explains the particularly high number of 115 regenerated shoots/shoot clumps of this species in October 1994. Among these, there was low mortality, resulting in the fact that the lower amount

Table 3. *B. orientalis* seedling recruitment in disturbed and undisturbed plots in various habitats in 1995 (SD1 1994). Shown are mean total number of recruits and the percentage of seedlings that survived in October (undisturbed plots). Mean number of observed seedlings at census time is given for the disturbed plots. The number of seedlings is also given as a percentage of the number of seeds in the seed bank except for the soil deposition sites (SD1 and SD2), where the seed bank apparently was exhausted by the recruitment event after soil deposition. Mean leaf length was determined in October.

Type of habitat	Seedlings total [m^{-2}] (± SE)	Survived [%]	Proportion of seed bank[%]	Mean leaf length [cm]	Mean leaf length > 7 cm [%]
Undisturbed plots					
Old fallow (OF)	31 ± 16	6	1.5	2.1	0
Ruderal, nutrient rich (RNR)	55 ± 19	12	0.5-2	3.2	0
East- or southward exp. slopes (SES)	20 ± 13	14	2	2.5	0
Grassland vegetation mown once a year (GV)	40 ± 13	24	2	5.8	5
Disturbed plots					
Old fallow (OF)	190	-	9	3.3	2
Fertile Slope 1 (FS1)†	1000, 27	-	6.5, 1.5	3.6, 4.0	7
Fertile Slope 2 (FS2)	180	-	11	6.0	24
Wayside vegetation (WV)	32	-	3	?	6
Moist soil spoil (MSS)	140	-	5	3.8	10
Mown roadside vegetation (RVM)	300‡	-	~7.5	5.2	13
Soil deposition 1 (SD1)	120	-	-	30	>>50
Soil deposition 2 (SD2)	270	-	-	15	> 50

†Values are separately shown for the plot at the bottom of the slope (left numbers) and for the plot at the top of the slope (right numbers), ‡data from plots placed in dense stands of *B. orientalis* within the population

Table 4. Number of shoot clumps regenerated from the dumped root systems of six perennial ruderal species.

Species	Number of regenerated shoot clumps	
	October 1994	August 1995
Bunias orientalis	115	63*
Rumex obtusifolius	54	38
Arctium tomentosum	22	12
Urtica dioica	2	2
Artemisia vulgaris	1	0
Cirsium vulgare	0	0

*The lower value reflects merger of some of the shoot clumps separately recorded in October 1994

of shoot clumps recorded in 1995 simply reflects a merge of previously separated shoot clumps due to rosette growth and further shoot regeneration.

The regenerating shoots/shoot clumps of *A. tomentosum* and *R. obtusifolius* grew rapidly and attained diameters of 30 to 100 cm as early as October 1994. The plants were about the same size in August 1995. In contrast, regenerated shoots of *B. orientalis* were small in October 1994 (76% of these had a mean leaf length of 10 cm or below,

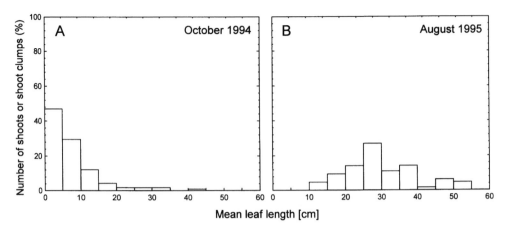

Fig 3. Frequency distribution of mean leaf length of regenerated shoots or shoot clumps of *B. orientalis.*

Fig. 3). In 1995, rosette growth increased considerably and, in August 1995, mean leaf length of most rosettes was in the range of 20 to 40 cm.

Stage transitions

There was low mortality of generative plants in all habitats, with most of the plants remaining in the generative stage in 1995 (Fig. 4). The proportion between those and generative plants which died or reverted to the vegetative stage in 1995 was not significantly different between the undisturbed (OF, RNR) and the disturbed (GV/R) habitats (P > 0.3, Fisher's exact test). Small generative plants at OF, and large generative plants at RNR showed a slightly reduced rosette diameter, but a trend for increased shoot number, in 1995. At the disturbed sites (GV/R), however, rosette diameter of generative plants on average was greater in 1995 than in the previous year. With one exception (generative plants reverting to the vegetative stage at OF), plants which were already generative in 1994 had a consistently higher mean shoot number than the 1995 generative plants originating from vegetative plants or seedlings.

Sizes and fates were more distinct among vegetative plants, both within and between habitats (Fig. 4). The fates of vegetative plants were significantly different between the undisturbed and the disturbed habitats (P < 0.001, Pearson's χ^2 test), with a higher proportion of transitions to the generative stage at GV/R and higher mortality at OF, RNR. In contrast to the situation for the generative plants, vegetative plants at OF were, on average, about as large as they were at RNR and GV/R. At OF, one fifth of the vegetative plants died, 70% remained vegetative and the rest were generative in 1995. At RNR, 54% of the vegetative plants died; however, these plants were particularly small (mean rosette diameter 8 cm). The fraction of plants remaining vegetative in 1995 (28%) were smaller than the vegetative plants which didn't survive at OF. Only particularly large plants (19%) reached the generative stage in 1995 at the cost of reduced size. At GV/R, most of the vegetative plants (54%) switched to the generative stage in 1995 and attained a similar size as they had in 1994 (mean rosette diameter = 31 cm). Most of the smaller vegetative plants remained vegetative in 1995 while few (8%) died.

Differences between the undisturbed and the disturbed habitats were most pronounced

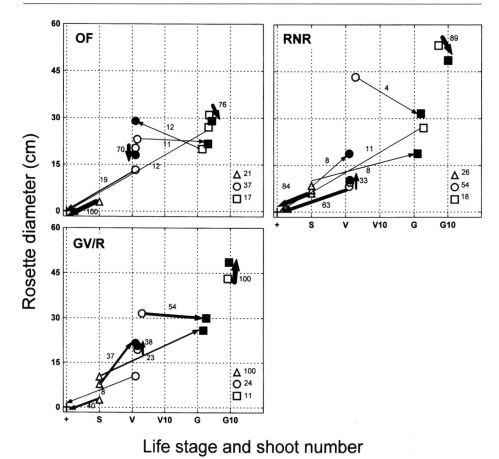

Life stage and shoot number

Fig 4. Mean stage-specific fates of *B. orientalis* plants marked in different habitats in 1994. Open symbols denote stages present in 1994. From these, arrows point to filled symbols (or crosses) representing the plant's stage in 1995. Numbers assigned to the arrows indicate the proportion of individuals of a given stage changing to one of the three possible stages in 1995. The location of the symbols in the graphs reflects mean stage-specific plant size in terms of rosette diameter (ordinate) and number of shoots (abscissa). The abscissa is subdivided into three sections, one for each of the different stages: S, seedling (position invariable); V, vegetative plant; G, generative plant (for both position within the sections vary with respect to number of shoots ranging from 0 to 20). In case there were both spring and autumn data available for a plant in 1994 or 1995, these were lumped together by averaging the size values. For seedlings, only rosette diameter at first census was considered. The values in the lower right corner of each graph denote the absolute number of individuals of the respective stage. OF, old fallow; RNR, ruderal, nutrient rich habitat; GV, grassland vegetation (data combined with those of the regenerating vegetation patches of RNR (R). Triangles, seedling stage; circles, vegetative stage; squares, generative stage; crosses, plants dead in autumn 1995.

in the case of seedling fate (P < 0.001, Pearson's χ^2 test, Fig. 4). Besides a few exceptions at RNR, all seedlings died by autumn 1995 in the undisturbed habitats. In contrast, 60% of the seedlings marked at GV/R were vegetative (37%) or generative (23%) in 1995. There was a consistent rank order of mean seedling size at GV/R, with the largest plants (rosette diameter 10.5 cm) switching to the generative stage, intermediate-sized plants (8 cm rosette diameter) becoming vegetative, and with small seedlings (rosette diameter 2.5 cm) dying. This correlation between size and fate also holds for the other stages and habitats.

Discussion

In accordance with previous studies (Steinlein *et al.* 1996), our results confirm *B. orientalis* to be (pre)adapted to anthropogenic disturbance as the key factor fostering establishment, development and persistence of its stands, as well as its spread. It is most striking that *all* life stages profit from disturbance regimes in one or more respects, i.e. the colonizing success of *B. orientalis* seems to be due to the joint promotion of crucial growth and propagation processes within its life-cycle.

Seed bank

Our results indicate that seed bank density in unmown habitats may be higher than in mown habitats of the same population density of *B. orientalis*, as partial seed loss due to litter removal and strong recruitment events following mowing can add considerably to seed bank depletion (cf. Rice 1989). In later years, however, these relations may be reversed, as *B. orientalis* stands tend to degenerate when undisturbed, but prosper at moderately mown habitats (see below). Furthermore, transport of litter containing seeds of the species may serve as an additional vector for its spread.

Despite complete inflorescence removal at RVM1, no exact data can be given on seed bank depletion, probably because the area was affected by the litter collection process from adjacent areas where *B. orientalis* fruit ripening was not prevented. Nevertheless, in the mown habitats, the seed bank of *B. orientalis* seems to decrease exponentially with a half-life of roughly half a year, if no further seed input occurs (Fig. 1 A and B). Accordingly, the seed bank may drop below 10% of its starting density after two years without replenishment, although there may be considerable year-to-year variation. A permanent seed bank, often described for pioneer or alien plants, is regarded generally as a prerequisite for a rapid reestablishemt after disturbance events (Carey and Watkinson 1993). Therefore, delayed seed movement to the soil by litter layers, along with reduced seed bank loss by low germination rates, presumably extending the persistence of the total seed pool in undisturbed stands of *B. orientalis*, may increase the probability of successful stand regeneration by seedling recruitment following disturbance once adult *B. orientalis* plants are largely displaced by taller competitors in later successional phases.

Recruitment

Seedling recruitment proved to be strongly dependent on disturbance events. In undisturbed vegetation, seedling recruitment was largely restricted to springtime, when elevated nutrient availability and pronounced temperature fluctuations may trigger seed germination (Fenner 1985). Except for GV, unfavorable site conditions (closed canopy at OF and RNR, soil dryness at SES) in the following months prevented considerable seedling growth, leading to high seedling mortality; seedlings surviving to the end of the growing period were too small to survive the winter. The few successful establishments at RNR represent seedlings which grew to larger sizes in gaps within the vegetation (Fig. 4, RNR). However, under conditions of higher gap size or -frequency, a considerable proportion of seedlings may reach sufficient size to become established, either as vegetative or generative plants in the next year, even if the disturbance occurred several months in the past (see Fig. 4, GV/R).

When soil disturbance occurred in summer, soil and light conditions were favorable for a longer time period. Thus, at least a small amount of the massively recruited seedlings after soil disturbance attained a sufficient size (mean leaf length > 7.5 cm, cf. stage transitions results) to survive the winter, and probably become established the next year. Two particular adaptations to soil disturbance were apparent for *B. orientalis* seedlings. First, large seed size (5-11 mg, Steinlein *unpublished*) and, correspondingly, the large size of newly recruited seedlings, resulted in greater seedling survival with respect to possible drought stress during the first few weeks after germination (cf. Baker 1972). Secondly, there was a high exploitation capability of favorable soil conditions (elevated nutrient availability and altered physical conditions due to soil surface disintegration, Hobbs 1989) by immediate mass recruitment after disturbance, with rapid seedling growth, particularly under conditions of intense disturbance, as indicated by SD1 and SD2. At these sites, dominant stands of *B. orientalis* developed the year after soil deposition, with 40% of all plants being in the reproductive stage.

Adult plants

Beneficial effects of disturbance were more complex for the adult stages of *B. orientalis*. Belowground mechanical disturbance severing, or even fragmenting, rootstocks clearly causes regeneration of *B. orientalis* stands due to the plants' ability to laterally regenerate multiple shoot clusters. Since regenerating shoots of *B. orientalis* cover a larger area than those of species regenerating only punctually from the root crown region, they preempt space and are able to use increased resource levels (e.g. light) after disturbance more efficiently. High lateral regeneration capacity also seems to particularly assist rapid stand development at new sites, where severed root stocks were introduced by soil depositions. The results of our study probably underestimate regeneration performance of severed *B. orientalis* root stocks in the field, as the root stocks used in our experiment originated from rather small individuals, which suffered probably from strong resource competition (Dietz *et al.* 1998).

Generally, the life stage transitions of vegetative or generative plants were strongly size dependent at all habitats surveyed, with the smallest plants having the highest risk of mortality and the largest plants having the best chance to become or stay generative (cf. Gross 1981). However, despite the relatively small sample size of marked plants, there were some important and significant differences between the undisturbed (OF, RNR) and disturbed habitats (GV/R). Several effects associated with mowing at GV (i.e. a higher proportion of grasses, rapid rosette regrowth in times of low vegetation height and a longer leaf duration of rosettes (Steinlein *et al.* 1996; Dietz and Ullmann 1997)) resulted in the growth of large vegetative rosettes of *B. orientalis*, of which more than half switched to the generative stage at GV/R. In contrast, only a few larger-sized individuals of the generally smaller vegetative plants at OF and RNR reached the reproductive stage. The strikingly larger size of most of the vegetative plants at OF, as compared to RNR, seems to be inconsistent with the particularly unfavorable site conditions at OF. However, as reproductive plants were observed to revert to the vegetative stage only at OF, the comparatively large size of vegetative plants may be explained by degeneration of previously generative plants, being indicative of regressive stand development. The vegetative plants at RNR, on the other hand, apparently all grew from the seedling stage. The fates of generative plants also reflect the favorable site conditions at GV/R, based on the same factors discussed above for the vegetative plants. Only at GV/R did generative plants on average increase in both rosette diameter and shoot number, whereas rosette size decreased at the undisturbed habitats, and a few generative plants even degenerated or died at OF. These results may also be applied to reproductive output as rosette size was strongly correlated with number of fruits produced per plant (at OF generative plants had only about 100 fruits per plant in 1994 compared to 500-1000 fruits per plant at GV/R and > 1000 fruits per plant at RNR; Dietz *unpublished*).

Synopsis

B. orientalis can be characterized as a plant especially adapted to opportunistically exploiting elevated resource levels created by (anthropogenic) disturbance or phenological phenomena, such as a reduced matrix vegetation cover in late summer (Dietz and Ullmann 1997). Both regeneration and development of existing stands, as well as further spread by rapid establishment of new stands, are facilitated. As *B. orientalis* shows a highly positive reaction norm to elevated nutrients (Steinlein *et al.* 1996), regeneration processes are particularly vigorous at nutrient rich disturbed sites. There, rapid establishment leads to pre-occupation of resources (cf. Steinlein *et al.* 1993) and, therefore, may partly compensate for its competitive inferiority to taller nitrophytes.

Many of the life-cycle traits of *B. orientalis,* including high plasticity (cf. Dietz *et al.* 1998), rapid reaction to resource pulses following disturbance, high relative growth rates (Steinlein *et al.* 1996), and space-filling radial shoot morphology, are generally known for plants adapted to disturbed habitats (Bazzaz 1983, Grubb 1985). Other ruderal rosette plants, such as *Oenothera biennis* L. and *Verbascum thapsus* L., may have longer-lived permanent seed banks (cf. Gross and Werner 1982) than *B. orientalis*. However, those ruderals are typically characterized as 'biennial' or 'short-lived perennial' (Grubb 1985), whereas *B. orientalis* is polycarpic and may attain ages of ten years or more (Dietz, *unpublished*). Thus, it may be the combination of life-cycle adaptions to disturbance, and the potential of individuals to persist in the reproductive stage for several years, even without further disturbance, which underlies the current success of the species.

Variation in type and frequency of disturbance seems to have a pronounced impact on the population development of *B. orientalis*. Field observations indicate that frequent mowing (> 2 times a year) may prevent seed set at all and, thus, sustains a static population of recurring regenerating rosettes over several years, with a complete lack of regeneration by seedling recruitment. At the other extreme, at rarely mown sites (e.g. less than once in three years), increased competition by taller successional species may displace *B. orientalis* in the long run. The same effect may be observed in habitats with varying frequencies of soil disturbance, whereas very frequently soil-disturbed sites support therophyte communities (e.g. Bazzaz 1983). Intermediate disturbance levels, however, seem to interfere most favorably with the life-cycle traits of the species.

A further spread of *B. orientalis* by translocation of root fragments or seed banks has to be expected as well as a continuing expansion of *B. orientalis* populations at sites with a moderate frequency of disturbance. Among these are meadows and grasslands, which are categorized as seminatural habitats, from where the species is still absent in some other regions (Pyšek *et al.* 1995). However, according to the results of our study, *B. orientalis* should not be able to invade undisturbed vegetation.

Acknowledgements

We are grateful to Birgit Baumann of the University of Würzburg and C. Jaekel of the University of Bielefeld for their help with data collection. We acknowledge the staff of the Botanical Garden for technical assistance. We thank Wolfram Beyschlag for valuable comments on the manuscript. Helpful comments on the manuscript by Ingo Kowarik and an anonymous reviewer are gratefully acknowledged. Parts of this work were supported by a PhD-fellowship of the Deutsche Forschungsgemeinschaft to H. Dietz (Graduierten Kolleg University of Würzburg).

References

Baker, H.G. 1972. Seed weight in relation to environmental conditions in California. Ecology 53: 997-1010.

Bazzaz, F.A. 1983. Characteristics of populations in relation to disturbance in natural and man-modified ecosystems. In: Mooney H.A. and Gordon M. (eds.), Disturbance and ecosystems. pp. 259-275. Ecological Studies Vol. 44. Springer, New York.

Brock, J.H., Child, L.E., de Waal, L.C. and Wade, P.M. 1995. The invasive nature of Fallopia japonica is enhanced by vegetative regeneration from stem tissues. In: P. Pyšek, K. Prach, M. Rejmánek and P.M. Wade (eds.) Plant invasions: general aspects and special problems. pp. 131-139. SPB Academic Publishing, Amsterdam.

Burke, M.J.W. and Grime J.P. 1996. An experimental study of plant community invasibility. Ecology 77: 776-790.

Carey, P.D. and Watkinson A.R. 1993. The dispersal and fates of seeds of the winter annual grass *Vulpia ciliata*. Journal of Ecology 81: 759-767.

di Castri, F. 1989. History of biological invasions with special emphasis on the old world. In: J.A. Drake, H.A. Mooney, F. Di Castri, R.H. Groves, F.J. Kruger, M. Rejmanek and M. Williamson (eds.) Biological invasions: a global perspective. pp 1-30. Scope 37. Wiley, Chicester.

Dietz, H. 1996. Etablierung und Bestandsentwicklung des nicht-klonalen Neophyten *Bunias orientalis* L. (Brassicaceae). Ph. D. Thesis, Universität Würzburg.

Dietz, H., Steinlein T. and Ullmann I. 1998. The role of growth form and correlated traits in competitive ranking of six perennial ruderal plant species grown in unbalanced mixtures. Acta Oecologia 19: 25-36.

Dietz, H. and Ullmann, I. 1997. Phenological shifts of the alien colonizer plant species *Bunias orientalis* L: Image-based analysis of temporal niche separation. Journal of Vegetation Science 8: 839-846.

Drake, J.A., Mooney, H.A., di Castri, F., Groves, R.H., Kruger, F.J., Rejmanek, M. and Williamson M. 1989. Biological invasions: a global perspective. Scope 37. Wiley, Chicester.

Ellenberg, H. 1988. Vegetation ecology of Central Europe. Cambridge University Press, Cambridge.

Fenner, M. 1985. Seed Ecology. Chapman & Hall, London.

Gross, K.L. 1981. Predictions of fate from rosette size in four 'biennial' plant species: *Verbascum thapsus, Oenothera biennis, Daucus carota,* and *Tragopogon dubius*. Oecologia 48: 209-213.

Gross, K.L. and Werner, P.A. 1982. Colonizing abilities of 'biennial' plant species in relation to ground cover: implications for their distributions in a successional sere. Ecology 63: 921-931.

Grubb, P.J. 1985. Plant populations and vegetation in relation to habitat, disturbance and competition: problems of generalization. In: White, J. (ed.), Handbook of Vegetation Science Vol. 3. pp. 595-621. Dr.W Junk Publishers, Dordrecht.

Hahn, W. 1993. Regenerationsverhalten von *Bunias orientalis* Wurzelstücken. Diploma Thesis, University of Würzburg.

Hobbs, R.J. and Atkins, L. 1988. Effect of disturbance and nutrient addition on native and introduced annuals in plant communities in the Western Australien wheatbelt. Australian Journal of Ecology 13: 171-179.

Hobbs, R.J. 1989. The nature and effects of disturbance relative to invasions. In: J.A. Drake, H.A. Mooney, F. Di Castri, R.H. Groves, F.J. Kruger, M. Rejmanek and M. Williamson (eds.), Biological invasions: a global perspective. pp. 389- 405. Scope 37. Wiley, Chicester.

Hobbs, R.J. and Huenneke L.F. 1992. Disturbance, diversity and invasion: implications for conservation. Conservation Biology 6: 324-337.

Mac Kay, D.B. 1972. The measurement of viability. In: Roberts E.H. (ed.), Viability of seeds. Syracuse University Press.

Manly, B.F J. 1991. Randomization and Monte Carlo Methods in Biology. Chapman & Hall, London.

Mooney, H.A. & Drake, J.A. 1986. Ecology of biological invasions of North America and Hawaii. Ecological Studies Vol. 58. Springer, New York.

Orians, G.H. 1986. Site characteristics favoring invasions. In: Mooney, H.A. and Drake, J.A. (eds.), Ecology of biological invasions of North America and Hawaii. pp. 133-148. Ecological Studies Vol. 58. Springer, New York.

Pyšek, P. 1995. On the terminology used in plant invasion studies. In: P. Pyšek, K. Prach, M. Rejmánek and P.M. Wade (eds.), Plant invasions: general aspects and special problems. pp. 131-139. SPB Academic Publishing, Amsterdam.

Pyšek, P., Prach, K. and Šmilauer, P. 1995. Relating invasion success to plant traits: an analysis of the Czech alien flora. In: Pyšek, P., Prach, K., Rejmánek, M and Wade M. (eds.), Plant invasions – General aspects and special problems. pp. 39-60. SPB Academic Publishing, Amsterdam.

Rejmánek, M. 1995. What makes a species invasive? In: Pyšek, P., Prach, K., Rejmánek, M and Wade, M. (eds.), Plant invasions – General aspects and special problems. pp. 3-13. SPB Academic Publishing, Amsterdam.

Rejmánek, M. and Richardson, D.M. 1996. What attributes make some plant species more invasive? Ecology 77: 1655-1661.

Rice, K.J. 1989. Impacts of seed banks on grassland community structure and population dynamics. In: Leck, M.A., Parker, V.T. and Simpson R.L. (eds.), Ecology of soil seed banks. pp. 211-230. Academic Press, San Diego.

Steinlein, T., Heilmeier, H. and Schulze, E.D. 1993. Nitrogen and carbohydrate storage in biennials originating from habitats of different resource availability. Oecologia 93: 374-382.

Steinlein, T., Dietz, H. and Ullmann, I. 1996. Growth patterns of *Bunias orientalis* (Brassicaceae) underlying its rising dominance in some native plant assemblages. Vegetatio 125: 73-82.

Tutin, T.G., Heywood, V.H., Burges, N.A., Chater, A.O., Edmondson, J.R., Moore, D.M., Valentine, D.H., Walters, S.M. and Webb, D.A. 1993. Flora Europaea. 2nd ed., Vol. 1. Cambridge University Press, Cambridge.

Ullmann, I., Heindl, B., Fleckenstein, M. and Mengling, I. 1988. Die straßenbegleitende Vegetation des Mainfränkischen Wärmegebietes. Berichte ANL 12: 141-187.

Young, J.A., Palmquist, D.E. and Wotring, S.O. 1997. The invasive nature of *Lepidium latifolium*: a review. In: J.H. Brock, M. Wade, P. Pyšek and D. Green (eds.), Plant invasions: studies from North America and Europe. pp. 59-68. Backhuys Publishers, Leiden.

A COMPARATIVE STUDY OF *PINUS STROBUS L.* AND *PINUS SYLVESTRIS L.*: GROWTH AT DIFFERENT SOIL ACIDITIES AND NUTRIENT LEVELS

Dáša Hanzélyová

University of South Bohemia, Faculty of Biological Sciences, Branišovská 31, 370 05 České Budějovice, Czech Republic; e-mail: hanzely@tix.bf.jcu.cz

Abstract

The invasion of *Pinus strobus* into the forest ecosystems of northern Bohemia, Czech Republic, during the last 20 years has had several reasons. The most important reason may be changed habitat conditions at some sites, as a consequence of high nitrogen deposition and soil acidification. The main aim of this study is a comparative assessment of conditions affecting seedling establishment in *Pinus strobus*, an invasive species originating from N. America, and *Pinus sylvestris*, a native European species.
   Seedlings of both species were grown under different conditions of pH and nutrient supply. The seedlings were cultivated in a glasshouse in solutions with pH 2.7, 4.0 and 6.2 respectively, at the nutrient level of 2x basic nutrient solution (BNS, Ingestad and Kähr 1985), and in solutions with different nutrient levels: distilled water, BNS, 2xBNS, 4xBNS and 6xBNS all at pH = 6.2. I investigated the effect of the treatments on seedling establishment, the dry weight of roots and shoots of the seedlings, seedling survival, number of lateral branches and dormancy of apical buds. There were significant differences between the establishment, growth and development of *P. strobus* and *P. sylvestris* seedlings. *P. strobus* seedlings were less sensitive both to low pH (pH = 2.7) and high nutrient levels (4xBNS and 6xBNS). Seedling biomass and RGR were significantly higher in *P. sylvestris* than in *P. strobus* grown under favorable conditions (pH = 4.0 or 6.2 and 2xBNS). Under unfavorable conditions (pH = 2.7, distilled water or 6xBNS) total dry weight and R/S were similar in both species, but *P. sylvestris* seedlings had a significantly higher mortality. The conditions of cultivation influenced the onset of dormancy in the apical buds and the number of branches formed by both species. Seedlings of both species grown at pH = 2.7 or in distilled water formed either no branches at all or (seldom) only one branch; their apical buds developed later than in the other treatments. The average number of branches formed by one seedling was one in *P. strobus* and two to three in *P. sylvestris*. The results of this study show that, under unfavorable conditions (high soil acidity and/or very high or very low concentrations of nutrients in the soil), seedling establishment is greater and subsequent seedling growth faster in *P. strobus* seedlings than in those of *P. sylvestris*.

Introduction

Large quantities of chemical compounds are deposited in terrestrial ecosystems from atmospheric emission of pollutants. Deposition is a significant pathway for entry of several compounds to the forests in Europe (Lindberg *et al.* 1990; Lovett and Lindberg 1993). In soil, acid deposition may have three effects: 1) a fertilizer effect caused by the deposition of N and probably also S (e.g., Abrahamsen 1981); 2) an acidification effect caused by deposition of H^+ ions and increased leaching of base cations (Likens and Butler 1981; Reich *et al.* 1987b); and 3) a toxicity effect (e.g., Al^{3+} ions) as a result of increased soil acidity (Abrahamsen 1984; Eldhuset *et al.* 1987). On plants,

Plant Invasions: Ecological Mechanisms and Human Responses, pp. 185–194
edited by U. Starfinger, K. Edwards, I. Kowarik and M. Williamson
© *1998 Backhuys Publishers, Leiden, The Netherlands*

deposition has two important impacts: 1) direct damage to plants (to their roots and/ or leaves) (Bäck et al. 1995); and 2) indirect damage, by changing the state of ecosystems (changes in nutrient ratios in the soil, increased nutrient inputs). Fertilization may have a positive effect on forest ecosystems by promoting growth and net photosynthesis of the trees (McLaughlin et al. 1982, Reich et al. 1987a, Bäck et al. 1993, Sverdrup et al. 1994). On the other hand, increased inputs of H^+ ions influence negatively the growth of the trees (Abrahamsen 1980).

In the last 20 years, invasive white pine (Pinus strobus), an American species, has been spreading in many places of the Czech Republic. Rapid invasion was observed also in forests dominated by the native Scotch pine (Pinus sylvestris) with indigenous plant species (e.g. Ledum palustre) in the undergrowth in sandstone regions. Some of these areas are Protected Landscape Areas (e.g. Labské pískovce, Český Ráj). Even aged seedlings of white pine often form dense monospecific patches offering a reduced possibility for reestablishment of new Scotch pine seedlings (but not only them). Because of the higher growth rate of white pine than Scotch pine (Hanzélyová, unpublished data) the invasion of white pine brings about changes in species composition and a reduction of species diversity in the invaded plant communities.

The invasion of Pinus strobus has occurred first of all in North Bohemia. This area is exposed to most air pollution in the Czech republic, which involves an increased deposition of H^+, and of S and N compounds (Abrahamsen 1980). In 1993, the average wet deposition of pure N (mainly in the form of NH_4^+ and NO_3^-) was 156 kg ha^{-1} $year^{-1}$ in this part of the country (Air pollution in Czech Republic in 1994 1995). One of the main reasons for the invasion of Pinus strobus may thus be the increased atmospheric nitrogen deposition in the last 30 years (Rejmánek, personal communication), changing the forest habitats. The observed increase in the number of even-aged Pinus strobus, while the number of seedlings of the other tree species have remained unchanged (Hanzélyová, 1993) suggests that the responses of the trees to the environmental changes are species dependent (Ågren 1983; Pan and Raynal 1995; Turner and Knapp 1996).

It may be hypothesized, that:

1) the growth rate of Pinus strobus responds more positively to an increased deposition of N (and probably other nutrients) than that of Pinus sylvestris;
2) Pinus sylvestris is more sensitive than Pinus strobus to stress (caused by soil acidification, high nutrient concentrations).

To test these hypotheses, an experiment was set up testing the influence of pH and nutrient supply on seedlings of Pinus sylvestris and Pinus strobus.

Material and methods

The seeds of P. strobus and P. sylvestris were kept for four months (January – April) in the refrigerator between layers of wet cotton at 4 °C. Then, the seeds were pregerminated at laboratory temperature (22 ± 2 °C). The germinated seeds (May) were transplanted into pots (one seed per pot) filled with perlite and hydroponic solution (see below) and placed in a glasshouse. The pots were assigned to each of the following treatments: three pH treatments (i.e., pH = 6.2, 4.0 and 2.7) at one nutrient level of 2 x BNS (Basic Nutrient Solution (Ingestad and Kähr 1985)), and five nutrients treatments (distilled water, BNS, 2xBNS, 4xBNS, 6xBNS – all at pH = 6.2), all of them for the two species (*P. sylvestris* and *P. strobus*). Each treatment comprised 82 seedlings and had two replicates. The hydroponic solutions were changed every four days in all treatments. pH of the solutions was adjusted with 1M HCl. One liter of BNS contains: 50 mg N (as nitrate and ammonium), 32.4 mg K^+, 7.43 mg P (as $H_2PO_4^{2-}$), 3.01 mg Ca^{2+}, 3.01 mg Mg^{2+} and 5.59 mg S (as SO_4^{2-}) + microelements.

During the whole experiment, the seedlings were kept in a glasshouse with regulated maximum temperature in summer. In winter, the temperature never fell below 0 °C.

Data collection

At three-week intervals, 5 to 10 randomly selected whole plants (including both roots and shoots) were sampled from each treatment. The plants were dried at 85 °C for 48 hours. The root and shoot dry weight of each plant was recorded. For the analysis of survival, the total number of seedlings of both species was scored monthly in each treatment during the growing season. The first observation was made ten days after the transplantation of germinated seeds into the pots. The number of living seedlings was determined as a percentage of initially established seedlings. After the growing season, in October, the number of lateral branches was recorded for all seedlings. Three times in the winter (on 4th December, 22nd January and 1st March), the length and width of apical buds were recorded in each remaining (unharvested) seedling in each treatment. The width and length of the apical buds were measured, but because of a high positive correlation between the two dimensions and a higher accuracy of the length measurements, only that was tested statistically.

The data were analyzed with general additive and linear models and analysis of variance (ANOVA), using S-plus (Statistical Sciences 1995a, b).

Results

The number of established seedlings differed significantly between the species and treatments (Table 1). *P. sylvestris* seedlings seem to be highly sensitive to pH = 2.7 and 4, as well as to nutrient concentrations of 4xBNS and 6xBNS, whereas the negative reaction of *P. strobus* seedlings to the same conditions is nonsignificant. The highest proportion of seedling establishment, 89% for *P. sylvestris* and 69% for *P. strobus*, occurred in soil with the basic nutrient level of BNS. Significant differences between the two species were found in the treatments with distilled water, BNS, 2xBNS and pH = 6.2.

Table 1. Percentages of established seedlings of *P. sylvestris* and *P. strobus* grown at different pH and nutrient levels. Between-species differences are significant (bold figures) at pH = 6.2, distilled water, BNS and 2xBNS (P<0.05). Differences between treatments were significant only for *Pinus sylvestris* (P<0.05).

		nutrient treatments				pH		
	distil.water	BNS	2xBNS	4xBNS	6xBNS	2.7	4.0	6.2
Pinus sylvestris	**87%**	**89%**	**87%**	67%	69%	56%	70%	**87%**
Pinus strobus	**64%**	**69%**	**62%**	60%	62%	61%	63%	**62%**

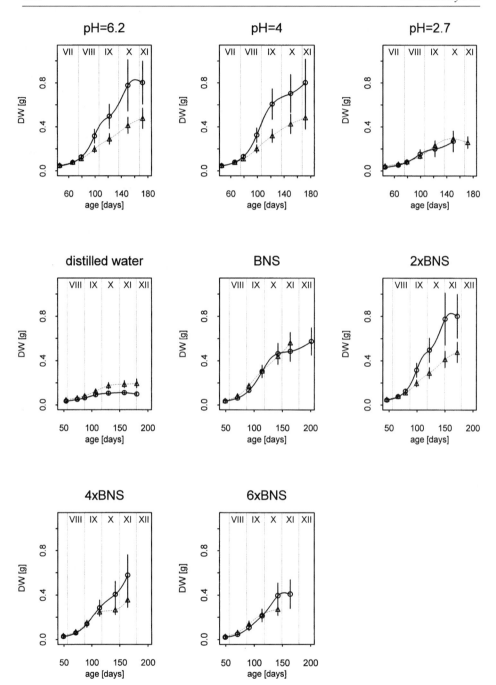

Fig. 1. Growth curves (with their 95% confindence intervals) of mean dry weight (DW) per plant in
Pinus sylvestris and *P. strobus* seedlings in treatments with different soil pH and nutrient levels.
The circles and full line are for *P. sylvestris*, the triangles and dashed line are for *P. strobus*.

Soil pH and nutrient levels influenced significantly the dry weight of *P. sylvestris* and *P. strobus* seedlings (P< 0.05) (Figure 1). Optimum conditions for the growth of *P. sylvestris* seedlings appeared to be at pH = 6.2 and 4.0 or at a nutrient concentration of 2xBNS. For the growth of *P. strobus* the nutrient concentration of BNS seemed to be optimal. The seedlings of *P. sylvestris* produced two times as much dry matter as those of *P. strobus* and also had a higher relative growth rate (RGR) in the treatments with pH = 6.2 and 4.0 and at the nutrient concentration of 2xBNS. On the other hand, seedling growth was strongly affected in a negative manner under conditions of low pH or when nutrients were either lacking or supplied at high concentrations. At pH = 2.7, reduction in seedling dry weight was by 34 % in *P. sylvestris* and by 54% in *P. strobus* at the end of the experiment (in November), in comparison with seedlings grown at pH = 6.2 or 4. In comparison to seedlings grown in the 2xBNS treatment, seedling growth was reduced, by 12% (*P. sylvestris*) and 41% (*P. strobus*) in the distilled water, and by 51% (*P. sylvestris*) and 57% (*P. strobus*) in the 6xBNS treatment.

In this experiment, adverse conditions affecting seedling growth produced an unhealthy appearance of their needles. Young needles of *P. sylvestris* seedlings were chlorotic or became copper-coloured in the treatments with pH = 2.7. Similar symptoms were found in the needles of *P. sylvestris* seedlings grown in hydroponic solutions with nutrient concentrations of 4xBNS or 6xBNS. These changes were probably caused by high concentrations of certain ions (H^+ and/or microelements) in the root zone (Mengel and Kirkby 1982; van den Driessche 1991) and were often followed by death of the *P. sylvestris* seedlings.

The winter measurements showed that the two species differed significantly in the size of dormant apical buds (P<0.05) and that the conditions of cultivation influenced the size, phenology and variability of buds in both species (P< 0.01) (Figure 2). Low pH and either very low or very high concentration (6xBNS) of nutrients influenced negatively the length of dormant buds in both species, more strongly in *P. sylvestris* than in *P. strobus*. In general, *P. sylvestris* seedlings form buds which are nearly two times as big as those of *P. strobus* seedlings.

Differences were found between the two species in the onset of dormancy. While *P. sylvestris* had formed dormant buds by the time of the first measurement (on 4th December), *P. strobus* seedlings were forming their dormant apical buds only in January.

The number of lateral branches differed significantly both between the species and treatments (P< 0.01) (Figure 3). In all treatments, except distilled water, *P. sylvestris* seedlings formed, on average, nearly two times as many lateral branches as *P. strobus* seedlings. The average number of branches formed by one seedling was one in *P. strobus* and two to three in *P. sylvestris*. Seedlings of both species grown at pH = 2.7, or in distilled water, did not produce any branches at all, or (seldom) only one branch per plant. The highest average number of branches was recorded in *P. sylvestris* seedlings grown at pH = 4.0 and the nutrient concentration of 2xBNS.

Discussion

The establishment of seedlings appears to be the most sensitive period of the tree life cycle, even more than germination (Abrahamsen 1976, Lee and Weber 1979 in Abrahamsen 1980). Many of the authors cited by Abrahamsen (1980) reported results

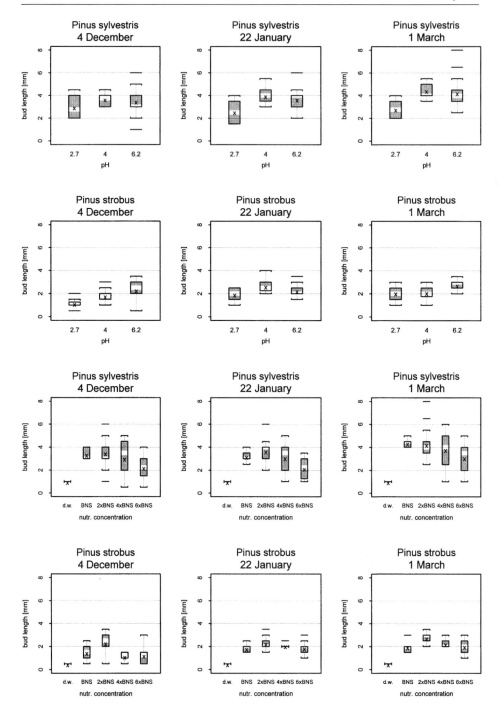

Fig. 2. Box plot showing the changes in the length of dormant apical buds of *Pinus sylvestris* and *Pinus strobus* seedlings. (x – mean, white stripe – median, d.w.-distilled water).

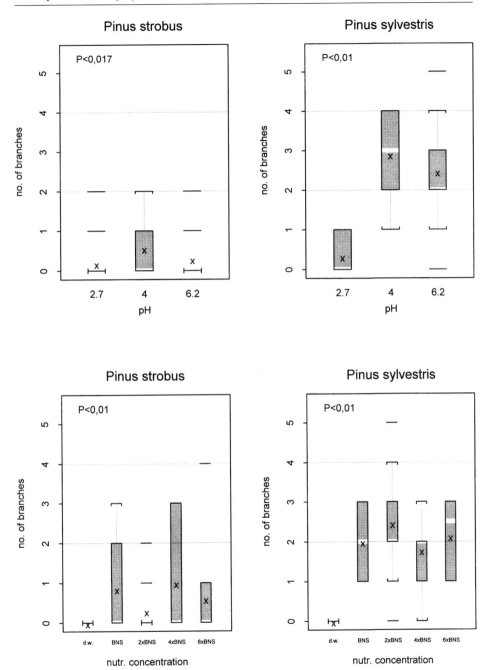

Fig. 3. Box plot showing the differences between the numbers of lateral branches in *Pinus sylvestris* and *Pinus strobus* seedlings in dependence on soil pH and nutrient concentration in the soil. (x – mean, white stripe – median, d.w.-distilled water).

similar to my results, namely that the percentage of established conifer seedlings rap-idly decreases at low soil pH. High soil acidity has also been reported as the main factor causing high seedling mortality in *Acer saccharum* (Raynal *et al.* 1982b). In my experiments, the main reason for the small percentage of established seedlings and increased mortality both in *Pinus strobus* and *P. sylvestris* at pH = 2.7 may have been an increased sensitivity of the roots to bacterial infections (Raynal *et al.* 1982a, b; Kozlowski *et al.* 1991).

Several experimental studies have recently been made on the effect of soil pH on tree growth. These experiments were accomplished in forests stands in which the soil conditions were apparently modified . Results have been summarized by Sverdrup *et al.* (1994). Enhancement of tree growth due to decreasing pH occurred in the first years of most of the experiments, but growth was retarded in the long term. Reich *et al.* (1987a) also reported an increase in the dry matter production of *Pinus strobus* seedlings with increasing acidity of rainwater in the first year of their experiment. Under experimental glasshouse conditions, significant reduction of conifer seedling growth was observed from the start of the experiments in treatments with low soil acidity (pH = 2.5 or less) (Abrahamsen 1980, 1984). This agrees with my results. The main reason for growth reduction may be the toxic effects of H^+ ions: high concentra-tions influence negatively the permeability of cell membranes and the gradient of pro-tons in the root cells (Fitter and Hay 1987, van den Driessche 1991).

Ingestad (1991) described the differences between cultivated birch (*Betula sp.*) seed-lings under nitrogen deficiency and under optimal conditions in the root zone. The seedlings cultivated under nitrogen deficiency had a higher root : shoot ratio (R/S). In my experiments, the seedlings grown at the nutrient concentrations of BNS and in distilled water had a higher R/S ratio, lower RGR (neither being presented in this paper), smaller apical buds, and reduced number of lateral branches (this in the treat-ment with distilled water only) as compared with seedlings exposed to the other treat-ments. These results shows that the nutrients contained in the BNS and, obviously, also distilled water, are insufficient for healthy growth both of white pine and Scotch pine seedlings. On the other hand, nutrient concentrations of 4xBNS and 6xBNS af-fect adversely the growth, survival and development of the seedlings. The changes in colour of their needles and other negative effects observed, such as a low percentage of seedling survival and reduced growth, confirm the toxicity of these concentrated hydroponic solutions (Mengel and Kirkby 1982; von Evers and Hüttl 1995; van den Driessche 1991).

The difference in total dry weight between the seedlings of *P. sylvestris* and *P. strobus* grown under near-optimum conditions was apparently due to physiological and probably also morphological differences between the two species. Net photosyn-thesis was found to be nearly two times as high in *P. sylvestris* seedlings as in *P. strobus* seedlings in the treatment with pH = 6.2 (D. Hanzélyová unpublished). The differences in photosynthesis match the differences in seedling dry weight between the two species in the treatments with pH = 6.2 and 4.0 and at the nutrient concentra-tion of 2xBNS.

The onset of dormancy in the apical buds appears to depend both on the species and the treatment. Although the conditions in the glasshouse differed somewhat from outdoor conditions, seedlings of both species formed dormant apical buds in winter. In *P. sylvestris* seedlings the first buds appeared at the end of August, in the treat-ments with pH = 6.2. But more than 50% of the seedlings in the experiment formed their apical buds one month later, at the end of September. As expected, the biggest buds were formed by seedlings in the treatments with pH = 6.2 and 4.0, and 2xBNS (see also Kozlowski *et al*. 1991). These seedlings, grown under near-optimum conditions, have sufficient reserves to form big buds which ensure good shoot growth in the subsequent year. The timing of the onset of dormancy differs between *P. sylvestris* and *P. strobus*. For all experimental treatments of *P. sylvestris* seedlings, only small differences were found between the variability in average length of terminal buds observed on 4th December and 22nd January (Figure 2). This indicates that the apical buds were dormant at that time. On the other hand, great differences were found between the variability in average length of buds in *P. strobus* seedlings measured on 4th December and 22nd January, whereas small differences were found between the measurements taken on 22nd January and 1st March. This indicates that the buds of *P. strobus* were still growing at the start of winter and became dormant later than the buds of *P. sylvestris* seedlings.

Conclusions

The results of the experiments with *P. strobus* and *P. sylvestris* seedlings grown at different pH values and nutrient levels in the soil can be summarized as follows:
1) Under favorable conditions (pH = 4.0 or 6.2 and 2xBNS), total dry weight, growth rate, number of lateral branches and size of dormant apical buds are higher in *Pinus sylvestris* seedlings than in those of *Pinus strobus*.
2) Under unfavorable conditions (pH = 2.7 and either very low or very high nutrient contents in the soil), the establishment, growth and survival of seedlings is reduced in both species, but to a greater extent in *Pinus sylvestris* seedlings.
3) *Pinus strobus* seedlings are more tolerant of stress (brought about by pH = 2.7 and either a deficiency or extremely high concentration of nutrients) than are seedlings of *Pinus sylvestris*.
4) Seedlings of *Pinus sylvestris* and *Pinus strobus* differ in their phenology, which is most apparent from the different timing of the onset of dormancy of their terminal buds.

Acknowledgments

I am very grateful for valuable criticism made by Jan Květ and Frideta Seidlová. The language of the manuscript was revised by Jan Květ and Keith Raymond Edwards. The work was financed by the Academy of the Czech Republic and Faculty of Biological Sciences, České Budějovice.

References

Abrahamsen, G. 1980. Acid precipitation, plant nutrients and forest growth. Proc., Int. conf. ecol. impact acid precip., SNSF project, Norway: 58-63.
Abrahamsen, G. 1981. In: Fazzolare, R. A. and Smith, C. B. (eds.), Beyond the energy crisis opportunity and challenge. pp. 433-446 . Pergamon Press, Oxford and New York.
Abrahamsen, G. 1984. Effects of acid deposition on forest soil and vegetation. Phil. Trans. R. Soc. Lond. B 305: 369-382.

Ågren, G. I. 1983. Nitrogen productivity of some conifers. Can. J. For. Res. 13: 494-500.

Air pollution in Czech Republic in 1994. REPORT CHI 1995, Air Quality Protection Department, Praha.

Bäck, J., Huttunen, S. and Roitto, M. 1993. Pollutant response and ecophysiology of conifer in open-top chamber experiments. Aquilo Ser. Bot. 32: 9-19.

Bäck, J., Huttunen, S., Turunen, M. and Lamppu, J. 1995. Effects of acid rain on growth and nutrient concentrations in Scots pine and Norway spruce seedlings grown in a nutrient-rich soil. Environmental Pollution, 89, No. 2: 177-187.

Eldhuset, T., Göranson, A. and Ingestad, T. 1987. In: Hutchinson, T. C. and Meema, K. M. (eds.), Effects of Atmospheric Pollutants on Forest, Wetlands and Agricultural Ecosystems. pp. 401-409. Springer-Verlag, Berlin, Heidelberg.

Fitter, A. H. and Hay, R. K. M. 1987. Environmental Physiology of Plants. Academic Press, London.

Hanzélyová, D. 1993. The dependence of selected morphological characteristics on age in *Pinus strobus.* BSc. Thesis. Faculty of Biological Sciences, University of South Bohemia, České Budějovice.

Ingestad, T. 1991. Nutrition and growth of forest trees. Tappi Journal, Vol. 74 (1): 55-62.

Ingestad, T. and Kähr, M. 1985. Nutrition and growth of coniferous seedlings at varied relative nitrogen addition rate. Physiologia Plantarum 65: 109-116.

Kozlowski, T. T., Kramer, P. J. and Pallardy, S. G. (eds.) 1991. The Physiological Ecology of Woody Plants. Academic Press. California.

Likens, G. E. and Butler, T. J. 1981. Recent acidification of precipitation in North America. Atmospheric Environment 15: 1103 – 1109.

Lindberg, S. E., Bredemeier, M., Schaefer, D. A. and Qi, L. 1990. Atmospheric concentrations and deposition and major ions in conifer forests in the United States and Federal Republic of Germany. Atmospheric Environment Vol. 24A, No.8: 2207-2220.

Lovett, G. M. and Lindberg, S. E. 1993. Atmospheric deposition and canopy interactions of nitrogen in forests. Can. J. For. Res. 23: 1603-1616.

McLaughlin, S. B., McConathy, R. K., Duvick, D., and Mann, L. K. 1982. Effects of chronic air pollution stress on photosynthesis, carbon allocation, and growth of white pine trees. For. Sci. 28: 60-70.

Mengel, K. and Kirkby, E. A. (eds.) 1982. Principles of Plant Nutrition. International Potash Institute. "Der Bund" AG, Bern/Switzerland.

Pan, Y. and Raynal, D. J. 1995. Predicting growth of plantation conifers in the Adirondack Mountains in response to climate change. Can. J. For. Res. 25: 48-56.

Raynal, D. J., Roman, J. R. and Eichenlaub, W. 1982a. Response of tree seedlings to acid precipitation – I. Effect of substrate acidity on seed germination. Env. Exp. Bot. 22: 377-383.

Raynal, D. J., Roman, J. R. and Eichenlaub, W. 1982b. Response of tree seedlings to acid precipitation – II. Effect of simulated acidified canopy throughfall on sugar maple seedling growth. Env. Exp. Bot. 22: 385-392.

Reich, P. B., Schoettle, A. W., Stroo, H. F., Troiano, J. and Amundson, R. G. 1987a. Effects of ozone and acid rain on white pine (*Pinus strobus*) seedlings grown in five soils. I. Net photosynthesis and growth. Can. J. Bot. 65: 977-987.

Reich, P. B., Schoettle, A. W., Stroo, H. F., Troiano, J. and Amundson, R. G. 1987b. Effects of ozone and acid rain on white pine (*Pinus strobus*) seedlings grown in five soils. III. Nutrient relations. Can. J. Bot. 65: 977-987.

Statistical Sciences. 1995a. S-PLUS Guide to Statistical and Mathematical Analysis, Version 3.3 for Windows. StatSci, a division of Mathsoft, Inc., Seattle, Washington, USA.

Statistical Sciences. 1995b. User's manual, Version 3.3 for Windows. StatSci, a division of Mathsoft, Inc., Seattle, Washington, USA.

Sverdrup, H., Warfvinge, P. and Nihlgard, B. 1994. Assessment of soil acidification effects on forest growth in Sweden. Water, Air and Soil Pollution 78: 1-36.

Turner, C. L. and Knapp, A. K. 1996. Responses of a C_4 grass and the three C_3 forbs to variation in nitrogen and light in tallgrass prairie. Ecology 77(6): 1738-1749.

van den Driessche, R. 1987. Nursery growth of conifer seedlings using fertilisers of different solubilities and application time, and their forest growth. Can. J. For. Res., 18: 172-180.

van den Driessche, R. (ed.) 1991. Mineral nutrition of conifer seedlings. CRC Press. London.

von Evers, F. H. and Hüttl, R. F. 1995. Magnesium-, Calcium- und Kaliummangel bei Waldbäumen – Ursachen, Symptome, Behebung. Blackwell Wissenschafts-Verlag. Waldschutzmerkblatt 16.

EFFECT OF *GUNNERA TINCTORIA* (MOLINA) MIRBEL ON SEMI-NATURAL GRASSLAND HABITATS IN THE WEST OF IRELAND

Betsy Hickey and Bruce Osborne*
*Botany Department, University College Dublin, Belfield, Dublin 4, Ireland, Telephone: + 3531 706 2249. Fax: +3531 706 1153; * Address for correspondence; e-mail: BOsborne@Macollamh.ucd.ie*

Abstract

Areas colonised by *Gunnera tinctoria* were characterised by significant reductions in the number of species and alterations in community composition. Virtually all of the locally-common grassland species were replaced by a very sparse group of isolated plants, mainly shade-tolerant dicotyledons. The fragmentary community associated with *G. tinctoria* may have some similarities with the natural successional sequence often associated with these areas, where *G. tinctoria*, rather than *Salix cinerea*, becomes the major canopy forming species. Despite the possibility of direct inputs of atmospherically – fixed nitrogen and the significantly higher tissue N concentrations of the *G. tinctoria* plants, there was no evidence that the observed changes were associated with higher soil organic N concentrations. A plausible explanation for this is that only the petioles, leaves and inflorescences decay, whereas the rhizome, which persists, could retain ~70% of the total plant nitrogen above ground. Changes in hydrology, or the availability of NO_3^- or NH_4^+- N, were site-specific and unlikely to be associated with major alterations in species abundance or composition. However, the highest rates of mineralisation of N occurred earlier in the year in the colonised plots and this may provide an additional source of N, when there is a high demand, during the initial stages of re-growth of *G. tinctoria* plants. In the colonised plots plant productivities (March-August), were 4-fold higher than those found in the uncolonised plots. This, together with the development of an extensive and persistent canopy, is thought to be a major factor in the success of established plants in these nutrient deficient areas. Based on this work and an analysis of previous studies, nitrogen fixation *per se* is unlikely to be an important factor in the success of invasive species, such as *G. tinctoria*, unless this is accompanied by rapid growth, extensive canopy development and/or nitrogen enrichment. Given the restriction of *G. tinctoria* to regions of high rainfall, where frosts are infrequent, suggests that factors associated with the maintenance of plant water balance and survival at low temperatures will be of particular significance in determining the success of this invasive species.

Introduction

Persistent and often large colonies of the introduced species *Gunnera tinctoria* are now well established in parts of western Ireland, particularly in the region of Achill Island, Co. Mayo (Fig. 1; Osborne 1988). On Achill Island *G. tinctoria* can be found growing in several habitats, both semi-natural and man-made, including ditches, river banks, cliff faces, old quarries, and wet meadows. Based on the earliest record of naturalised plants in Ireland, the range of this species is increasing as colonies have now been recorded in a number of additional localities (Campbell and Osborne 1990; Osborne *et al.* 1991; Osborne, unpublished). There are also reports of scattered naturalised plants in the British Isles, particularly in the southwest of England, and the Channel Islands

Plant Invasions: Ecological Mechanisms and Human Responses, pp. 195–208
edited by U. Starfinger, K. Edwards, I. Kowarik and M. Williamson
© *1998 Backhuys Publishers, Leiden, The Netherlands*

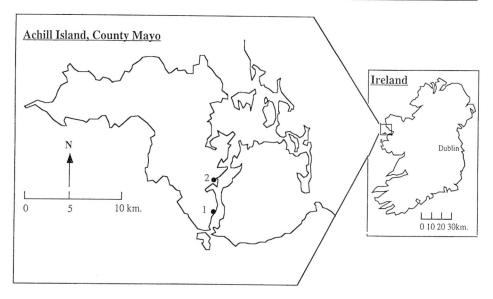

Fig. 1. Location of the two study sites on Achill Island, Co. Mayo, off the west coast of Ireland. Site 1, Grace O'Malleys Castle and site 2, Sraheens.

(see Osborne *et al.* 1991; Clement and Foster 1994), as well as north-western France and the Azores (Philips and Rix 1993). As far as we are aware, however, it is only in Ireland that the formation of large extensive colonies has been observed (Fig. 2). Based on local information colonisation can be rapid, with an almost monospecific stand of *G. tinctoria* occurring after ~ 20 years and individual plants can attain heights of 2m or more during one growing season. Although the origin of this material is uncertain, anecdotal evidence indicates that many are derived from garden escapes, although some may have been deliberately planted in the vicinity of local communities. It is thought that most of the original material is derived from plants brought back from South America by Victorian plant collectors during the nineteenth century (Osborne *et al.* 1991).

In its native habitat in South America, *G. tinctoria* is normally found growing in leached soils in areas of high rainfall and high humidity, where freezing temperatures are uncommon (Jarzen 1980; Osborne 1989). To some extent similar conditions are found in the west of Ireland (Osborne, 1989; Osborne *et al.* 1991). In localities occupied by *G. tinctoria* the annual rainfall is >1200 mm and frosts are infrequent (Osborne 1989). However, plant productivities comparable to sites in the west of Ireland were achieved in an experimental garden in Dublin, where freezing temperatures are more common, when water was supplied in amounts that matched those found in naturalised sites (Osborne 1988; Osborne *et al.* 1991; Campbell and Osborne 1993; Campbell 1994). This suggests a more complex relationship between plant distribution, temperature and rainfall. There is, however, little evidence for eastward colonisation and the pronounced westerly distribution of this species in Ireland and the United Kingdom is similar to a number of European plants that are intolerant of low temperatures (Grace 1987).

Among the factors that may contribute to the success of *G. tinctoria*, is its potential for achieving high productivities in nutrient deficient habitats. The symbiotic relationship which *G. tinctoria* has with the nitrogen-fixing cyanobacterium *Nostoc punctiforme* (Osborne *et al.* 1991), could be important in these situations. All plants

Fig. 2. Naturalised colonies of Gunnera tinctoria at site 2, Sraheens, Achill Island, Co. Mayo. The photograph was taken looking almost due south on the 20[th.] of July 1997.

have been shown to possess the symbiotic cyanobacterium, irrespective of the environmental conditions or soil water/nutrient status (Osborne *et al.* 1991), and N-fixation by the symbiosis is thought to be capable of providing all the nitrogen required for vegetative growth (Osborne *et al.* 1992). A capability for atmospheric nitrogen fixation, particularly in severely nutrient-nitrogen deficient habitats, may be a key feature of some alien invaders (Vitousek and Walker 1989; Vitousek 1990). In turn, other features associated with invasive species, such as nitrogen enrichment (Vitousek 1990), rapid early season growth and extended leaf longevity (Harrington *et al.* 1989a, b), may also be important.

   In order to investigate the reasons for the success of *G. tinctoria* in western Ireland we have initiated a programme of studies on plant performance and nutrient cycling in habitats dominated by this species on Achill Island. Seasonal assessments of plant productivity and nitrogen mineralisation are currently being made at two sites containing extensive, almost monospecific, stands of *G. tinctoria*. Comparisons are also being made with closely adjacent uncolonised areas and this will provide information on the likely impact of this species on ecosystem dynamics. The results obtained in this study will also enable us to explore the possibility of whether we can make generalisations about the characteristics associated with all invasive species.

Materials and Methods

Study sites

Two coastal sites, site 1 at Grace O'Malley's castle (53° 52' N, 9° 57'W), and site 2 at Sraheens, Achill Island (53° 55' N, 9° 57' W), separated by approximately 4 km, were used in this study (Fig. 1). Site 1 is relatively freely draining and shows some drying out during the summer, whilst site 2 is predominantly waterlogged. In both sites there are extensive colonies of *G. tinctoria* (Fig. 2). At each site, two plots (20 by 20 m^2), were established which contained mature stands of *G. tinctoria*, whilst a third adjacent uncolonised plot (20 by 20 m^2), was used for comparison. Four sampling points were randomly chosen within each plot along a gradient that extended down to the sea. In the plots containing *G. tinctoria* the sampling locations for mineralization cores and total N measurements were adjacent to *G. tinctoria* plants.

Soil profiles were examined at three locations by digging pits down through the substrate until bedrock was reached (35-70 cm). One profile was taken in both a colonised and an uncolonised plot at site 1 and another profile was examined in a colonised plot at site 2.

The annual maximum and minimum temperatures for this area are approximately 13°C and 7°C, respectively and the mean annual temperature is about 10°C, with a rainfall of 1100-1200 mm (Rohan 1986). Frosts are unusual during the major part of the growing season (April-October, see Rohan 1986). Under these climatic conditions the *G. tinctoria* plants are deciduous, dying back in November-December, with re-growth from the rhizomes occurring the following March-April (Osborne 1989).

Soil and plant analyses

Soil samples were collected in March 1997 and then on a monthly basis from May onwards. Initially in March two cores (length 25 cm, diameter 5 cm), were obtained by inserting modified sections of commercial plastic piping, to a depth of 25 cm at each sampling point (n=6), at sites located within each plot. The cores were covered with tape on either end. One core was returned to the ground for incubation, whilst the other was brought back for analysis. Two more cores were collected and a sub-sample used for the measurement of N (n=6). In May the incubated core was removed for laboratory analysis, and a new core inserted into the ground. The same sampling technique was used on subsequent trips.

Measurements of soil moisture content (SMC) were made on sub – samples collected for the organic and inorganic N determinations, by drying fresh samples at 80 °C until a constant weight was obtained. SMC was then determined from fresh weight – dry weight measurements. Soil pH was measured on another sub-sample, by adding 25 ml of distilled water to 10 g of dry soil. This was stirred to make a slurry and left for an hour with occasional stirring using a glass rod. The pH of the soil slurry was measured using a Philips PW 9418 meter with a glass electrode.

Organic N determinations were made on the leaves and rhizomes of *G. tinctoria*, as well as on randomly selected shoot material from the uncolonised plots, which had been used for the standing biomass determinations. The total N content of the soils and vegetative material was analysed colorimetrically, after a micro-Kjeldahl digestion, using the technique described by Hendershot (1985). For analyses of nitrate and ammonium, fresh soil was extracted in 100 ml of 2 M KCl. Soil samples were agitated for 1 hr and then left to settle for 24 hours, after which they were filtered. Both NH_4^+ and NO_3^- - N were determined colorimetrically, using an automated flow injection system (Tecator, FIAstar autoanalyzer). Ammonium N was analysed directly and NO_3^- - N was determined from measurements of NO_2^- – N, following reduction with a Cd catalyst (Ruzicka and Hansen 1988). Mineralization was calculated as the change in NH_4^+ - N or NO_3^- - N from one sampling period to the next.

Plant biomass determinations

At the start of the growing season (early March), living above ground rhizomes of *G. tinctoria* (n=6) were collected from a known area of ground ($\sim$4 m^2), and oven-dried for biomass determinations. (No petioles or leaves were present at this time). Any remaining dead material following the winter was also dried and weighed. Harvests were also taken from the uncolonised plots by removing all standing material, within an area of 0.25 m^2 (n=4), to 5 cm above soil level. A second harvest of all plant material was carried out in both sites in August 1997 and a third in site 1 in November 1997. For biomass determinations on *G. tinctoria* the above-ground structures were separated into their component parts (leaves, petioles and rhizomes), before being oven-dried. In August 1997 an assessment of root and dead plant biomass was also made in both sites.

Statistical analysis

All the data was analysed using the ANOVA or GLM procedures of the SAS statistical package (Statistical Analysis System, Cary, North Carolina, USA).

Results

Site description and plant species

Both sites were on gently sloping ground (<10 °), with elevations of <5 m above sea level. All of the plots were south or southeast facing. The soil in site 1 of the colonised plots was classified as a shallow, ground water mineral gley of marine origin, overlying schist. The soil in site 1 of the uncolonised plots was similar, except that it was a surface water mineral gley, whilst in site 2 it was a surface water minerotrophic peaty gley. There was no evidence of any recent attempt to drain either site, although the colonised plot in site 1 had some drainage *via* a small stream that ran through it.

In total there were 22 higher plant species present in the uncolonised plots at site 1 and 15 at site 2. Eight species, 3 of which were grasses, were common to both uncolonised sites (*Agrostis capillaris, Anthoxanthum odoratum, Epilobium palustre, Holcus lanatus, Juncus articulatus, Juncus effusus, Potentilla erecta* and *Rumex acetosa*; Table 1). In the colonised plots the number of species was reduced to 17 and 10 species at sites 1 and 2, respectively. Three species were common to both colonised sites *(Cardamine hirsuta, Galium aparine* and *Rubus fruticosus* agg.). In contrast to the uncolonised plots, only one grass species (*Poa pratensis*, at site 1), was found in a colonised plot (Table 1). Of the species specifically found in association with *G. tinctoria*, there were only 6

Table 1. List of higher plant species growing in the two sites. Species shown in bold were restricted to the colonised plots

Site 1		Site 2	
Colonised	Uncolonised	Colonised	Uncolonised
Apium nodiflorum	*Agrostis capillaris*	*Angelica sylvestris*	*Agrostis capillaris*
Cardamine hirsuta	*Anthoxanthum odoratum*	**Cardamine flexuosa**	*A. stolonifera*
Chrysosplenium			
oppositifolium	*Calluna vulgaris*	**C. hirstua**	*Angelica sylvestris*
Digitalis purpurea	*Cerastium fontanum*	**Cirsium arvense**	*Anthoxanthum odoratum*
Filipendula ulmaria	*Cirsium palustre*	*Epilobium palustre*	*Epilobium palustre*
Galium aparine	*Epilobium palustre*	**Galium aparine**	*Holcus lanatus*
Geranium robertianum	*Festuca rubra*	*Juncus effusus*	*Juncus acutiflorus*
Hedera helix	*Holcus lanatus*	**Lemna minor**	*J. articulatus*
Iris pseudacorus	*Juncus articulatus*	**Ranunculus repens**	*J. effusus*
Poa pratensis	*J. effusus*	*Rubus fructicosus agg.*	*Lythrum salicaria*
Rubus fruticosus agg.	*Leontodon autumnalis*		*Potentilla erecta*
Rumex acetosa	*Lotus corniculatus*		*Pteridium aquilinum*
R. crispus	*Lychnis flos-cuculi*		*Rubus fruticosus agg.*
Taraxacum officinalis agg.	*Orchis mascula*		*Rumex acetosa*
Umbilicus rupestris	*Phragmites australis*		*Salix viminalis*
Urtica dioica	*Potentilla erecta*		
Veronica beccabunga	*Ranunculus acris*		
	R. ficaria		
	R. repens		
	Rumex acetosa		
	Succisa pratensis		
	Vicia sepium		

in site 2, the wetter site, whilst there were 16 species in site 1, the drier site (Table 1).

The above ground material in the uncolonised plots comprised a large proportion of partially decayed dead shoot material that accumulates in these plant communities, presumably because of the wet anaerobic conditions.

Soil pH and moisture content

The soil pH was always <5.3 and significantly (P<0.05) higher in the colonised plots at site 1, compared to the uncolonised ones (Table 2). In contrast, the SMC was consistently lower in the colonised plots (P<0.01) at this site, throughout the sampling period (Table 2). Values for SMC in the colonised plots ranged from 40–47% in site 1 and from 72–77% in site 2. In the uncolonised plots, the SMC ranged from 47–76% and from 73–84%, in sites 1 and 2, respectively (Table 2).

Soil nitrogen concentrations

There were relatively small differences in soil organic N content between colonised and uncolonised plots at either site (Fig. 3a, c), although, at site 2, significantly higher values were found in the colonised plots during June and July (P<0.05). Site 1 had the lower and less variable N concentrations (5-10 mg g^{-1}), compared to site 2 (7.7-19.6

Table 2. Soil moisture content (SMC), and pH in colonised and uncolonised plots.

| | SMC (%) | | | | pH | | | |
| | Colonised | | Uncolonised | | Colonised | | Uncolonised | |
	1	2	1	2	1	2	1	2
March	41.5 ± 4.6	76.0 ± 7.9	75.7 ± 7.0	83.9 ± 2.5	5.5 ± 0.49	5.2 ± 0.27	4.5 ± 0.44	4.9 ± 0.20
May	46.7 ± 4.2	77.2 ± 11.9	47.1 ± 7.0	84.1 ± 6.4	4.7 ± 2.40	4.9 ± 0.37	4.6 ± 0.06	4.9 ± 0.21
June	45.9 ± 23.2	75.9 ± 11.3	66.5 ± 14.3	83.1 ± 7.3	5.5 ± 0.73	5.1 ± 0.45	5.0 ± 0.06	5.1 ± 0.12
July	40.0 ± 8.6	72.8 ± 9.3	59.6 ± 11.8	73.4 ± 3.7	5.3 ± 0.64	5.1 ± 0.44	4.7 ± 0.06	4.9 ± 0.42
Mean	43.5 ± 12.3	75.3 ± 9.4	62.2 ± 13.1	81.1 ± 6.5	5.3 ± 1.26	5.1 ± 0.20	4.7 ± 0.28	5.0 ± 0.21

Table 3. Rates of mineralization of NO_3^- and NH_4^+ in colonised (n=6), and in uncolonised (n=3), plots.

Rates of mineralization ($\mu g\ g^{-1}\ day^{-1}$)

| | Site 1 | | | | Site 2 | | | |
| | Colonised | | Uncolonised | | Colonised | | Uncolonised | |
	NO_3^-	NH_4^+	NO_3^-	NH_4^+	NO_3^-	NH_4^+	NO_3^-	NH_4
March-May	0.91 ± 0.67	0.17 ± 0.08	-0.10 ± 0.05	-0.57 ± 0.58	0.34 ± 0.26	0.08 ± 0.29	0.01 ± 0.04	-0.77 ± 0.74
May-June	-1.77 ± 1.56	-0.15 ± 0.53	0.09 ± 0.13	-0.05 ± 0.44	-0.06 ± 0.63	0.16 ± 0.53	0.57 ± 0.50	0.47 ± 1.06
June-July	1.48 ± 0.72	-0.31 ± 0.49	0.16 ± 0.14	0.00 ± 0.00	-0.17 ± 0.98	-0.57 ± 0.80	-0.18 ± 0.74	-1.07 ± 0.71

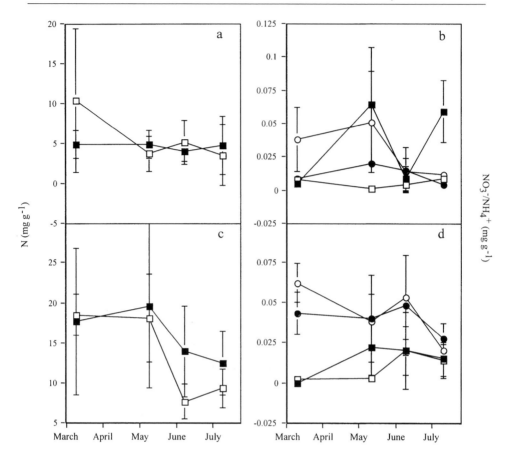

Fig. 3. Seasonal variation in soil organic N (a, c), NO_3^- - N and NH_4^+ - N (b, d) concentrations in plots on Achill Island, Co. Mayo. Open symbols refer to uncolonised sites, closed symbols refer to colonised sites (a, b = site 1: c, d = site 2). In b, d the squares refer to NO_3^- - N, whilst the circles refer to NH_4^+ - N.

mg g^{-1}). Little seasonal trend in organic N was found at site 1. In site 2 there was evidence for a significant decline in N during May to July (P<0.01 for the colonised plots and P<0.05 for the uncolonised plots), to a level comparable with the average values found in site 1 (Fig. 3a, c).

In general, soil NO_3^- - N concentrations were higher (P<0.05) in the colonised plots in site 1 (Fig. 3b), where it tended to be the dominant N-form. There was little difference in NO_3^- - N between the colonised and uncolonised plots in site 2, where NH_4^+ was the major available form of N (Fig. 3d). In site 1, NH_4^+ - N was highest in the uncolonised plots (P<0.05), whereas in site 2 there was no difference in NH_4^+ - N between the colonised and uncolonised plots. In the uncolonised plots in site 1 and in both the uncolonised and colonised plots in site 2, there was a general decline in NH_4^+ - N during the growing season (Fig. 3b, d).

Table 4. Seasonal variation in the concentration of N in *Gunnera tinctoria* leaves and rhizomes (colonised), and in shoot material from uncolonised plots (n=4).

	N (mg g^{-1})					
	Site 1			Site 2		
	Colonised		Uncolonised	Colonised		Uncolonised
	Leaf	Rhizome		Leaf	Rhizome	
March	–	6.18 ± 2.99	12.34 ± 1.89	–	4.99 ± 0.03	5.00 ± 0.22
May	25.79 ± 3.75	17.73 ± 3.01	15.82 ± -	22.07 ± 0.89	–	14.93 ± 1.45
June	21.36 ± 1.58	–	10.02 ± 1.63	22.08 ± 1.26	–	–
July	24.01 ± 5.22	9.39 ± 3.42	–	18.80 ± 1.48	7.22 ± 1.94	11.80 ± 5.92
Mean	23.72 ± 2.23	11.10 ± 5.82	12.72 ± 2.92	20.98 ± 1.90	6.26 ± 1.83	10.58 ± 5.08

Mineralisation of nitrogen

There was some evidence that NO_3^- mineralisation was greater than NH_4^+ mineralisation, although there was considerable variation (Table 3). Rates of mineralisation of NO_3^- – N were high at both colonised sites during March-May, although the highest rates of mineralisation of NO_3^- - N (1.48 ± 0.72 µg g^{-1} day^{-1}), occurred in the drier colonised plot (P<0.05), during June to July (Table 3). In the uncolonised plots, rates of mineralisation of NO_3^- - N were low and the highest values (0.57 µg g^{-1} day^{-1}), were found in the wetter site during May to June (Table 3). The highest rates of mineralisation of NH_4^+ - N were also found at this time (Table 3).

Tissue nitrogen concentrations

In general, the N values for leaf material were greater in site 1 (P<0.05), compared to site 2 (Table 4), for both the colonised and uncolonised plots. Leaf material of *G. tinctoria* generally had N concentrations ~50% higher (P<0.01) than the shoot (mainly leaf) material found in the uncolonised plots (Table 4). There was little evidence for any seasonal variations in leaf or shoot N during the sampling period. From March to May the N concentration in the rhizome more than doubled from 6.2 to 17.7 mg g^1, but decreased to 9. 4 mg g^{-1} in July (Table 4).

Standing above ground biomass

Standing above-ground biomass in the colonised plots in August 1997 at site 1 (1.37 ± 0.41 kg m^{-2}) and site 2 (1.17 ± 0.08 kg m^{-2}), was somewhat lower than in the uncolonised plots at site 1 (1.61 ± 0.47 kg m^{-2}) and site 2 (1.99 ± 0.29 kg m^{-2}). Simi-lar values for standing biomass were obtained in the colonised plots in November at site 1 (1.31 ± 0.22 kg m^{-2}), indicating that this is maintained into the early winter months. As the colonised plots are essentially monospecific stands, all of this biomass was due to the presence of *G. tinctoria*. Rhizome production dominated the above ground pro-portion of plant mass in both sites, with the rhizome comprising 60-70% of the total plant biomass (Fig. 4). Maximum standing biomass peaked around August, but a high above ground biomass, together with a significant canopy (leaf + petiole), biomass was maintained until November (Fig. 4). The proportion of dead material associated with

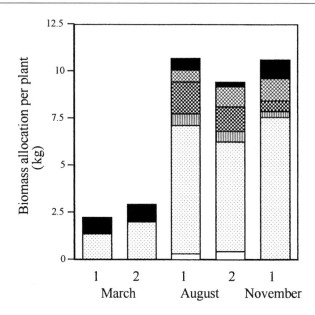

Fig. 4. Allocation of biomass between roots, rhizomes, petioles, leaves, inflorescences and litter for *G. tinctoria* plants at both site 1 and site 2. □ = roots, ▦ = rhizome, ▥ = leaf petiole, ▨ = leaf, ▓ = inflorescence, ■ = dead material.

mature plants in the colonised plots was always <9% of the total above ground biomass (Fig. 4).

Discussion

There can be several effects on ecosystem dynamics following invasion by alien plant species. Among these is a potential to alter soil chemistry, hydrology and ecosystem productivity. Other changes include the introduction of new, or more productive, life forms and a change in species composition, often resulting in a predominantly monospecific stand and a reduction in diversity (Williamson 1996; Cronk and Fuller 1995; Pyšek and Prach 1994). From the results to date it would appear that *G. tinctoria* may be fulfilling some, if not all, of the above criteria. Clearly, a more productive species has been introduced into the habitat. In addition, there was a 22-33% reduction in the number of species (Table 1), and a change in community composition, with the replacement of nearly all the native or semi-native grassland species by a number of new dicotyledonous species which were not found in closely adjacent uncolonised areas.

The plant communities invaded by *G. tinctoria* at both sites fall largely into the phytosociological class Molinio-Arrhenathereatea, order Molinietalia. In Ireland these are typically unfertilised wet meadows, which may exhibit some reduction in moisture content during the summer (White and Doyle 1982). This classification also has some affinities with the *Holcus lanatus-Juncus effusus* rush pasture described in Rodwell (1992). Although the most extensive stands are found in hydromorphic soils, which formerly contained these communities (Hickey and Osborne, unpublished) and may be the most prone to invasion by *G. tinctoria*, naturalised colonies can be found in a wide range of habitats in the west of Ireland (Osborne *et al.* 1991). The plant community specifically associated with the colonised areas is sparse but, unusually, includes species (e.g. *Galium aparine, Hedera helix, Rubus fruticosus, Digitalis purpurea*), which can be found in shaded environments. This community has a remarkable similarity to that found in *Salix cinerea-Galium palustre* woodlands (the Salicetalia auritae of White and Doyle 1982), a pioneer, often fragmentary, vegetation which develops on wet mineral soils, where willow is the dominant canopy forming species (see White and Doyle 1982; Rodwell 1992). This raises the possibility that *G. tinctoria* is replacing willow as the major canopy forming species and that the establishment of this new community could be largely a consequence of a shading effect. The rapid early growth of *G. tinctoria* and the development of a large and persistent canopy would support this proposal. As *S. cinerea* can invade similar sites in the west of Ireland it is possible that the introduction of *G. tinctoria* is simply advancing the natural successional sequence. Confirmation of this will require a more detailed long-term assessment of community structure and composition. Interestingly, fewer species were found in the understory in the wetter colonised site, suggesting that an ability to tolerate shade in combination with waterlogging is less common.

Overall, soil N contents in these sites were somewhat higher than a number of published values (Úlehová 1993; Raison *et al.* 1987; Allen *et al.* 1974), particularly in site 2, presumably because of a high organic matter content. In wet areas mineralisation may be reduced leading to higher levels of organic N and this is supported by the data on inorganic N concentrations (Fig. 4b and 4d). These are at the lower end of the range for unfertilised meadows. Ulehová (1993), recorded values of 0.02 ± 0.02 mg g^{-1} to 0.10 ± 0.09 mg g^{-1} for NO_3^- - N and 0.04 ± 0.007 mg g^{-1} to 0.08 ± 0.02 mg g^{-1} for NH_4^+ - N in a non-fertilised meadow. In the colonised sites mineralization of NH_4^+ - N and NO_3^- - N occurred earlier in the year and this nitrogen would be available to support the early growth of shoots of *G. tinctoria*, prior to significant canopy development. Mature plants may actually have a limited capacity to utilize soil inorganic N and be almost entirely reliant on atmospheric sources, irrespective of soil nutrient-nitrogen availability (Osborne *et al.* 1992). In the current study, leaf N concentrations were independent of variations in the mineralisation of NO_3^- or NH_4^+ (Table 4).

The higher levels of NO_3^- - N in site 1 were associated with a lower SMC in the colonised plots. Also, the highest rates of mineralisation of NO_3^- in colonised plots at both sites were generally associated with the lowest soil moisture contents. This is presumably due to increased water uptake and canopy transpiration by the *G. tinctoria* plants. The impact of this in the wetter site was, however, much smaller and, under these conditions, the soil was likely to have still been anaerobic, thus accounting for the dominance of NH_4^+ - N (Mitsch and Gosselink 1986).

Invasion of these habitats by *G. tinctoria* was, therefore, also associated with some modifications in soil chemistry and hydrology. The presence of ruderal species, such as *Rumex crispus, Taraxacum officinalis* and *Urticia dioica*, in the colonised plots, may be associated with higher soil nitrate concentrations (site 1), and /or small inputs of atmospherically-fixed N in the close vicinity of individual *G. tinctoria* plants. However, soil organic N concentrations were largely unaffected by invasion and ruderal species were not a common feature of the colonised plots. Given the possibility of nitrogen being available, either directly *via* fixation and/or the decay of a large part of the above-ground standing biomass, little of which is left in the following year (Fig.

4), this is somewhat surprising. However, an enhanced soil N status associated with N fixing species may only occur in long established stands (Boring and Swank 1984). Although we have no specific information on the age of the stands that we examined, the evidence is that they are long-lived (see Osborne *et al.* 1991), so that this is unlikely to be a major consideration. A possible reason for the absence of any nitrogen enrichment could be that a large proportion of plant N is retained within the rhizome. Based on the biomass data shown in Fig. 4 and measurements of the N contents of individual plant parts, ~70% of the N could be retained within these persistent above ground structures (see also Osborne *et al.* 1992). As the leaves and petioles could still make a contribution to the soil N status, this may also require significant transfer of nitrogen from the leaves, petioles and inflorescences to the rhizome at the end of the growing season, before these structures decay. Although the fluctuations in rhizome nitrogen concentration reported in the results provide some evidence for movement of N from the leaves to the rhizomes during early growth, further analyses are required to confirm that movement occurs in the opposite direction at the end of the year. Some loss of nitrogen could also occur due to birds, which feed on the fleshy fruits/seeds but this would be difficult to assess. In any case the seeds are often retained on the inflorescences until the following spring. In these high rainfall areas leaching of N (as NO_3^-), could also contribute to a low soil N status. Whatever the reasons for these results, the often-quoted suggestion that alterations in species composition, due to invasion by nitrogen-fixing species, are primarily a consequence of a significantly enhanced soil nitrogen status (see Vitousek 1990; Cronk and Fuller 1995), does not apply to *G. tinctoria*-dominated ecosystems in these habitats.

Values for the above ground standing biomass obtained in this work (1.3 kg m^{-2}), are similar to those recorded by Campbell and Osborne (1992), and Campbell (1994), for *G. tinctoria* plants grown under the simulated rainfall conditions found in the west of Ireland. These are within the range of values normally associated with many temperate ecosystems (Beadle *et al.* 1985). In contrast, the standing biomass for the uncolonised plots (1.8 ± 0.38 kg m^{-2}), was higher than typical estimates (0.25 kg m^{-2}), for temperate grasslands (Beadle *et al.* 1985). However, we did not separate living from dead biomass, which comprises a major fraction of the plant material in the uncolonised plots under these conditions. This would have led to an overestimate of current production. Based on the differences in biomass between March and August we estimate plant productivities of 2.4 ± 0.82 kg m^{-2} yr^{-1} for the colonised plots and 0.6 ± 0.66 kg m^{-2} yr^{-1} for the uncolonised plots. The values for the uncolonised plots are comparable to the *primary* productivities of many natural grassland communities (0.25-1.6 kg m^{-2} yr^{-1}; Caldwell 1975; Beadle *et al.* 1985), whilst the values for the colonised plots are within the higher range of reported values (1-3.5 kg m^{-2} yr^{-1}; Beadle *et al.* 1985). These are also likely to be underestimates of true above ground *primary* productivity, as no assessment of decomposition processes or grazing was made. Based on a ~75% decrease (Fig. 4), in the proportion of dead material in the colonised plot, the true primary productivity may be significantly greater than the values estimated in this study. The ability of *G. tinctoria* to achieve such a high productivity in hydromorphic soils with low concentrations of mineral N suggests that this may be of major significance in the success of this species in these habitats. Changes in hydrology and associated alterations in N transformations were not consistent between the two sites, although the productivities were similar, suggesting that these modifications are not directly associated with the success of this species under these conditions.

It has often been suggested that an ability to fix atmospheric nitrogen may play a key role in determining invasive ability (Vitousek and Walker 1989). However, of the detailed case studies reported by Cronk and Fuller (1995), only 4 out of 17 plant invaders listed can be regarded as putative N-fixers. In the list of invasive species (111) given in the same publication only approximately 24% could be considered to have some capacity for N fixation. Similarly, an ability to fix atmospheric N alone does not appear to be contributing to the success of native plant species in the N deficient habitats reported in this work, as only two potential N-fixing species were identified in the uncolonised plots. Although it may be difficult to generalise about the particular attributes associated with invasive species, most appear to be trees or have arborescent life forms and all have a capacity for rapid growth, often combined with the formation of dense stands (Harrington *et al.* 1989a, b; Cronk and Fuller 1995). These characteristics are certainly a feature of the *G. tinctoria* colonies described, although a high productivity and an ability to form an extensive canopy may be more important than the formation of dense stands with a high standing biomass. On the basis of this evidence an ability to fix atmospheric N may only be an important attribute of an invader when combined with large size and/or a potential for rapid growth. For *G. tinctoria*, part of its success may also be determined by the timing of the rapid increase in shoot re-growth as this is initiated in the early spring some weeks prior to that of most of the species found in the uncolonised plots. Extended leaf and petiole longevity may also be important as an almost intact canopy can be retained into the autumn/winter months after the growth of native species has declined. Similar conclusions have been proposed to explain the success of exotic invaders in Southern Wisconsin (Harrington *et al.* 1989a, b).

Although, at maturity, the development of an extensive canopy may have a major role in eliminating many species, it is still unclear how individual *G. tinctoria* plants become established, or how these largely single-species stands are maintained when there is little shoot development. This may involve suppression of the growth of co-occurring species (see Cronk and Fuller 1995), and is currently being investigated. During the initial phases of invasion factors that influence rhizome growth may also be important, as this structure will provide reserves to support the early development of the canopy. It is also clear that, for many invasive species, including *G. tinctoria*, the environmental conditions may also play an important role. Restriction of the range (Osborne 1989; Osborne *et al.* 1991), or performance (Campbell and Osborne 1992; Campbell 1994), of *G. tinctoria*, by low rainfall, soil water deficits, or freezing temperatures indicates that these environmental factors will have a significant impact on the success of this species as an invader.

Acknowledgements

We thank Jim White and Gerry Clabby for discussions on plant communities.

References

Allen, S.E., Grimshaw, M., Parkinson, J.A. and Quarmby, C. 1974. Chemical Analysis of Ecological Materials. 565pp. Blackwell Scientific Publications, Oxford.

Beadle, C.L., Long, S.P., Imbamba, S.K., Hall, D.O. and Olembo, R.J. 1985. Photosynthesis in Relation to Plant Production in Terrestrial Environments. 156pp. Tycooley Publishing Limited, Oxford.

Boring, L.R. and Swank, W.T. 1984. The role of black locust (*Robinia pseudo-acacia*), in forest succession. Journal of Ecology 72: 749-766.

Caldwell, M.M. 1975. Primary production of grazing lands. In: Cooper, J.P. (ed), Photosynthesis and Productivity in Different Environments, pp. 41-73. Cambridge University Press, Cambridge.

Campbell, G.J. 1994. Water Supply, Plant Productivity and Gas Exchange Responses *of Gunnera tinctoria* (Molina) Mirbel (Gunneraceae). 137pp. Ph.D. thesis, University College Dublin.

Campbell, G.J. and Osborne, B.A. 1990. *Gunnera tinctoria* (Molina) Mirbel in Western Ireland. Irish Naturalists Journal 23: 222-223.

Campbell, G.J. and Osborne, B.A. 1993. Watering regime and photosynthetic performance of *Gunnera tinctoria* (Molina) Mirbel. In: Borghetti, M., Grace, J. and Raschi, A. (eds), Water Transport in Plants Under Climatic Stress, pp. 247-255. Cambridge University Press, Cambridge.

Clement, E.J. and Foster, M.C. 1994. Alien Plants of the British Isles. 590pp. Botanical Society of the British Isles, London.

Cronk, Q.C.B. and Fuller, J.L. 1995. Plant Invaders. 241pp. Chapman and Hall, London.

Grace, J. 1987. Climatic tolerance and the distribution of plants. New Phytologist. 106: 113-130.

Harrington, R.A., Brown, B.J. and Reich, P.B. 1989a. Ecophysiology of exotic and native shrubs in Southern Wisconsin. I. Relationship of leaf characteristics, resource availability and phenology to seasonal patterns of carbon gain. Oecologia 80: 356-367.

Harrington, R.A., Brown, B. J., Reich, P.B. and Fownes, J. H. 1989b. Ecophysiology of exotic and native shrubs in Southern Wisconsin. II. Annual growth and carbon gain. Oecologia 80: 368-373.

Hendershot, W. H. 1985. An inexpensive block digester for nitrogen determinations in soil samples. Communication in Soil Science and Plant Analysis 16: 1271-1278.

Jarzen, D.M. 1980. The occurrence of *Gunnera* pollen in the fossil record. Biotropica 12: 117-123.

Mitsch, W.J. and Gosselink, J.G. 1986. Biogeochemistry of Wetland. In: Mitsch, W.J. and Gosselink, J.G. (eds), Wetlands, pp. 89-125. Van Nostrand Rheinhold, New York.

Osborne, B.A. 1988. Photosynthetic characteristics of *Gunnera tinctoria* (Molina) Mirbel. Photosynthetica 22: 168-178.

Osborne, B.A. 1989. Effect of temperature on photosynthetic O_2-exchange and slow fluorescence characteristics of *Gunnera tinctoria* (Molina) Mirbel. Photosynthetica 23: 77-88.

Osborne, B.A., Cullen, A., Doris, F., McDonald, R., Campbell, G.J. and Steer, M. 1991. *Gunnera tinctoria*: an unusual nitrogen fixing invader. Bioscience. 41: 224-234

Osborne, B.A., Cullen, A., Jones, P.W. and Campbell, G.J. 1992. Use of nitrogen by the *Nostoc-Gunnera tinctoria* (Molina) Mirbel symbiosis. New Phytologist. 120: 481-487.

Philips, R. and Rix, M. 1993. Perennials, vol 2. Late Perrenials. 182pp. Pan Books, London.

Pyšek, P. and Prach, K. 1994. How important are rivers in supporting plant invasions? In: de Waal, L.C., Child, L.E., Wade, P.M. and Brock, J.H. (eds), Ecology and Management of Invasive Riverside Plants, pp. 19-26. John Wiley and Sons, London.

Raison, R.J., Connell, M.J. and Khanna, P.K. 1987. Methodology for studying fluxes of soil mineral-N *in situ*. Soil Biology and Biochemistry 19: 521-530.

Rodwell, J.S. 1991. British Plant Communities I. Woodlands and Scrub. 395pp. Cambridge University Press, Cambridge.

Rodwell, J.S. 1992. British Plant Communities III. Grasslands and Montane Communities. 540pp. Cambridge University Press, Cambridge.

Rohan, P.K. 1986. The Climate of Ireland. 146pp. Meteorological Service, Dublin.

Ruzicka, J. and Hansen, E. H. 1988. Flow Injection Analyses. 498pp. John Wiley, New York.

Úlehová, B. 1993. Fertility of grasslands soils. In: Rychnovská M. (ed), Developments in Agricultural and Managed-Forest Ecology 27. Structure and Functioning of Seminatural Meadows, pp. 307-321. Elsevier, Amsterdam.

Vitousek, P. M. 1990. Biological invasions and ecosystem processes: towards an integration of population biology and ecosystem studies. Oikos 57: 7-13.

Vitousek, P.M. and Walker, L.R. 1989. Biological invasion by *Myrica faya* in Hawaii: plant demography, nitrogen fixation and ecosysten effects. Ecological Monographs 59: 247-265.

White, J. and Doyle, G. 1982. The vegetation of Ireland: a catalogue raisonne. In: White, J. (ed), Studies on Irish Vegetation, pp. 289-367. Royal Dublin Society, Dublin.

Williamson, M. 1996. Biological Invasions. 244pp. Chapman and Hall, London.

HISTORY OF THE SPREAD AND HABITAT PREFERENCES OF *ATRIPLEX SAGITTATA* (CHENOPODIACEAE) IN THE CZECH REPUBLIC

Bohumil Mandák and Petr Pyšek
Institute of Botany, Academy of Sciences of the Czech Republic, CZ-252 43 Průhonice, Czech Republic; e-mail: mandak@ibot.cas.cz, pysek@ibot.cas.cz

Abstract

In the Czech Republic, an alien species *Atriplex sagittata* has been spreading rapidly recently. The paper analyses the rate of its spread and pattern of habitat preferences in the course of more than 150 years. The ancient distribution is inferred from archaeological records and compared with recent situation. The oldest record from the study area comes from the Bronze Age. The first recent report is dated 1810. *A. sagittata* is known from 64.6 % of the map squares of the European phytogeographical grid. An exponential regression fits the increase in the cumulative number of localities. The curve obtained can be divided into four parts corresponding to periods in which the number of localities increased abruptly. The most remarkable increase follows the Second World War, and can be explained by species autoecology and creation of suitable habitats. The species is closely confined to ruderal sites and habitats facilitating transport. It prefers warm and moderately warm climatic regions. The main determinants of success of *Atriplex sagittata* in Central Europe are probably a combination of (a) special adaptation mechanisms, such as heterocarpy and salt-tolerance which favour the species survival in disturbed habitats, and (b) increasing frequency of suitable habitats, related to frequent disturbances of the landscape and human-related transport activities. These habitats are similar, in terms of temporal and spatial variability, to those in which the species occurs in the native distribution area (i.e. salt steppe and riparian habitats).

Introduction

The *Chenopodiaceae* is known as a family containing many extremely specialized species, adapted to dry (deserts and semideserts) and saline (temperate salt marshes and sea shore) situations, often occupying habitats disturbed by human activities and/ or stressed (McArthur and Sanderson 1984). The occurrence in extreme ecological situations is made possible by a number of adaptive modifications such as the presence of different photosynthetic pathways, seed dormancy mechanisms, heterocarpy, xeromorphy, and leaf area reduction. The distribution of *Chenopodiaceae* is largely conditioned by another very important factor, their ability to grow in saline habitats (see e.g. Osmond *et al.* 1980; Flowers *et al.* 1986; Breckle 1995). The species of *Atriplex* are often facultative halophytes (Kelley *et al.* 1982). It is generally suggested that majority of halophytes are restricted to salt habitats because of their limited ability to compete with other species in other habitats (Kelley *et al.* 1982). *Atriplex* is a very heteromorphic genus, including various life forms.

Globally, the *Chenopodiaceae* is amongst the most invasive plant families. Possible

Plant Invasions: Ecological Mechanisms and Human Responses, pp. 209–224
edited by U. Starfinger, K. Edwards, I. Kowarik and M. Williamson
© *1998 Backhuys Publishers, Leiden, The Netherlands*

clues to the success of the family could be high reproductive rate, long viability of seeds, salt tolerance, heterocarpy, and C4 photosynthetic pathway (Pyšek 1998).

Within the *Chenopodiaceae*, it seems useful to distinguish between two main ways of translocations, between-continents and within-continent. The former way is determined by the fact that semi-arid regions in Australia, Central Asia and North America dominated by endemic species have been used for sheep grazing. Consequently, in the last 100 years species were introduced, particularly into Europe, by to the wool trade. For example Probst (1949) reported 15 Australasian species of *Atriplex* in the "Wolladventiv" flora of Europe and a few South American and Central Asian species have found their way into Europe in the same way (Aellen 1960; Aellen and Akeroyd 1993). These species are usually classified as neophytes (i.e. introduced after 1500 A. D., see e.g. Holub and Jirásek 1967 and Pyšek 1995a for terminology); the Australian species *A. semilunaris* Aellen is an example in the Czech Republic. Relatively large number of *Atriplex* species, now in Central Europe, were involved in the within-continent movements. They are not native to this region and their introduction happened during the ancient colonization of Europe. These species are archeophytes, i.e. introduced before 1500: *A. oblongifolia*, *A. patula*, *A. rosea*, *A. sagittata* or *A. tatarica*. Others in the genus are, from the geographical point of view, native to the region; from their natural habitats, i.e. small-scale salt marshes, expanded into other habitats; *A. prostrata* subsp. *latifolia* is an example. These species probably occurred in warmer regions of the Czech Republic and now have expanded and increased their distribution due to the human activities creating suitable habitats beyond the limits of their native distribution. Hence, in case of this group of *Atriplex* species, the native distribution is sometimes rather obscure and very difficult to identify.

The present paper aims at describing the invasion history of an alien species *Atriplex sagittata* in the Czech Republic, demonstrating changes in habitat preferences in the course of the invasion process and characterizing the main community types in which it occurs. This alien species was chosen for the study because of the following reasons: (i) it is remarkably successful in the present landscape, (ii) it produces three different types of fruit which makes it very special not only among its congeners but among heterocarpic species in general, and this fact (iii) makes it possible to infer on the role of heterocarpy in the process of invasion.

Study species

Description and ecological characteristics

Atriplex sagittata Borkh. (syn. *A. acuminata* W. et K., *A. hortensis* L. subsp. *nitens* (Schkuhr) Pons, *A. nitens* Schkuhr, see Kirschner 1984 for notes on nomenclature) belongs to the section *Dichosperma* Dumort. of the *Chenopodiaceae*. This section contains three annual species each possessing three types of dimorphic achenes (Iljin 1936; Aellen 1960). The species of this section bear on the same plant female flowers subtended in bracts and ebracteolate flowers with perianth. The latter flowers are very similar to the perfect flowers of *Chenopodium* and most other chenopods (Stutz *et al.* 1990, 1993). Phylogenetically, these species seem to be an intermediate evolutionary link between types with perianth flowers and "true" *Atriplex* lacking perianth and having female flowers subtended in bracts extremely variable in shape and size (Osmond *et al.* 1980).

A more primitive species of this section is a Central Asiatic species *A. aucheri* which inhabits salt steppes, deserts and semideserts. Aellen (1960) considered *A. sagittata* a cultivated species which evolved from *A. aucheri* and then escaped from cultivation.

A. sagittata is an annual herb reaching the height of 1–2 m. The species is monoecious with non-Krantz anatomy, 2n = 18. Leaves are usually over 10 cm long, often irregularly and coarsely dentate. In the Czech Republic, flowering starts in July and seeds mature from the middle of October. Flowers in terminal or axillary spicate inflorescences are dimorphic and produce three types of fruits (Kirschner and Tomšovic 1990; Kopecký and Lhotská 1990): The first type (further termed A) originates from female or bisexual flowers and contains small, dormant, black lens-shaped seeds with glossy, smooth testa and 5-lobed perianth. The second type (B) is produced by female flowers and contains medium-size dormant seeds of similar appearance as the previous type; the seed is covered by extended bracteoles. Finally, the third type (C) is produced by female flowers, contains rather big, brown non-dormant seeds, that are covered by extended bracteoles (Mandák 1998).

The species of *Atriplex* have been often studied for salt tolerance (Black 1958; Moore *et al.* 1972; Osmond *et al.* 1980; Kelley *et al.* 1982; Schirmer and Breckle 1982; Freitas and Breckle 1993, 1994; Breckle 1995). Of the two principal groups reducing the salt concentration in the plant body, i.e. salt excluders and salt absorbers (Schirmer and Breckle 1982), all the species of genus may be classified as salt excluders. This happens through bladder hairs located on the leaf surface (Freitas and Breckle 1993, 1994; Breckle 1995). *A. sagittata* is common along roads, in habitats rich in salt (KCl and NaCl) due to the winter treatment of roads.

Distribution and dispersal

At present, *A. sagittata* is widely distributed from western Europe through central and southeastern Europe to Central Asia, Asia Minor and western Siberia; it is classified as an Irano-Turanian floristic element (Meusel *et al.* 1965; Hultén and Fries 1986). The westernmost localities are in western Germany on the border with Netherlands and in south-eastern France.

Jalas and Suominen (1987) published a map of the European distribution where they made an attempt to distinguish between native and adventive occurrence. However, what they consider as native distribution probably corresponds to the territory the species inhabited before 1492 (i.e. the discovery of America). In this area, the species should be considered as an archaeophyte. The western part of the European distribution is probably of a more recent origin and the species can be viewed as a neophyte there. The problem is, however, that the boundary between the archaeophytic and neophytic distribution follows perfectly the border between the former eastern block and the rest of Europe. Hence this border reflects rather different concepts of understanding the native vs. adventive status than the real situation. There is no doubt that the occurrence of *A. sagittata* in some areas is of very recent character but the map of Jalas and Suominens (1987) is over-generalized.

The native distribution area of the species covers Central Asia, Asia Minor and eastern Europe from where the species has spread across southeastern and eastern Europe to western Europe (see Meusel *et al.* 1956; Aellen 1960). A remarkable spread was re-corded after The Second World War, especially on ruins and ruderal places of Ger-man cities (Gebhardt 1954; Schreier 1955; Fröde 1956; Korneck 1956; Ullman 1977; Brandes 1982).

Materials and methods

Present distribution and history of spread

The data on the distribution of *A. sagittata* in the the Czech Republic were taken from the following sources: (a) major Czech herbaria – BRNU, HR, MJ, MP, PB, PL, PR, PRC, ZMT, (b) floristic literature, (c) unpublished floristic data obtained from personal communication, (d) our own data from recent years. The information about habitat, altitude, and number of inhabitants was recorded for each locality. The distribution was mapped using a grid of approximately 12×11 km which is commonly used in phytogeographical mapping (Schönfelder and Bresinsky 1990). The probability of being occupied (i.e. the proportion of occupied squares) was calculated separately for squares in warm, moderate and cold regions. Classification of climatic districts was taken from Quitt (1971). The distinction between warm and cooler regions was the –5°C January isotherm.

Previous papers (Pyšek 1991; Pyšek and Prach 1993, 1995) demonstrated how flo-ristic data, systematically gathered over an area for a long time, may be used to recon-struct the pattern of invasion of a species on a large geographical scale. There are, however, some limitations to the floristics data which should be emphasised. A suffi-cient intensity of floristic research within an area is necessary for a successful retro-spective analysis of species spread. This is possible because of the strong, long-term floristic tradition in the Czech Republic. If systematic recording of the flora is carried out, one can assume that the more common a species is, the more often it is recorded. The species itself should be currently (1) worthy of note, i.e. rare enough or otherwise interesting from the point of view of ecology, spreading dynamics etc., (2) conspicu-ous in order not to be overlooked and (3) taxonomically unproblematic, i.e. easily rec-ognisable by amateur botanists who are the main producers of floristic data (Pyšek 1991). These points may be considered reasonably fulfilled by *Atriplex sagittata*.

The spread of the species on an historical time scale was expressed by constructing a plot of the cumulative number of localities reported or squares occupied (i.e. the number reported/occupied up to the given year) against time (Trewick and Wade 1986; Pyšek 1991). An exponential regression model best fitted the data. The rate of invasion was expressed as the value of the slope *b* of the linear regression of the log-transformed cumulative number of localities on time: Log (CUMULATIVE NUMBER OF LO-CALITIES) = a + b × YEAR. (see Pyšek 1991; Pyšek and Prach 1993, 1995 for details).

The curve describing the increase in the cumulative number of localities (or "inva-sion curve") was divided into particular sections. The division was made in a year in which the dynamics of the invasion curve changes. Regressions were fitted to the par-ticular sections and the best fit was found for the log-linear model in each part of the curve. Four distinct periods were found (Fig. 2), and analysed separately.

Although the calculations of the rate of spread were performed on log-transformed data, the figures of invasion curves are shown in the real scale (i.e. untransformed) throughout the paper as it provides a more realistic picture of the dynamics of spread.

Analysis of communities with Atriplex sagittata

In total, 97 phytosociological relevés of communities in which *A. sagittata* occurred were made. All vegetation types were sampled (i.e. including those with low representation of *A. sagittata*) to cover the variation of plant communities in which the species grows in the Czech Republic.

Canonical correspondence analysis (ter Braak 1987) with all species present in a relevé was used to treat the data. Habitat type was used as a nominal environmental variable with the following categories distinguished: dumps, arable land, dung heaps, road margins. Successional stage was coded as follows: 1 – initial, 2 – intermediate, 3 – later successional stages. These characteristics were recorded for each relevé in the field.

Results

The history of spread

Archaeobotanical records prove that the species has been present at the territory of the Czech Republic for about 4,000 years. The oldest report comes from the Bronze Age (Khün 1981a), and up to 15th century, there were at least 33 localities located in 18 mapping squares (i.e. 2.7 % of the total number of squares located at the territory of the Czech Republic were occupied then). The earliest floristic record is from 1810; a herbarium specimen collected in Prague by J. and C. Presl for the first Czech Flora (Presl and Presl 1819). Up to 1900, the distribution of the species was scattered and the localities known were concentrated mostly in warmer regions and close to big cities (Fig. 1b). At that time, the distribution very closely reflects that inferred from archaeobotanical data (compare Figs. 1a with 1b).

At present, *A. sagittata* is significantly over-represented in mapping squares located in warm and moderate climatic regions compared to those located in cold regions; it is rather rare in the latter (Table 1). The total number of localities reported up to

Table 1. The effect of climate on the present distribution of *Atriplex sagittata* in the Czech Republic. Difference between the observed number of squares occupied by the species and expected value derived from the total number of squares located in particular climatic regions was tested by using the χ^2 goodness-of-fit test.

Climatic region	Observed frequency	Expected frequency	Contributions to χ^2
Warm	126	129	0.07
Moderate	264	394	42.89
Cold	48	155	73.86
Total	438	678	116.82
Significance level P = 0.0001 (df 2)			

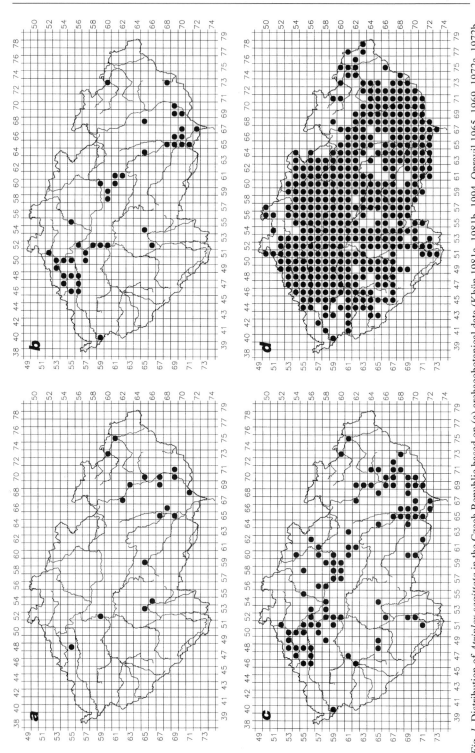

Fig. 1. Distribution of *Atriplex sagittata* in the Czech Republic based on (a) archaeobotanical data (Kühn 1981a, 1981b, 1994, Opravil 1965, 1969, 1972a, 1972b, 1976a, 1976b, 1979, 1980a, 1980b, 1980c, 1981, 1985a, 1987, 1990a, 1990b, 1990c, 1993a, 1993b, 1994, 1996, Šikulová and Opravil 1974), and floristic

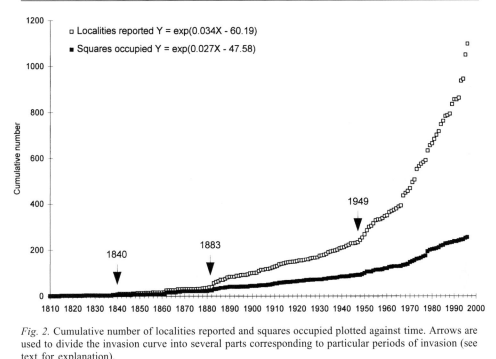

Fig. 2. Cumulative number of localities reported and squares occupied plotted against time. Arrows are used to divide the invasion curve into several parts corresponding to particular periods of invasion (see text for explanation).

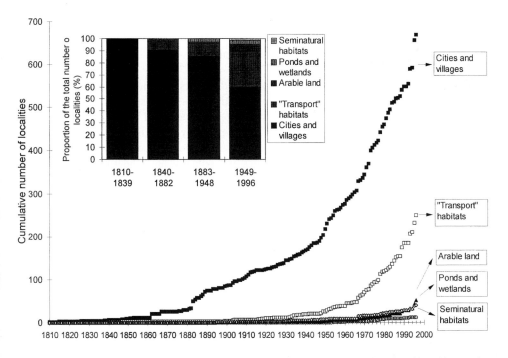

Fig. 3. Cumulative number of localities reported for particular habitats plotted against time. Changes in proportion of particular habitats expressed for particular periods of invasion (see text for details on estimation) are shown in the upper left part of the diagram.

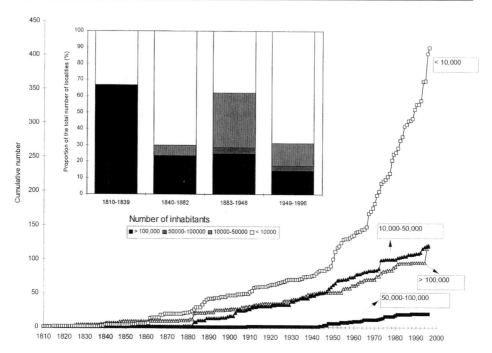

Fig. 4. Increase in the cumulative number of localities reported from cities and villages. Contribution of settlements of different size to the total number of localities is shown for particular periods of invasion (see text for details on estimation) in the upper left part of the diagram.

now is 1101 and 438 mapping squares are occupied, i.e. 64.6 % of the total number located in the area studied (Fig. 1d).

The cumulative number of localities reported and squares occupied has been increasing exponentially over time (Fig. 2). Four rather distinct phases of spread can be recognized. A remarkable increase in the number of localities occurred after 1840, 1883 and 1949 (Fig. 2). Assuming that different habitat types may play different parts in various period of the spreading process, the four phases were analysed separately (Fig. 3).

The spread was fastest ($b = 0.013$) in cities and villages; until the 1940s, the occurrence in habitat types other than these two was rather negligible. After the Second World War, the number of localities reported from the vicinity of roads, railways, paths etc. (or "transport habitats") began to increase rapidly. A remarkable increase in other habitat types, i.e. arable land, ponds and wetlands happened as late as the latest decades. The species is almost absent from semi-natural habitats (meadows, forests and their margins, scrub margins) where only 11 localities were recorded (Fig. 3).

The spread in cities and villages was analysed with respect to the number of inhabitants. In 1839, 66.6 % of localities (out of 6) were reported from cities with more than 100,000 inhabitants; the proportional contribution of these big cities to the total number of localities gradually decreased over time (Fig. 4). At present, the species is present in settlements of any population size; the highest number of localities is from cities and villages up to 10,000 inhabitants (Fig. 4).

The "transport habitats" have gradually become more important; this probably reflects their dispersal role in the spreading of the species from cities and villages into

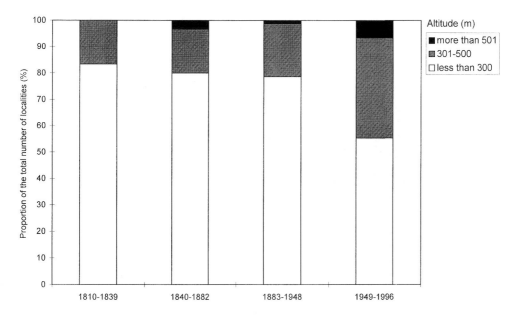

Fig. 5. Distribution of localities with respect to altitude. Proportional contribution of localities found in particular altitudinal ranges is shown for particular periods of invasion.

the open landscape ($b = 0.013$) (Fig. 3). Analysing separately particular "transport habitats", the localities along roads contributed most to the total number of "transport" localities (i.e. 46 %) and the rate of spread was faster in this habitat ($b = 0.011$) compared to railways ($b = 0.008$, 16 % of localities), path margins ($b = 0.007$, 20 % of localities) and water courses ($b = 0.009$, 15 % of localities).

A consistent pattern was found when the spread of the species was related to the altitude (Fig. 5). Up to 1948, 78.7 % of localities reported were located in lower altitudes (below 300 m a. s. l.), and the invasion into higher altitudes started after that. From 1949, there were more than 40 % of localities reported from altitudes above 300 m a. s. l. and the proportion of those occurring above 500 m a. s. l. reached 6.6 %, compared to 1.2 % up to 1948 (Fig. 5). The altitudinal maximum recorded at the territory of the Czech Republic is 763 m a. s .l. (herbarium specimen from the Šumava Mts).

In a similar vein, the probability of a square being occupied was highest, over the whole period of spread, in warm regions (Fig. 6). A remarkable increase in moderate regions was observed about a century later and its timing corresponds to the post-war changes in the landscape. In the cold climatic regions, only 30.9 % of the total number of squares are occupied (Table 1).

Analysis of communities with Atriplex sagittata

Fig. 7 displays the CCA ordination of relevés. The first axis ($\lambda = 0.18$) is seen to separate the stands with respect to the successional status (r = 0.75) with initial stages located in the right part of the diagram. The first axis accounted for 36.2 % variance in the data set and the Monte Carlo test for this axis was highly significant (P < 0.01). The

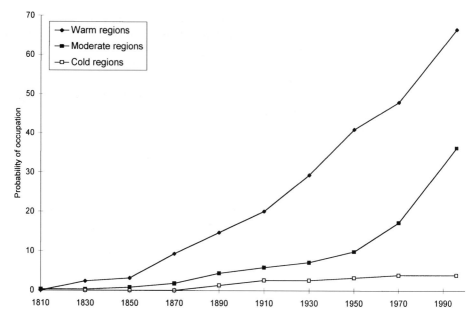

Fig. 6. Probability of being occupied (expressed as a proportion of occupied squares) shown for squares located in warm, moderate and cold climatic regions. Classification of climatic district was taken from Quitt (1971).

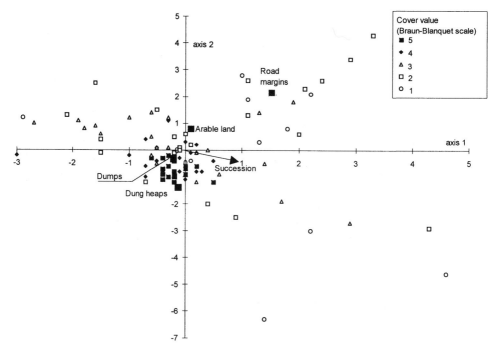

Fig. 7. Correspondence analysis ordination diagram of the relevés of plant communities with *Atriplex sagittata* in the area of the Czech Republic. The performance of *Atriplex sagittata* in particular stands is expressed by using different symbols according to the cover/abundance value the species reached in the respective relevés. Centroids for nominal variables (habitats) are hatched.

second axis ($\lambda = 0.15$) was best correlated with roads ($r = 0.47$). *A. sagitatta* reaches higher dominance (degrees 4 and 5 of the Braun-Blanquet scale) in the post-initial successional stages, and has its optimum on dumps and dung heaps (Fig. 7). The occurrence in the other habitats and communities is more accidental or represents remnants from previous successional stages.

Twenty-three species were present in more than 20 % of the total number of relevés (i.e. in 97 relevés): *Achillea millefolium* agg., *Amaranthus retroflexus*, *Artemisia vulgaris, Atriplex oblongifolia, A. patula, Ballota nigra* subsp. *nigra, Carduus acanthoides, Chenopodium album, Cirsium arvense, Convolvulus arvensis, Echinochloa crus-galli, Elymus repens, Galium aparine, Lactuca serriola, Lolium perenne, Plantago major* subsp. *major, Polygonum aviculare* agg., *Sisymbrium loeselii, Sonchus oleraceus, Stellaria media, Taraxacum* sect. *Ruderalia, Tripleurospermum inodorum, Urtica dioica*.

In total, 236 species were recorded in the complete data set, indicating a high variation in species composition of communities with *A. sagittata*. Even in those stands where *A. sagittata* dominated (value 4–5 in the Braun-Blanquet scale), that are usually treated as a single association *Atriplicetum nitentis* Knapp 1945 (1948), there were still 145 species recorded.

Discussion

History of spread

Studying the spread of an alien species which arrived in the area long ago before herbaria and floristic records were available is extremely difficult. Fortunately, much archaeobotanical research have been done in the Czech Republic (see caption to Fig. 1a). These data make it possible to outline the approximate distribution of some archaeophytes, i.e. those possessing fruits easily recognizable in archaeological excavations. For *A. sagittata*, the picture of the distribution before the 15th century obtained by this method corresponds very closely to the distribution at the end of the last century. Despite some limitations imposed by using floristic records to analyse the spread of an alien species in the last two centuries (see e.g. Pyšek 1991), the rapid spread of *A. sagittata* in the Czech landscape is evident. It may be argued the spread expressed by using the cumulative number of reported localities is partly an artefact of increasing floristic activity. However, a simple comparison of the present occurrence of this very common species with the situation 50 years ago, when it was considered very rare, and up to 1962, when it was missing from relatively large regions (Skalický *et al.* 1962; Kopecký and Lhotská 1990) shows without doubt its remarkable spread.

As in many other invading species (Pyšek *et al.*, this volume), ruderal habitats played the most important role in the spread of *A. sagittata*. These habitats are usually characterized by (1) a high level of disturbance, (2) low moisture, (3) high concentration of nitrogen and other nutrients, (4) high habitat heterogeneity, and (5) specific climatic conditions (many of them lie in big cities in the so-called "heat islands" – Gilbert 1989; Wittig 1991). The rapid spread in these habitats and capability to persist in a site for relatively long time (> 3 years) is probably due to the adaptive mechanism of the species, i.e. (1) heterocarpy and (2) salt-tolerance, and other features determining its successional position (e.g. annual life cycle, vigorous growth, high production of biomass, high fecundity).

A remarkable biological feature of the species, heterocarpy, is one of the typical attributes of plants in arid and semiarid regions (Mandák 1997). The species which produce two or more fruit types represent groups where divergent strategies usually found in different taxa are combined by one individual. In such groups, there is a tendency for fruit functions to diverge, each type specializing on some aspect of environmental variation to which it is predisposed while being buffered by the other seed type (Venable *et al.* 1987, 1995). Variation in dispersal and dormancy strategies of diaspores have been suggested to represent an adaptive response to desert conditions (Venable and Lawlor 1980). Species with special mechanisms such as heterocarpy are at an advantage under high levels of disturbance and unpredictable occurrence of suitable habitats, and their chance of survival is increased. A high level of disturbance is typical of habitats harbouring *A. sagittata*. In spite of this, the species can persist in a site for more than three years (B. Mandák and P. Pyšek, personal observation). The success of *A. sagittata* in Central Europe can be thus explained as an adaptation to such habitats that are similar, in terms of temporal and spatial variability, to those in which it occurs in the native distribution area (i.e. salt steppe and riparian habitats).

In *A. sagittata*, the particular fruit types differ in their ecological functions (Mandák and Pyšek, in preparation). The A type is undispersed, deeply dormant, with low germinability, forming a Type IV seed bank (Thompson and Grime 1979). The B type is easily dispersed, dormant, with germinability intermediate between the A and C types. It forms a Type IV seed bank. The fruit type C is easily dispersed, non-dormant and with Type II seed bank. In general, A-type fruits represent behaviour which favours later germination and restricted dispersal with less survival risk, in contrast to the C type which favours earlier germination and more efficient dispersal with associated survival risk. The intermediate ecological position of the B type helps create an ecological continuum between the two contrasting strategies.

The role of salt tolerance in the species spread probably increased in the last three decades. The winter treatment of roads by salt (mainly NaCl and KCl) favours species with physiological predisposition to growth in such habitats, consequently keeping them free from competition of native plants the majority of which are not salt tolerant. This phenomenon is well documented by the massive spread of the halophytic grass *Puccinelia distans* (B. Mandák and P. Pyšek, personal observation), which is common along roads exposed to winter salt treatment and does not occur in other habitats. Human activities thus created a very specialized niche which may be used for the spread and persistence of species with specific ecology in modern countryside which was relatively poor in salt affected habitats in the past. Salinity-tolerance with an increase of human-induced transport activity have been probably important determinants of the fast spread of the study species.

Heterocarpy and salt tolerance are ecological properties which, along with others (such as annual life cycle, vigorous growth, high production of biomass, high fecundity fruits) determine the species early successional position (Harper 1977). Most of the habitats occupied by *A. sagittata* are in human settlements and the occurence of the species is therefore associated with building activities, creating specific "urban" niches broadly characterized by mechanical perturbation of soil. The establishment of *A. sagittata* in these habitats is enabled by creating open space and reduced competition from other species (Kowarik 1995).

The preference for bigger cities in the early stages of expansion was due to their warmer climate in comparison to surrounding countryside (Wittig 1991; Pyšek 1995b) *A. sagittata* is a thermophilous and heliophilous species (indicator value for temperature = 7 on 9 degree scale, that for light = 9 on 9 degree scale – Ellenberg *et al.* 1991) and the presence of habitats meeting these demands played a significant role in the process of naturalization in the present-day countryside.

In general, two factors or their combination could have caused the invasion of *A. sagittata*, i.e. (1) genetic adaptation, or (2) increasing number of suitable habitats. A high level of genetic plasticity is commonly mentioned as one of the properties of colonizing species (see Baker and Stebbins 1965; Bazzaz 1986). A plant which occurred in a region for a long period has had time to select genetically successful types which are better predisposed to local conditions. On the other hand, the increase of building activity in the second half of the 20th century was remarkable and the rapid spread could have been only a response to this fact and to increased dispersal possibilities. A North American invasive species *Conyza canadensis* is an example of another alien with similar history and determinants of spread.

The main determinants of success of *Atriplex sagittata* in Central Europe are probably a combination of (a) special adaptation mechanisms and (b) increasing frequency of suitable habitats which are similar, in terms of temporal and spatial variability, to those in which the species occurs in its native distribution area (i.e. salt steppe and riparian habitats).

Communities with Atriplex sagittata

A. sagittata is an early successional species (Pyšek and Pyšek 1991). However, it can persist in a site for more than three years (B. Mandák and P. Pyšek, personal observation) and occassionally appears even later in succession as a "successional relic" from preceding stages.

Communities dominated by *A. sagittata* are usually species-poor due to the close canopy, strong competition for water, nutrients, and shading. This community type is common on dumps and dung heaps where *A. sagitata* has its ecological optimum. Pyšek (1977) described two different successional pathways following the stands dominated by *A. sagittata,* depending on the nutrient status of the site: (1) On soils with moderate contents of nutrients, *A. sagittata* populations are followed by the community of tall perennial forbs *Tanacetum vulgare* and *Artemisia vulgaris*; later on a mixture of ruderal grasslands and woodlands is formed. (2) On sites very rich in nutrients, the sequence proceeds from *A. sagittata* community to those dominated by *Ballota nigra* and *Chenopodium bonus-henricus*, later on followed by the stands dominated by *Agropyron repens* and *Aegopodium podagraria*, and the communities with *Sambucus nigra*.

Acknowledgments

Our thanks are due to E. Opravil for providing us with archaeobotanical data, V. Chán for unpublished floristic records, V. Grulich and J. Hadinec for help with studying herbarium data. We also thank late K. Kopecký for inspiring discussion on the biology of the species. Our thanks are also to Mark Williamson and Uwe Starfinger for their comments on the manuscript. Mark Williamson kindly improved the English. We are grateful to I. Ostrý for technical support.

References

Aellen, P. 1960. *Atriplex*. In: Hegi, G. (eds.), Illustrierte Flora von Mitteleuropa. 3/2: 664–693. München.

Aellen, P. and Akeroyd, J.R. 1993. *Atriplex*. In: Tutin, T.G. *et al.* (eds.), Flora Europaea. Vol. 1. pp. 115–117. Cambridge University Press, Cambridge.

Baker, H.G. and Stebbins, G.L. (eds.) 1965. The genetics of colonizing species. Academic Press, London.

Bazzaz, F.A. 1986. Life history of colonizing plants: some demographic, genetic, and physiological features. In: Mooney, H.A. and Drake, J.A. (eds.), Ecology of biological invasions of North America and Hawaii. pp. 96–110. Springer Verlag, New York.

Black, R.F. 1958. Effect of sodium chloride on leaf succulence and area of *Atriplex hastata* L. Austr. J. Bot. 6: 306–321.

Brandes, D. 1982. Das *Atriplicetum nitentis* Knapp 1945 in Mitteleuropa - insbesondere in Südost-Niedersachsen. Documents Phytosociol. 6: 131–153.

Breckle, S.-W. 1995. How do halophytes overcome salinity? In: Khan, M.A. and Ungar, I.A. (eds.), Biology of salt tolerant plants. pp. 199–213. Dept. of Botany, University of Karachi, Pakistan.

Ellenberg, H., Weber, H.E., Düll, R., Wirth, V., Werner, W. and Paulißen, D. 1991. Zeigerwerte von Pflanzen in Mitteleuropa. Scripta Geobot. 18: 1–248.

Flowers, T.J., Hajibagheri, M.A. and Clipson, N.J.W. 1986. Halophytes. Quart. Rev. Biol. 61: 313–337.

Freitas, H. and Breckle, S.-W. 1993. Accumulation of nitrate in bladder hairs of *Atriplex* species. Plant Physiol. Biochem. 31: 887–892.

Freitas, H. and Breckle, S.-W. 1994. Importance of bladder hairs for seedling of some *Atriplex* species. Mésogée 53: 47–54.

Fröde, E. 1956. Zur Frage der Versteppung im Braunschweiger Raum. Braunschw. Heimat 42: 65–69.

Gebhardt, E. 1954. Notizen über die Trümmerflora in Franken. Hess. Flor. Briefe 3(34): 3.

Gilbert, O. 1989. Ecology of urban habitats. Chapman and Hall, London.

Harper, J.L. 1977. Population biology of plants. Academic Press, London, New York and San Francisco.

Holub, J. and Jirásek, V. 1967. Zur Vereinheitlichung der Terminologie in der Phytogeographie. Folia Geobot. et Phytotax. 2: 69–113.

Hultén, E. and Fries, M. 1986. Atlas of North European vascular plants North of the Topics of Cancer. Vol. 1. Koeltz Scientific Books, Königstein.

Iljin, M.M. 1936. *Atriplex*. In: Komarov, V.L. and Šiškin, B.K. (eds.), Flora URSS. Vol. 6. pp. 77–116. Academiae Scientarium URSS, Moskva and Leningrad.

Jalas, J. and Suominen, J. 1987. Atlas Florae Europaeae. Distribution of vascular plants in Europe. II. map 516. Cambridge University Press, Cambridge.

Kelley, D.B., Goodin, J.R. and Miller, D.R. 1982. Biology of *Atriplex*. In: Sen, D.N. and Rajpurohit, K.S. (eds.), Tasks for vegetation science. Contributions to the ecology of halophytes. Vol. 2. pp. 79–107. Dr. W. Junk Publishers, The Hague.

Khün, F. 1981a. Crops and weeds in Šlapanice near Brno from early Bronze Age to now. Zeitschrift f. Archäol. 15: 191–198.

Khün, F. 1981b. Rozbory nálezů polních plodin. Přehled Výzkumů (Archeol. Ústav ČSAV Brno) 1979: 75–79.

Khün, F. 1994. Pěstované rostliny v Brně v době hradištní a ve středověku. Sborník Prací Filos. Fak. Brněn. Univ., Brno, E39: 83–91.

Kirschner, J. 1984. *Atriplex sagittata* Borkhausen. A nomenclature note. Preslia 56: 159–160.

Kirschner, J. and Tomšovic, P. 1990. *Atriplex*. In: Hejný, S. and Slavík, B. (eds.), Flora of the Czech Republic. Vol. 2. pp. 266–280. Academia, Praha.

Kopecký, K. and Lhotská, M. 1990. K šíření druhu *Atriplex sagittata*. Preslia 62: 337–349.

Korneck, D. 1956. Beiträge zur Ruderal- und Adventivflora von Mainz und Umgebung. Hess. Flor. Briefe 5(60): 1–6.

Kowarik, I. 1995. On the role of alien species in urban floras and vegetation. In: Pyšek, P., Prach, K., Rejmánek, M. and Wade, M. (eds.), Plant invasions – general aspects and special problems. pp. 85–103. SPB Academic Publishing, Amsterdam, The Netherlands.

Mandák, B. 1997. Seed heteromorphism and the life cycle of plants: a literature review. Preslia 69: 129–159.

Mandák, B. 1998. Reproductive ecology of *Atriplex sagittata*. PhD Thesis, Czech Agricultural University Praha.

McArthur, E.D. and Sanderson, S.C. 1984. Distribution, systematics, and evolution of *Chenopodiaceae*: an overview. In: Tiedmann, A.R., McArthur, E.D., Stutz, H.C., Stevens, R. and Johnson, K.L. (eds.), Proceedings - symposium on the biology of *Atriplex* and related chenopods. Provo, UT. General Technical Report INT-172.

Meusel, H., Jäger, E. and Weinert, E. 1965. Vergleichende Chorologie der zentraleuropäischen Flora. Gustav Fischer Verlag, Jena.

Moore, R.T., Breckle, S.-W and Caldwell, M.M. 1972. Mineral ion composition and osmotic relations of *Atriplex confertifolia* and *Eurotia lanata*. Oecologia 11: 67–78.

Opravil, E. 1965. Rostlinné nálezy z archeologického výzkumu středověké Opavy prováděného v roce 1962. Čas. Slez. Muz., ser. A, Opava, 14: 77–83.

Opravil, E. 1969. Synantropní rostliny dvou středověkých objektů ze SZ Čech. Preslia 41: 248–257.

Opravil, E. 1972a. Rostliny z velkomoravského hradiště v Mikulčicích. Stud. Archeol. Úst. Čs. Akad. Věd, Brno, 1/2: 6–31.

Opravil, E. 1972b. Synantropní rostliny ze středověku Sezimova Ústí (jižní Čechy). Preslia 44: 37–45.

Opravil, E. 1976a. Archeobotanické nálezy z městského jádra Uherského Brodu. Stud. Archeol. Úst. Čs. Akad. Věd, Brno, Praha, 3/4: 1–60.

Opravil, E. 1976b. Z nejmladší historie luhu řeky Moravy u Kvasic (okres Kroměříž). Zpr. Vlastiv. Úst. Olomouc, 181: 5–11.

Opravil, E. 1979. Rostlinné zbytky z Mohelnice 1. et 2. – Čas. Slez. Muz., ser. A, Opava, 28: 1–13 [1.] et 97–109 [2.].

Opravil, E. 1980a. Rostlinné nálezy ze středověku Starého Města (okres Uherské Hradiště). Přehl. Výzk. Archeol. Úst. ČSAV Brno 1977: 103–105.

Opravil, E. 1980b. Rostlinné zbytky z pravěkého sídliště a zaniklé středověké vsi. In: Unger, J. *et al.*, Pohořelice-Klášterka, 8/2: 96–101, Stud. Archeol. Úst. Čs. Akad. Věd Brno, Praha.

Opravil, E. 1980c. Z historie synantropní vegetace 1–6. Živa 28 (66): 4–5, 53–55, 88–90, 130–161, 167–168, 206–207.

Opravil, E. 1981. Rostlinné zbytky z archeologické výzkumu v Jihlavě. Přehl. Výzk. Archeol. Úst. ČSAV Brno 1979: 62–65.

Opravil, E. 1985a. Rostliny z mladší doby hradištní z Olomouce. Přehl. Výzk. Archeol. Úst. ČSAV Brno 1983: 51–54.

Opravil, E. 1985b. Výsledky archeobotanických analýz z historického jádra města Uherské Hradiště. Přehl. Výzk. Archeol. Úst. ČSAV Brno 1983: 74–82.

Opravil, E. 1987. Rostlinné zbytky z historického jádra Prahy. Archeol. Prag., Praha, 7(1986): 237–271.

Opravil, E. 1990a. Die Vegetation in der jüngeren Burgwallzeit in Přerov. Čas. Slez. Muz., ser. A, Opava, 39: 1–32.

Opravil, E. 1990b. Postmediévální archeobotanické nálezy z Olomouce. Stud. Postmediaev. Archeol., Praha, 1: 231–248.

Opravil, E. 1990c. Archeobotanické nálezy z Kolářské ulice v Opavě. Archeol. Hist., Brno, 15/90: 491–509.

Opravil, E. 1993a. Archeobotanické nálezy z Hrnčířské ulice v Opavě (hotel Orient – dostavba). Čas. Slez. Muz., Opava, A42: 193–214.

Opravil, E. 1993b. Rostliny ze středověku Uherského Brodu – Soukenická ulice a Lidový dům. Přehl. Výzk. Archeol. Úst. ČSAV Brno 1989: 135–143.

Opravil, E. 1994. Příspěvek k poznání rostlinných makrozbytků ze staré Prahy. Archeol. Rozhledy, Praha, 46: 105–114.

Opravil, E. 1996. Archeobotanické nálezy z historického jádra Opavy z výzkumné sezony 1993–1994. Čas. Slez. Muz., Opava, A45: 1–15.

Osmond, C.B., Björkman, O. and Anderson, D.J. 1980. Physiological processes in plant ecology – towards a synthesis with *Atriplex*. Springer Verlag, Berlin, Heidelberg and New York.

Presl, J.S. and Presl, C.B. 1819. Flora Čechica. Květena-Česká. Pragae.

Probst, R. 1949. Wolladventivflora Mitteleuropas. Vogt. Schild., Solothurn.

Pyšek, A. 1977. Sukzession der Ruderalpflanzendgesellschaften von Gross-Plzeň. Preslia 49: 161–179.

Pyšek, P. 1991. *Heracleum mantegazzianum* in the Czech Republic: dynamics of spreading from the historical perspective. Folia Geobot. et Phytotax. 26: 439–454.

Pyšek, P. 1995a. On the terminology used in plant invasion studies. In: Pyšek, P., Prach, K., Rejmánek, M. and Wade, M. (eds.), Plant invasion – General aspects and special problems. pp. 71–81. SPB Academic Publishing, Amsterdam, The Netherlands.

Pyšek, P. 1995b. Approaches to studying spontaneous settlement flora and vegetation in central Europe: a review. In: Sukopp, H., Numata, M. and Huber, A. (eds.), Urban ecology as the basis of urban planning. pp. 23–39, SPB Academic Publishing, Amsterdam.

Pyšek, P. 1998. Is there a taxonomic pattern to plant invasions? Oikos 82: 282–294.

Pyšek, P. and Prach, K. 1993. Plant invasion and the role of riparian habitats: a comparison of four species alien to central Europe. J. Biogeogr. 20: 413–420.

Pyšek, P. and Prach, K. 1995. Invasion dynamics of *Impatiens glandulifera* – a century of spreading reconstructed. Biol. Conserv. 74: 41–48.

Pyšek, P. and Pyšek, A. 1991. Succession in urban habitats: an analysis of phytosociological data. Preslia 63: 125–128

Pyšek, P., Prach, K. and Mandák, B. 1998. Invasion of alien plants into habitats of Central European landscape: an historical pattern. (this volume)

Quitt, E. 1971. Klimatische Gebiete der Tschechoslowakei. Stud. Geogr. 16: 1–73.

Schirmer, U. and Breckle, S.-W. 1982. The role of bladders for removal in some *Chenopodiaceae* (mainly *Atriplex* species). In: Sen, D.N. and Rajpurohit, K.S. (eds.), Tasks for vegetation science. Contributions to the ecology of halophytes. Vol. 2. pp. 215–231. Dr. W. Junk Publishers, The Hague.

Schönfelder, P. and Bresinsky, A. 1990. Verbreitungsatlas der Farn- und Blütenpflanzen Bayerns. Eugen Ulmer, Stuttgart.

Schreier, K. 1955. Die Vegetation auf Trümmer-Schuttzerstörten Stadtteile in Darmstadt und ihre Entwicklung in pflanzensoziologischer Betrachtung. Schriftenr. Naturschutzstelle Darmstadt, 3 (1).

Skalický, V. 1962. Příspěvek ke květeně Rychnovska. Acta Mus. Reginaehradec. et Pardub., ser. A, 47: 63–81.

Stutz, H.C., Chu, G-L. and Sanderson, S.C. 1990. Evolutionary studies of *Atriplex*: Phylogenetic relationships of *Atriplex pleiantha*. Amer. J. Bot. 77: 364-369.

Stutz, H.C., Chu, G-L. and Sanderson, S.C. 1993. Resurrection of the genus *Endolepis* and clarification of *Atriplex phyllostegia* (*Chenopodiaceae*). Amer. J. Bot. 80: 592-597.

Šikulová, V. and Opravil, E. 1974. Nález zvířecích kostí a rostlinných zbytků ve Vávrovicích. Přehl. Výzk. Archeol. Úst. ČSAV Brno 1973: 122–124.

ter Braak, C.J.F. (1987): CANOCO – a FORTRAN program for canonical community ordination by [partial] [detrended] [canonical] correspondence analysis, principal component analysis and redundancy analysis (version 2.1). Agricultural Mathematics Group, Wageningen.

Thompson, K. and Grime, J.P. 1979. Seasonal variation in the seed banks of herbaceous species in ten contrasting habitats. J. Ecol. 67: 893–921.

Trewick, S. and Wade, P.M. 1986. The distribution and dispersal of two alien species of *Impatiens*, waterway weeds in the British Isles. – Proceedings EWRS/AAB Symposium on Aquatic Weeds 1986: 351–356.

Ullman, I. 1977. Die Vegetation des südlichen Maindreiecks. Hoppea, Denkschr. Regensb. Bot. Ges. 36: 5–190.

Venable, D.L. and Lawlor, L. 1980. Delayed germination and dispersal in desert annuals: escape in space and time. Oecologia 46: 272–282.

Venable, D.L., Búrquez, A., Corral, G., Morales, E. and Espinosa, F. 1987. The ecology of seed heteromorphism in *Heterosperma pinnatum* in Central Mexico. Ecology 68: 65–76.

Venable, D.L., Dyreson, E. and Morales, E. 1995. Population dynamics consequences and evolution of seed traits of *Heterosperma pinnatum* (*Asteraceae*). Amer. J. Bot. 82: 410–420.

Wittig, R. 1991. Ökologie der Großstadtflora. Gustav Fischer Verlag, Stuttgart.

INVASION OF THE ACCIDENTALLY INTRODUCED TROPICAL ALGA *CAULERPA TAXIFOLIA* IN THE MEDITERRANEAN SEA

Ulrike Meyer[1,*], Alexandre Meinesz and Jean de Vaugelas
Laboratoire Environnement Marin Littoral, Université de Nice-Sophia Antipolis, Faculté des Sciences, Parc Valrose, F - 06108 Nice Cedex 02, France; e-mail: meinesz@hermes.unice.fr, http://www.unice.fr/LEML
[1]*present address: Ökologie-Zentrum, Christian-Albrechts-Universität, Schauenburgerstr. 112, D - 24118 Kiel, Germany; e-mail: ulrikem@pz-oekosys.uni-kiel.de*

Abstract

Caulerpa taxifolia, a green alga of tropical origin, first colonized in the Mediterranean Sea in 1984 and has been spreading rapidly since. At the end of 1996, 77 stations colonized by this alien plant have been recorded along the coasts of France, Spain, Italy, Monaco and Croatia. The rapid expansion is due to its efficient vegetative reproduction and facilitated by the transport of fragments over long distances by boat anchors or fishing nets. Outstanding characteristics, such as its capacity to colonize all kinds of substratum at depths from 0,5 to 100 meters, its property to form dense, durable covers with very long fronds and stolons, and particularly its feature to contain anti-epiphytic and anti-grazing toxins, which proved to be harmful to many organisms, make *C. taxifolia* a strong competitor in its newly colonized marine environment. The introduction of *C. taxifolia* in the Mediterranean Sea has led to serious ecological consequences, such as an impoverishment of the species diversity in algal communities, a decline in sea urchin populations, and an impact on the ichthyofauna. Another major ecological risk is the invasion of the *Posidonia oceanica* seagrass beds by *C. taxifolia*, which leads to an obstruction and a clogging up of the meadow. *C. taxifolia* blocks the seagrass interstices and reduces light penetration, thereby altering the functions of this fragile and important ecosystem.

Introduction

The littoral ecosystems of the north-western Mediterranean Sea have been subjected for the last decade to the invasion of a tropical green alga, which differs from the original form in several important ecological features. Observations have shown, that the rapid spread of this dominant alien plant leads to severe consequences for the indigenous communities. This paper gives an overview over this recent ecological problem, the investigations undertaken and the perspectives in controlling the spread.

The species

The green alga *Caulerpa taxifolia* (Vahl) C. Agardh (Chlorophyta, Ulvophycea) originally occurs in tropical waters of the Red Sea, the Pacific, the Atlantic and the Indian Oceans. It has a world-wide pantropical distribution and its sparse occurrence is natu-

* corresponding author

Plant Invasions: Ecological Mechanisms and Human Responses, pp. 225–234
edited by U. Starfinger, K. Edwards, I. Kowarik and M. Williamson
© 1998 Backhuys Publishers, Leiden, The Netherlands

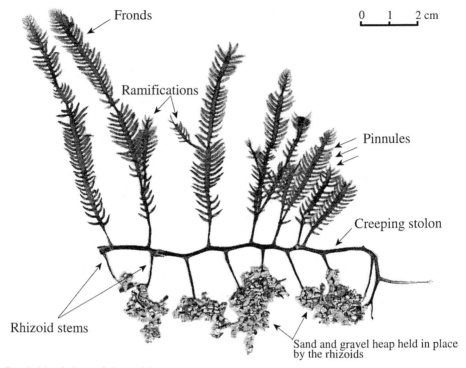

Fig. 1. Morphology of C. taxifolia.

rally limited by water temperatures of below 20°C mean winter temperatures (Meinesz *et al.* 1994). The alga develops on dead coral blocks, sand or mud at depths of between 0 and 30 meters, and never forms dense and large colonies. *C. taxifolia* is a siphonous alga of coenocytic habit which forms erect feathered fronds with numerous pinnules originating from horizontal creeping stolons which are attached to the substratum by rhizoid pillars (Fig. 1).

History of the invasion (arrival and spread in the Mediterranean Sea)

Since the 1970's, individuals of *C. taxifolia* have been used as decorative plants in tropical fish aquariums in the Aquarium of the Wilhelma Zoologisch-Botanischer Garten of Stuttgart, Germany (Meinesz *et al.* 1995). The origin of these individuals is not known. Fragments of *C. taxifolia* were given to the Oceanographic Museum of Monaco (France). In 1984, individuals of *C. taxifolia* were observed in the Mediterranean Sea off Monaco occupying an area of 1 m² (Meinesz and Hesse 1991). Five years later, the colony covered a surface of one hectare. The spread of the invasive alien alga has been progressing rapidly since its accidental introduction (Meinesz and Belsher 1993) and the invaded sites have extended along the northern parts of the Mediterranean Sea. The total area more or less strongly affected by this alga at the end of 1996 was estimated at more than 3096 ha (Fig. 2), with 77 recorded stations along the coasts of France, Spain, Italy, Monaco and Croatia (Meinesz *et al.* 1997).

The rapid spread of *C. taxifolia* in the Mediterranean Sea is due to both natural as well as human modes of expansion. First, vegetative reproduction occurs by fragmentation of various parts of the thallus and local dissemination of the fragments, which do not float but tend to sink and then build new colonies at their arrival. Branching and growth of the stolons is very efficient; total stolon growth measured over a 10 month period was 1860 mm (Komatsu *et al.* 1994). Sexual reproduction does not seem to be a dispersal factor, since only the release of male gametes has been observed so far (Meinesz *et al.* 1994). Second, human factors contribute to the dissemination of the alga at a wider range. The transport of fragments over long distances by boat anchors and fishing material is responsible for the invasion of sites far away from the already existing colonies. Sant *et al.* (1996) showed that *C. taxifolia* has a high resistance to desiccation and tolerates conditions of darkness, 18°C and 85-90% air humidity for eight days, which is comparable to conditions of anchor chests and heaped fishing nets.

Distribution and occurrence in the Mediterranean Sea

The distribution of the introduced seaweed in the Mediterranean ranges from the surface and rock pools near the shore to depths of 100 meters (Meinesz and Belsher 1993, Belsher and Meinesz 1995). At its optimal depth of 2.5 through 50 meters, *C. taxifolia* forms very dense and durable covers on the sea bed with more than 8000 fronds per square meter (Meinesz and Hesse 1991). All kinds of substratum of the infralittoral zone can be colonized by the alien plant, such as rock, sand, mud, gravel, seagrass beds or dead seagrass matte. It can be encountered in exposed sites, as well as in sheltered locations; it has been observed in clean waters, as well as in harbours or polluted environments.

Comparison of the Mediterranean and the tropical form of *C. taxifolia*

Mediterranean *C. taxifolia* differs in some important characteristics, compared to the tropical form of the alga (Table 1). Individuals observed in the Mediterranean have much longer fronds and grow much faster than the tropical counterpart (Meinesz and Hesse 1991). Mediterranean individuals are able to withstand water temperatures as low as 10°C for three months under controlled temperature and light conditions (Komatsu *et al.* 1997). In contrast to the sparse occurrence of *C. taxifolia* in its tropical habitat, the Mediterranean *C. taxifolia* forms very dense covers on the sea bed. There is a lower concentration of toxic substances in the tropical form compared to the Mediterranean form. A comparative study of three geographically different strains has shown large genetic differences, indicating a high degree of instability in the *C. taxifolia* genome (Dini *et al.* 1996).

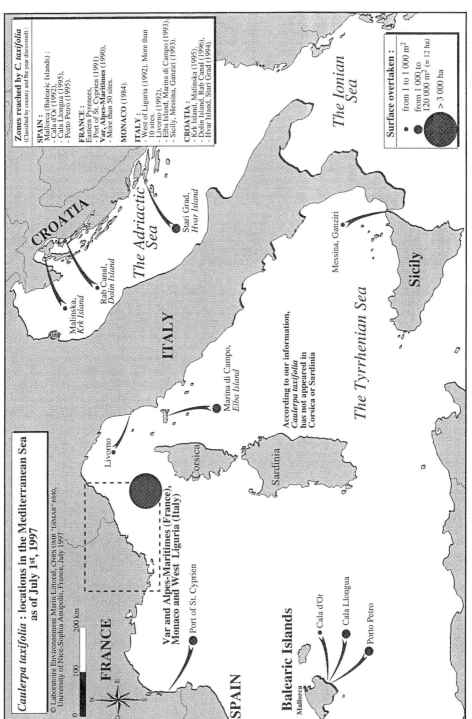

Fig. 2. Locations of invaded zones by *Caulerpa taxifolia* as of July 1st, 1997.

Table 1. Comparison of the characteristics of the tropical and the Mediterranean form of *C. taxifolia.*

Characteristics	Tropical *C. taxifolia*	Mediterranean *C. taxifolia*
Max. frond length	25 cm	83 cm: "gigantism"
Growth	fronds: 0.93 cm/2 weeks at 28°C	fronds: 2.95 cm/2 weeks at 30°C
Occurrence	tropical waters (> 20°C)	temperate waters (10-26°C)
Lethal temperature after 3 months	20°C	10°C
Density	Sparse occurrence	Very dense meadows (up to 8000 fronds/m^2)
Sexual reproduction	Male and female gametes, monoecious	Only male gametes have been observed
Caulerpenyne production	less than Mediterranean form	more than tropical form

Toxicity

Several secondary metabolites have been isolated from the Mediterranean *C. taxifolia* that have proved to be harmful to living organisms: The sesquiterpene caulerpenyne, the taxifolial A, B C and D, taxifolione, caulerpenynol and 10,11-epoxycaulerpenyne. The production of these toxic substances varies depending on depth and season, and according to the seasonality of the vegetative development of *C. taxifolia*. The concentration of caulerpenyne in Mediterranean *C. taxifolia*, which is the major toxin produced by the alga, is highest in summer and at 5 meters of depth, and lowest in winter and at a depth of 30 meters (Amade *et al.* 1996). It has antimicrobial and cytotoxic properties, causing repulsion of herbivores (Lemée *et al.* 1996) and inhibiting fixation of sessile organisms. Even low doses of caulerpenyne affect the behaviour of ciliated protists (Dini *et al.* 1996).

The composition of seawater bacterial communities of environments colonized by *C. taxifolia* colonies is very different from non-invaded zones (Giannotti *et al.* 1996). Experiments showed that the presence of *C. taxifolia* reduces the productivity of some macroalgae (Ferrer *et al.* 1996). Furthermore, extracts of *C. taxifolia* can lead to a blow out of zygote membranes or to an inhibition of rhizoid formation in macroalgae (Gómez Garreta *et al.* 1996). The toxic substances act also on toxicological models, such as animal and human cell cultures, cancerous cells and mice, as well as on marine models like ciliates, microalgae (Lemée *et al.* 1997) and sea urchin eggs (Pedrotti *et al.* 1996). Extracts of *C. taxifolia* act also on the central nervous system of the leech (Garcia-Gil *et al.* 1996). Changes in the digestive gland cells of the mollusc *Aplysia punctata* (Taieb and Vicente 1996) and in the liver of the fish *Serranus cabrilla* (Jouhaud *et al.* 1996), both feeding on *C. taxifolia,* have been observed.

Competitiveness

The growth of *C. taxifolia* does not suffer from nutrient limitation throughout the year; productivity rates were not enhanced by nutrient enrichment. In addition, the rates of alkaline phosphatase activity (APA) are similar to those of characteristic P-sufficient algae (Delgado *et al.* 1995). The alga has the capacity to survive low light intensities and is characterized by a low compensation point all year round, which enables the alga to grow at greater depths (Rodríguez-Prieto *et al.* 1996).The optimal light intensity for its productivity occurs at depths of between 5 and 30 meters, but its modest irradiance requirements could explain the existence of individuals at greater depths in relatively clear waters.

Comparing growth characteristics of *C. taxifolia* and the native macrophyte *Posidonia oceanica*, the two plants are characterized by shifted seasonal development cycles depending on water temperature and light. *C. taxifolia* grows from late spring until late autumn following the rise of water temperatures. When its fronds are longest in autumn, the seagrass *P. oceanica* sheds its long leaves, and the young seagrass shoots start to grow until next summer (Meinesz *et al.* 1993). The growth of *P. oceanica* rhizomes is very slow (3-10 cm per year), whereas the growth of *C. taxifolia* stolons is fast (more than 1 m per year).

Impact and consequences

Serious changes in the Mediterranean ecosystems have occurred since the introduction of *C. taxifolia* in the Mediterranean Sea. *C. taxifolia* overgrows and therefore may be a better competitor than many indigenous algal species leading to a decreased species diversity in the affected algal communities (Verlaque and Fritayre 1994). There has also been a noted decline in sea urchin populations (Ruitton and Boudouresque 1994) and an impact on the ichthyofauna by reducing the fish biomass in zones colonized by the alga (Harmelin-Vivien *et al.* 1996). A modification of the sedimentary balance in zones invaded by *C. taxifolia* has been demonstrated, and thus a development of anoxic conditions leading to changes in species composition of the meiofauna in invaded stations (Poizat and Boudouresque 1996).

A major ecological risk constitutes the invasion of the *Posidonia oceanica* seagrass beds by *C. taxifolia*. This marine phanerogam forms dense and large meadows along the coasts, covering the ground of shallow waters in depths between 0 and 40 meters (Molinier and Picard 1952). *P. oceanica* constitutes a major ecosystem (Hartog 1970), since these seagrass meadows concentrate a great amount of biomass (Bay 1984; Pergent *et al.* 1994) and show the highest primary and oxygen productivity of the communities in the Mediterranean Sea (Romero 1989). This community plays an important role in the complex functioning of marine life, promoting a rich feeding ground for omnivorous and herbivorous animals. Furthermore, they provide shelter to larvae, juveniles and adults of fish and invertebrates (Mazzella *et al.* 1989) and settling substratum for epiphytic organisms (Buia *et al.* 1989), that perform an intensive photosynthetic activity which also contributes to the high oxygen production of the seagrass community. The seagrass meadows of *P. oceanica* stabilize the sedimentary bottoms by solid rhizomes, regulate the sedimentation process and diminish the turbidity of the water by trapping sedimentary particles. In opposition to this, *C. taxifolia* does not fundamentally stabilize the sea bottom, since it is only fixed to the sediment by its small rhizoids.

Villèle and Verlaque (1995) noted consistent changes in invaded *P. oceanica* beds correlated with the seasonal development of the alga: a change in the arrangements of tannin cells in the seagrass leaves, a decrease of number, width and longevity of leaves, chlorosis, necrosis and finally death of shoots. The invasive *C. taxifolia* acts not only on the physiology and the morphology of this fragile marine phanerogam but also on the functional processes of the seagrass ecosystem. By clogging up the interstices between the seagrass roots and leaves, it reduces considerably the space formerly available to organisms colonizing the seagrass beds (Meyer 1996). *C. taxifolia* obstructs the circulation of water currents and organisms between the seagrass leaves by its high amount of biomass and volume. Furthermore, there is strong competition for light, resulting in a serious threat to the durable existence of the sensitive seagrass, mainly at greater depths.

Monitoring of the spread

In order to keep track of the spread of *C. taxifolia* in the Mediterranean Sea, a campaign of sensitizing the public has been undertaken in the coastal areas of the Mediterranean riverine countries. Informative documents and the diffusion of communiqués through the media have helped to increase peoples' awareness of the phenomenon. Verification of reports of *C. taxifolia* colonies by divers, fishermen, and tourists, as well as the active search for probable new colonies and the observation of the progression of already invaded zones (Belsher and Meinesz 1995) contribute to an accurate mapping and following of the spread of *C. taxifolia* (Vaugelas *et al.* 1996*)*. A model of the spread of *C. taxifolia* has been developed; discrete event computer simulation, coupled with a Geographic Information System, will help to assess algal expansion strategies and to test various situations of the invasion and the spread of the alga (Hill *et al.* 1997).

Perspectives

In the struggle against the spread of *C. taxifolia* in the Mediterranean Sea, preventing invasion of unaffected sites, as well as the elimination of already existing colonies, are of great importance. In order to prevent new contaminations, necessary precautions have to be applied, such as cleaning of boat anchors and fishing nets after usage. There are legal bases in France and Catalonia prohibiting purchase, sale, transport and the keeping of *C. taxifolia* individuals. Different methods of eradication of the invasive plant have been investigated, such as manual eradication, electrolysis by copper electrodes, copper ions on plate, dry ice, ultrasound, opaque sails, hypochlorite and hot water. Considering different criteria (effectiveness, working speed, yield, involuntary dispersal of fragments, security of the divers, costs), the most effective way would be a combination of different strategies (Boudouresque *et al.* 1996).

A different kind of approach is the biological control of the alga using organisms feeding on *C. taxifolia*. Two ascoglossan mollusc species of tropical origin, *Elysia subornata* and *Oxynoe azuropunctata,* are morphologically specialized in grazing the pinnule-bearing alga by adaptations of the radula. Data about the consumption rate and their reproduction indicate that these organisms could constitute a helpful tool in the struggle against the alien plant (Meinesz *et al.* 1996). But numerous precautions have to be taken and profound research studies have to be performed before the final decision of an expert commission is made concerning the introduction of new organisms into the Mediterranean Sea.

Acknowledgments

We would like to thank the team of the Laboratoire Environnement Marin Littoral, University of Nice (France) for their helpful assistance. We also are very grateful to Dr. Uwe Starfinger for organizing the interesting Workshop on the Ecology of Invasive Alien Plants.

References

Amade, P., Lemée, R., Pesando, D., Valls, R. and Meinesz, A. 1996. Variations de la production de caulerpényne dans *Caulerpa taxifolia* de Méditerranée. In: Ribera, M.A., Ballesteros E., Boudouresque C.F., Gómez A. and Gravez V. (eds.), Second International Workshop on *Caulerpa taxifolia*. pp. 223-231. Publicacions Universitat Barcelona, Spain.

Bay, D. 1984. A field study of the growth dynamics and productivity of *Posidonia oceanica* in Calvi Bay, Corsica. Aquatic Botany 20: 43-64.

Belsher, T. and Meinesz, A. 1995. Deep-water dispersal of the tropical alga *Caulerpa taxifolia* introduced into the Mediterranean. Aquatic Botany 51: 163-169.

Boudouresque, C.F., Ballesteros, E., Cinelli, F., Henocque, Y., Meinesz, A., Pesando, D., Pietra, F., Riebera, M.A. and Tripaldi, G. 1996. Synthèse des résultats du programme CCE-LIFE "Expansion de l'algue verte tropicale *Caulerpa taxifolia* en Méditerranée". In: Ribera, M.A., Ballesteros, E., Boudouresque, C.F., Gómez A. and Gravez, V. (eds.), Second International Workshop on *Caulerpa taxifolia*. pp. 11-57. Publicacions Universitat Barcelona, Spain.

Delgado, O., Rodríguez-Prieto, C., Gacia, E. and Ballesteros, E. 1995. Lack of severe nutrient limitation of *Caulerpa taxifolia* (Vahl) C. Agardh, an introduced seaweed spreading over the oligotrophic north-western Mediterranean. Botanica Marina 38: 61-67.

Dini, F., Capovani, C., Durante, M., Pighini, M., Ricci, N., Tomei, A. and Pietra, F. 1996. Principles of operation of the toxic system of *Caulerpa taxifolia* that undertook a genetically conditioned adaptation to the Mediterranean Sea. In: Ribera, M.A., Ballesteros, E., Boudouresque, C.F., Gómez, A. and Gravez, V. (eds.), Second International Workshop on *Caulerpa taxifolia*. pp. 247-254. Publicacions Universitat Barcelona, Spain.

Ferrer, E., Gómez Garreta, A. and Ribera, M.A. 1996. Effects of *Caulerpa taxifolia* on two Mediterranean Macrophytes. In: Ribera, M.A., Ballesteros, E., Boudouresque, C.F., Gómez, A. and Gravez, V. (eds.), Second International Workshop on *Caulerpa taxifolia*. pp. 271-276. Publicacions Universitat Barcelona, Spain.

Garcia-Gil, M., Romando, A., Bottai, D., Della Pietà, F. and Brunello, M. 1996. Crude extracts of *Caulerpa taxifolia* increase protein phosphorylation in the leech central nervous system. In: Ribera, M.A., Ballesteros, E., Boudouresque, C.F., Gómez, A. and Gravez, V. (eds.), Second International Workshop on *Caulerpa taxifolia*. pp. 349-354. Publicacions Universitat Barcelona, Spain.

Giannotti, A., Ghelardi, E., Dini, F., Pietra, F. and Senesi, S. 1996. Progressive modification of mediterranean bacterial communities along with the spreading of *Caulerpa taxifolia*. In: Ribera, M.A., Ballesteros, E., Boudouresque, C.F., Gómez A. and Gravez, V. (eds.), Second International Workshop on *Caulerpa taxifolia*. pp. 255-260. Publicacions Universitat Barcelona, Spain.

Gómez Garreta, A., Ferrer, E. and Ribera, M.A. 1996. Impact de *Caulerpa taxifolia* sur les zygotes de *Cystoseira mediterranea* (Fucales). In: Ribera, M.A., Ballesteros, E., Boudouresque, C.F., Gómez, A. and Gravez, V. (eds.), Second International Workshop on *Caulerpa taxifolia*. pp. 277-280. Publicacions Universitat Barcelona, Spain.

Harmelin-Vivien, M., Harmelin, J.G. and Francour, P. 1996. A 3-year study of the littoral fish fauna of sites colonized by *Caulerpa taxifolia* in the N.W. Mediterranean (Menton, France). In: Ribera, M.A., Ballesteros, E., Boudouresque, C.F., Gómez, A. and Gravez, V. (eds.), Second International Workshop on *Caulerpa taxifolia*. pp. 391-397. Publicacions Universitat Barcelona, Spain.

Hartog, C. den. 1970. The seagrasses of the world. North-Holland publ. Co., Elsevier publ., Amsterdam, 275 pp.

Hill, D., Coquillard, P., Vaugelas, J. de and Meinesz, A. 1995. A stochastic model with spatial constraints: simulation of *Caulerpa taxifolia* development in the north Mediterranean Sea. EUROSIM '95: 999-1005.

Hill, D., Coquillard, P., Vaugelas, J de and Meinesz, A. 1997. A computer model for invasive species

– application to *Caulerpa taxifoli*a (Vahl) C. Agardh development in the north-western Mediterranean Sea. Ecological Modelling (accepted).

Jouhaud, R., Valls, R., Fourcault, B. and Brusle, J. 1996. Expériences préliminaires d'intoxication alimentaire du serran *Serranus cabrilla* par *Caulerpa taxifolia*: Induction de modifications hépathiques. In: Ribera, M.A., Ballesteros, E., Boudouresque, C.F., Gómez A. and Gravez, V. (eds.), Second International Workshop on *Caulerpa taxifolia*. pp. 323-328. Publicacions Universitat Barcelona, Spain.

Komatsu, T., Molenaar, H., Blachier, J., Buckles, D., Lemée, R. and Meinesz, A. 1994. Premières données sur la croissance saisonnière des stolons de *Caulerpa taxifolia* en Méditerranée. In: Boudouresque, C.F., Meinesz, A. and Gravez, V. (eds.), First International Workshop on *Caulerpa taxifolia*. pp. 279-283. GIS Posidonie publ., Marseille, France.

Komatsu, T., Meinesz, A. and Buckles, D. 1997. Temperature and light responses of alga *Caulerpa taxifolia* introduced into the Mediterranean Sea. Marine Ecology Progress Series 146: 145-153.

Lemée, R., Boudouresque, C.F., Gobert, J., Malestroit, P., Mari, X., Meinesz, A., Menager, V. and Ruitton, S. 1996. Feeding behaviour of *Paracentrotus lividus* in presence of *Caulerpa taxifolia* introduced in the Mediterranean. Oceanologica Acta 19 (3-4): 245-253.

Lemée, R., Pesando, D., Issanchou, C. and Amade, P. 1997. Microalgae: a model to investigate the ecotoxicity of the green alga *Caulerpa taxifolia* from the Mediterranean Sea. Marine Environmental Research 44 (1): 13-25.

Mazzella, L., Scipione, M.B. and Buia, M.C. 1989. Spatio-temporal distribution of algal and animal communities in a *Posidonia oceanica* meadow. P.S.Z.N.I: Marine Ecology 10(2): 107-129.

Meinesz, A. and Hesse, B. 1991. Introduction et invasion de l'algue tropicale *Caulerpa taxifolia* en Méditerranée nord-occidentale. Oceanologica Acta 14 (4): 415-426.

Meinesz, A. and Belsher, T. 1993. Observations en sous-marin de *Caulerpa taxifolia* dans l'étage circalittoral de l'Est des Alpes Maritimes. Rapport IFREMER, Centre de Brest, France.

Meinesz, A., Vaugelas, J. de, Hesse, B., Mari, X. 1993. Spread of the introduced tropical green alga *Caulerpa taxifolia* in northern Mediterranean waters. Journal of Applied Phycology 5: 141-147.

Meinesz, A., Pietkiewicz, D., Komatsu, T., Caye, G., Blachier, J., Lemée, R. and Renoux-Meunier, A. 1994. Notes taxonomiques préliminaires sur *Caulerpa taxifolia* et *Caulerpa mexicana*. In: Boudouresque, C.F., Meinesz, A. and Gravez, V. (eds.), First International Workshop on *Caulerpa taxifolia*. pp. 105-114. GIS Posidonie publ., Marseille, France.

Meinesz, A., Benichou, L., Blachier, J., Komatsu, T., Lemée, R., Molenaar, H. and Mari, X. 1995. Variations in the structure, morphology and biomass of *Caulerpa taxifolia* in the Mediterranean Sea. Botanica Marina 38: 499-508.

Meinesz, A., Melnyk, J., Blachier, J. and Charrier, S. 1996. Etude préliminaire, en aquarium, de deux Ascoglosses tropicaux consommant *Caulerpa taxifolia*: une voie de recherche pour la lutte biologique. In: Ribera, M.A., Ballesteros, E., Boudouresque, C.F., Gómez, A. and Gravez, V. (eds.), Second International Workshop on *Caulerpa taxifolia*. pp. 157-161. Publicacions Universitat Barcelona, Spain.

Meinesz A., Cottalorda, J.-M., Chiaverini, D., Braun, M., Carvalho, N., Febvre, M., Ierardi, S., Mangialajo, L., Passeron-Seitre, G. 1997. Suivi de l'invasion de l'algue tropicale *Caulerpa taxifolia* devant les côtes françaises de la Méditerranée: Situation au 31 décembre 1996. Ed. Laboratoire Environnement Marin Littoral, Université de Nice – Sophia Antipolis, 190 pp.

Meyer, U. 1996. Clogging up of a *Posidonia oceanica* meadow due to the invasion of *Caulerpa taxifolia* in the Mediterranean Sea. Diploma thesis, Freie Universität Berlin, 97 pp.

Molinier, R. and Picard, J. 1952. Recherches sur les phanérogames marines du littoral méditerranéen français. Annales de l'Institut Océanographique, Paris, France 27 (3): 15-34.

Pedrotti, M.L., Marchi, B. and Lemée, R. 1996. Effects of *Caulerpa taxifolia* secondary metabolites on the embryogenesis, larval development and metamorphosis of the sea urchin *Paracentrotus lividus*. Oceanologica Acta 19 (3-4): 255-262.

Pergent, G., Romero, J., Pergent-Martini, C., Mateo, M.A. and Boudouresque, C.F. 1994. Primary production, stocks and fluxes in the Mediterranean seagrass *Posidonia oceanica*. Marine Ecology Progress Series 106: 139-146.

Poizat, C. and Boudouresque, C.F. 1996. Méiofaune du sédiment dans des peuplements à *Caulerpa taxifolia* du Cap Martin (Alpes-Maritimes, France). In: Ribera, M.A., Ballesteros, E., Boudouresque, C.F., Gómez, A. and Gravez, V. (eds.), Second International Workshop on *Caulerpa taxifolia*. pp. 375-386. Publicacions Universitat Barcelona, Spain.

Rodríguez-Prieto, C., Gacia, E., Delgado, O., Sant, N. and Ballesteros, E. 1996. Seasonality in the productivity of Mediterranean *Caulerpa taxifolia* (Vahl) C. Agardh in relation to light and temperature: January and April 1994. In: Ribera, M.A., Ballesteros, E., Boudouresque, C.F., Gómez, A. and Gravez, V. (eds.), Second International Workshop on *Caulerpa taxifolia*. pp. 197-201. Publicacions Universitat Barcelona, Spain.

Romero, J. 1989. Seasonal pattern of *Posidonia oceanica* production: growth, age and renewal of leaves. In: Boudouresque, C.F., Meinesz, A., Fresi, E. and Gravez, V. (eds.), International Workshop on *Posidonia* Beds 2. pp. 63-67. GIS Posidonie publ., Marseille, France.

Ruitton, S. and Boudouresque, C.F. 1994. Impact de *Caulerpa taxifolia* sur une population de l'oursin *Paracentrotus lividus* à Roquebrune Cap-Martin (Alpes Maritimes, France). In: Boudouresque, C.F., Meinesz, A. and Gravez, V. (eds.), First International Workshop on *Caulerpa taxifolia*. pp. 371-378. GIS Posidonie publ., Marseille, France.

Sant, N., Delgado, O., Rodriguez-Prieto, C. and Ballesteros, E. 1996. The spreading of the introduced seaweed *Caulerpa taxifolia* (Vahl) C. Agardh in the Mediterranean Sea: testing the boat transportation hypothesis. Botanica Marina 39: 427-430.

Taieb, N. and Vicente, N. 1996. Ultrastrucutre des cellules de la glande digestive de *Aplysia punctata* nourrie avec *Caulerpa taxifolia*. In: Ribera, M.A., Ballesteros, E., Boudouresque, C.F., Gómez, A. and Gravez, V. (eds.), Second International Workshop on *Caulerpa taxifolia*. pp. 265-270. Publicacions Universitat Barcelona, Spain.

Vaugelas, J. de, Cottalorda, J.M., Charrier, S., Commeau, T., Delahaye, L., Jaffrennou, F., Lemée, R., Meinesz, A. and Molenaar, H. 1996. Cartographie de l'invasion de *Caulerpa taxifolia*. Situation sur les côtes françaises de la Méditerranée à la fin de 1994. In: Ribera M.A., Ballesteros E., Boudouresque C.F., Gómez A., Gravez V. (eds.), Second International Workshop on *Caulerpa taxifolia*. pp. 91-97. Publicacions Universitat Barcelona, Spain.

Verlaque, M. and Fritayre, P. 1994. Modification des communautés algales méditerranéennes en présence de l'algue envahissante *Caulerpa taxifolia* (Vahl) C. Agardh. Oceanologica Acta 17: 659-672.

Villèle, X. de and Verlaque, M. 1995. Changes and degradation in a *Posidonia oceanica* bed invaded by the introduced tropical alga *Caulerpa taxifolia* in the north western Mediterranean. Botanica Marina 38: 79-87.

WHAT MAKES A TRANSGENIC PLANT AN INVASIVE ALIEN? – GENETICALLY MODIFIED SUGAR BEET AND THEIR POTENTIAL IMPACT ON POPULATIONS OF THE WILD BEET *BETA VULGARIS* SUBSPEC. *MARITIMA* ARCANG.

Matthias Pohl-Orf*[1], Ulrike Brand[2], Ingolf Schuphan[1] and Detlef Bartsch[3]

[1]*Lehrstuhl für Biologie V, Ökologie, Ökochemie, Ökotoxokologie, RWTH – Aachen, Worringerweg 1, 52056 Aachen, Germany, Tel.: +49(0)241/806676, Fax: +49(0)241/ 8888-182; e-mail: pohl-orf@rwth-aachen.de;* [2]*Institut für Entwicklungsbiologie, Universität zu Köln, Gyrhofstr. 17, 50923 Köln, Tel.: +49(0)221/470 3130, Fax: +49(0)221/470 5164;* [3]*present address: University of California, Riverside, Department of Botany and Plant Sciences, Riverside, California 92521, Tel.: +1 (909) 7875009, Fax: +1 (909) 7874437*

Abstract

The potential invasiveness of a transgenic plant is depending on the qualities of the newly introduced genes as well as the specific characteristics of the modified plant species and its communities. The present investigations aim to clarify the interactions between the transgenic plant and its relatives and to assess whether a transgenic plant is capable of becoming an invasive alien in these communities. Investigations were made to assess whether a transgenic sugar beet with different modifications is able to establish itself, or at least its new genes in different natural or disturbed plant communities.

The possibilities of outcrossing and hybridisation with wild and cultivated relatives were examined, as well as differences in competitiveness with and without virus infestation. The specific conditions in wild beet habitats were checked according to the occurrence of the virus and the susceptibility of different wild beet variants

According to our expectations, we found no barriers for crossing the transgenic sugar beets with cultivated relatives or wild beet variants. The competitiveness of the transgenic beet was enhanced under infestation conditions, while, in absence of the virus, no advantages in competition were detected. This was found with sugar beet and with hybrids between sugar beet and Swiss chard.

In the investigated natural habitats of *Beta vulgaris,* no virus infestation was detected, causing a lack of selection advantage due to virus resistance. Additionally, wild beets show a wide range of susceptibility to infection, ranging from low infection to nearly resistance.

These results lead to the following assessment: after releasing these modified sugar beet, transgenes will spread into natural beet populations, but ecological effects of rhizomania resistance on the degree of invasiveness of *Beta vulgaris* L. seem to be unlikely due to the absence of ecological advantage in natural habitats.

Introduction

Invasiveness of neophytes has often been used as an analogy to assess the risks towards the introduction of transgenic plants into the environment (Bartsch *et al.* 1993, Sukopp and Sukopp 1993). It is easier to determine the effects of transgenic plants,

to whom correspondence should be addressed;

Plant Invasions: Ecological Mechanisms and Human Responses, pp. 235–243
edited by U. Starfinger, K. Edwards, I. Kowarik and M. Williamson
© *1998 Backhuys Publishers, Leiden, The Netherlands*

because there is greater knowledge of the introduced genes, compared to neophytes, whose ecological effects and degree of success are often not predictable. Finding a proper habitat for monitoring transgenic plants is also easier, as it is mostly the same as that of their natural relatives.

Even though these theoretical considerations can help, these aspects alone are not enough for making a sound decision about the risk of an escape in the environment. Due to unspecific and uncontrolled integration of the transgenes into the plant genome, unexpected effects can occur which were not expected and, thus, not monitored. For this reason, risk assessment also needs a practical approach towards the characteristics that can make a transgenic plant become invasive.

The distribution of beet, *Beta vulgaris* L., is limited to supralittoral plant community *Beto-Atriplicetum laciniatae* (= *Atriplicetum sabulosae*) with its open, wind-resistant and nitrophilous characteristic (Runge 1990) and due to their low competitiveness and their tolerance of salt irrigation. Therefore, an escape of transgenic sugar beet attributes is more probable through gene flow into wild beet populations than the escape of the transgenic sugar beet plant itself. It is important to emphasise that all cultivated and wild varieties of beet (sugar beet, Swiss chard, red beet or fodder beet) belong to the same species and that there are no outcrossing barriers for transgenic traits within this species (Bartsch and Pohl-Orf 1996).

The aim of this work is to evaluate if this application of the models is appropriate and which parameters are decisive to determine invasive effects of a transgenic plant. In our studies, the potential ecological advantage of a coat protein mediated virus resistance (Beachy *et al.* 1990, Mannerlöf *et al.* 1996) against infection with the *Beet necrotic yellow vein virus* (BNYVV) was assessed under conditions similar to that in wild beet habitats (salt irrigation). This should give answers to the following questions
1. Are wild beets susceptible to infection with the rhizomania virus and if so, would the resistance of transgenic plants cause an ecological advantage ?
2. Does this advantage help the plant to become invasive in natural plant communities?

Investigations were made to localized wild beet populations in Italy near the center of seed production in the Po-valley (Bartsch and Schmidt 1997). In these habitats, the occurrence of BNYVV was checked. To assess the influence of salt irrigation, tests on salt tolerance and rate of infection with the rhizomania virus under mesohaline soil conditions were performed. Related to results from experiments about the competitiveness of transgenic and non-transgenic sugar beets, a clearer picture of the effects and the relevance of outcrossing and establishment of the transgenic virus resistance could be drawn.

Material and methods

Study site

From 1994 to 1997 wild beet populations of the Adriatic coast (see Fig. 1) from Trieste along the Gulf of Venice to the south till the Rimini area were observed. Notes about the number of plants per site and the characteristics of the habitat were taken and seeds and soil probes were sampled.

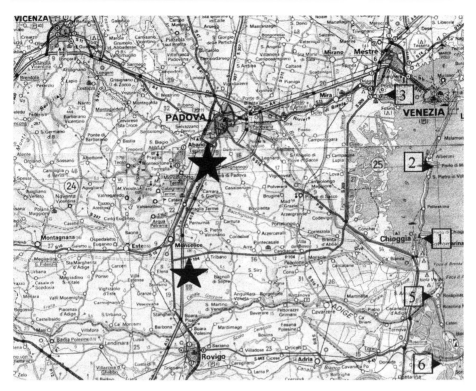

Fig 1. Examination area in the North Italian Po-valley. The observed populations, investigated in the virus infestation test, are numbered from 1 to 5. Population 4 was found in the Laguna di Grado near Trieste. The main breeding areas are marked with a star.

Plant material

The transgenic sugar beets we worked with were carrying the additional transgenic sequences of the c-DNA for the coat protein of BNYVV (Meulewater *et al.* 1989), the *nptII* (Beck *et al.* 1982) as a resistance marker against kanamycin and the *bar*-gene (Thompson *et al.* 1987) mediating resistance against the herbicide BASTA® / LIBERTY® with its active agent glufosinate-ammonium. The breeding lines, cultivars and transgenic varieties were made available to us by KWS/PLANTA, Einbeck, Germany. The samples of different wild beet populations were collected at the Adriatic coast in Italy.

Virus infestation

The seeds for the infection test (see Table 1) were germinated in the greenhouse in standard cultivation substrate under standard conditions (16 hours light, 20°/15°C day/ night rhythm). Twelve to sixteen days after germination, the seedlings were pricked out in pots containing BNYVV contaminated soil. This soil was composed of 50 % standard cultivation substrate and further 50 % of soil from a proven infestation site. Thirty-two plants of each treatment were watered twice a week with sea salt water at

Table 1. Different plant types and their characterization. The sugar beet breeding lines and cultivars were supplied by KWS, Einbeck/Germany, the wild genotypes were sampled at the coast in Italy from Venice about 50 km along the Adriatic coast to the south. The sites the wild beet populations were found are marked in Fig.1.

Name	Type	Characterization	Use
2K6481	Sugar beet	**transgenic** breeding line	Pollinator for production of **transgenic hybrids**
4B4857	Sugar beet, Breeding line	Cms-maintainer, multigerm, O-type	Breeding line for Transformation Pollinator for production of the control hybrids
Edda	Sugar beet, Cultivar	monogerm, diploid	Infection tests
Glatter Silber	Swiss chard	Cultivar diploid Control	Competitiveness trials
Ma/2K6481	*Beta*-Hybrid	Hybrid, diploid, transgen	Competitiveness trials
Ma/4B4857	*Beta*-Hybrid	Hybrid, diploid Control	Competitiveness trials
Wild beet pop. (Ital) 1 – (Ital) 6	Italy, Adriatic coast	wild, mostly annual	Infection tests

two different concentrations. The first treatment was 1 % salt, the second 0.5 % and the third was tap water for control. After cultivation for 95 to 106 days in the greenhouse, the roots were harvested and checked for the presence of BNYVV (Adlerliste and Van Euwijk 1992) with a specific antibody test (ELISA, *enzyme linked immunosorbent assay,* König *et al.* 1987, Kaufmann *et al.* 1992).

As positive control, transgenic plants were used, expressing the viral coat protein in a low amount. The extinction values of the samples had to been higher than 0.5 after 1 hour incubation. This value is as high as it is expectable in transgenic positive controls and multiple high as in the negative probe (Cultivar *Edda* in not infectious soil). Highly infected plants can reach extinction value of than 2.5 after the same time of incubation.

Experiments on competitiveness

Field test series were conducted at two sites with typical field soil (brown earth on calcareous subsoil) without additional fertilization with three different levels of inoculation:
(a)   low virus infection (Mainz, Rheinland-Pfalz),
(b)   high virus infection, due to pre–inocculation (Mainz, Rheinland-Pfalz) and
(c)   no virus infection (Aachen, Nordrhein-Westfalen).

Each test site comprised 180 subplots (3 genotypes x 3 competitive densities x 20 replicates), randomized into 20 blocks, respectively. Each subplot consisted of an area of 50 cm x 50 cm (0.25 m^2), each containing 4 *Beta* plants. Two hybrid genotypes (transgenic/non-transgenic) and the parental Swiss chard were planted when they reached the four leave stage into the field in April 1996 within three competitive densities of *Chenopodium album* L. var. *album* (without, with four and with sixteen *C. album* plants). C. album is a widespread and typical weed in the associations of the *Polygono-Chenopodietalia.* The plants for the high background virus infestation plots which were not pre-cultivated in BNYVV contaminated soil. The test sites at Mainz had only a low natural BNYVV infestation level, so at least a low background infestation was expected for individuals not pre-inoculated in the greenhouse.

C. album plants were planted when they were grown to 10 to 15 cm height and the sugar beets reached the six-leaf stage. To maintain perfect control over the intensity of competition in the different plots, natural weeds were removed from all subplots. The experiment was terminated when *C. album* began to die in July. Due to the large biomass productivity of the field test, only the fresh weight of the beets and the foliage were determined. Performance measurement was based on the absolute biomass production

Results and discussion

Study site

The conditions at all sites were characterized through stony underground and a low level of competitiveness. With immunological tests (ELISA, see above) the occurrence of BNYV-virus could not be detected in all investigated wild beet habitats.

Virus infection under different soil conditions

Extinction values of infected plants in the ELISA test were about 2.5 after 60 minutes of reaction time in the case of high infection. Extinction of 0.3 to 0.9 was assessed as low infection. Transgenic plants, expressing the virus coat protein and thus responding positively to the test showed values as a positive control of approximately 0.5. Negative controls were below 0.1.

There was a significant decrease in infection with increasing salinity in all populations (Two-Way-ANOVA with P < 0,05), except for wild beet population 1 (Fig. 2) an the high tolerant population 6. Population 1 showed only with 1 % salt watering a significant decrease compared with the control. The 0,5% treatment was not significantly different from, the control.

The different wild variants showed a different range of extinction values. One of the populations even proved to be completely tolerant against virus infection.

Especially in the case of the cultivar *Edda,* morphological changes due to salt irrigation were noticed. The plants developed succulent characteristics like thick leaves with a strong cuticle and more compact growth thus causing morphological similarity to their wild relatives.

Varying levels of infection between wild beet populations have been noted previously (Whitney 1989, Barr *et al.* 1995, Geyl *et al.* 1995), but not the strong dependence on soil salinity. One possible reason for this dependence may be the absence of the *Polymyxa betae* vector in the naturally mesohaline habitats (Ellenberg 1992) with conductivity corresponding to a salt concentration of 0.5% to 0.7% like in the infection experiments. At some sites it was not possible to check the salinity of the soil because of the too stony ground so that only spot checks were carried out (site 3: 0.5%, site 4: 0.1%). Based on our results, we can not determine why this virus is unable to infect the plants, but we can state that selective pressure for rhizomania resistance does

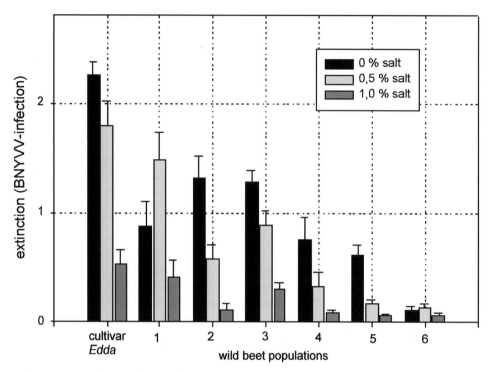

Fig. 2. Infection of one cultivar (*Edda*) and different wild populations of *Beta vulgaris* L. with BNYVV depending on salt solution watering (mean of 32 plants with standard error). The six different wild beet populations originated from the Adriatic coast in Italy. The occurrence of the virus in the samples (extinction/BNYVV-infection) is measured by an specific antibody interaction combined with an colorimetric detection (ELISA).

not appear to be the decisive factor for the potential invasiveness of the transgenic beet into the natural Italian populations at the Adriatic coast.

Competitiveness of sugar beet hybrids

Although we expected a higher performance of transgenic hybrids even at low background BNYVV infection, we found that the transgenic hybrids were superior to controls only at high background BNYVV infestation levels (Fig. 3). The superior performance amounted to approximately 20% over all weed competition levels. This decrease was in a T-test with a 5 % level of significance only positive without weed competition (p = 4,3%). With 4 and 16 C. album as competitors the increase was not significant on the 5 % level. In comparison with previous experiments (Bartsch *et al.* 1996), the superior performance of transgenic Swiss chard hybrids was within the reported range of 30% higher fresh weight of transgenic BNYVV-resistant sugar beet.

   The highly significant decrease in performance of transgenic hybrids compared with the non-transgenic genotypes in the low infestation plot (p < 0,1 % in all cases) is probably due to the double-selfing of the transgenic parent or in general a worse constitution of the transgenic line due to the transformation process. As described in the literature, inbreeding depression has a large effect on fitness in many plant species

Fresh weight [kg / 0.25 m²]

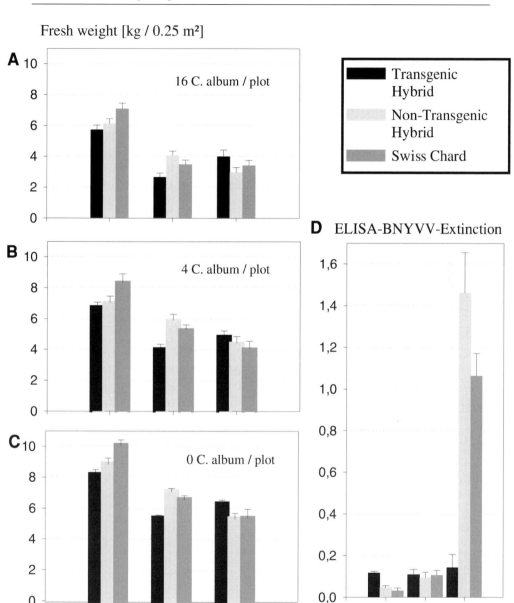

Fig. 3. Performance of transgenic and non-transgenic *Beta vulgaris*: The biomass production field test includes a transgenic hybrid between sugar beet and Swiss chard, a non-transgenic hybrid and Swiss chard. The plants were pre-cultivated in the greenhouse and planted in three different competition densities with *Chenopodium album* (Fig. 3A-C) Each density was tested at three different sites with different infestation level. BNYVV coat protein in roots representative for infection of the plants (Fig. 3D) was detected with ELISA. Bar lines represent the mean ± SE (Biomass test: n = 20 replicates, 4 plants each; ELISA: n = 5 for transgenic hybrid , n = 30 for non-transgenic hybrid, and n = 50 for swiss chard)

(Charleswoth and Charlesworth 1987, Damgaard and Loeschke 1993). In addition, the ecological advantage of virus resistance may have been weakened by planting stecklings, leading to a lower virus infection than sowing directly in contaminated soil.

Weed competition had a strong negative influence (significant on the 5 % level in all comparisons between the control and the 16 *C. album* treatment and in most 4 *C. album* cases, except Swiss chard on high infestation level with T= 0.9 and p=0.35) on hybrid performance within all genotypes and all infestation levels.

As shown in Fig. 3D, the virus coat protein is present in low amounts in the transgenic plants. The susceptible non-transgenic genotypes from the virus-free site showed no significant BNYVV infection. At the low infestation level, the detected virus coat protein of the susceptible genotypes were within the range of transgenic coat protein expression. At the high infestation site the susceptible plants were heavily infected, even at the beginning of the field trial.

Conclusion

To answer the questions stated in the introduction the results of this investigation confirm that in this special case the risk for the modified sugar beet to become invasive is expected to be low if introgression itself is not accepted as one possible mode of invasion. Ecological implications due to transgenic BNYVV-resistant hybrids might only be observable in natural beet habitats with high levels of BNYVV infestation and where susceptible *Beta vulgaris* genotypes grow. The possibility for outcrossing has been proved by investigations on outcrossing (Pohl-Orf and Bartsch 1996) and genetic markers (Santoni and Berville 1992, Boudry *et al.* 1993) but in our case the ecological relevance of the new traits is less important, because wild beets show a significantly lower rate of infection with the virus, and moreover, the infection is strongly reduced if the habitat is salt irrigated, as it is typical for wild beet habitats. As confirmed by the experiments on competitiveness, there is no ecological advantage for the transgenic plants when the virus is absent, which is the case under natural mesohaline soil conditions. On the other hand, the transgenic plants appear to be more competitive under high virus pressure. If the virus would be able to establish itself in these habitats, an increased fitness in wild habitats could be the consequence, regarding that this case has not been found yet and an occurrence of the virus is only common on agricultural areas.

References

Alderlieste, M.F.J. and Van Eeuwijk, F.A. 1992. Assessment of concentrations of beet necrotic yellow vein virus by enzyme-linked immunosorbent assay. Journal of Virological Methods 37: 163-176.

Barr, K.J., Asher, M.J.C. and Lewis, B.G. 1995. Resistance to *Polymyxa betae* in wild *Beta* species. Plant Pathology 44:301-307.

Bartsch, D. and Pohl-Orf, M. 1996. Ecological aspects of transgenic sugar beet: Transfer and expression of herbicide resistance in hybrids with wild beets. Euphytica 91: 55-58.

Bartsch, D., Schmidt, M., Pohl-Orf, M., Haag, C. and Schuphan, I. 1996. Competitiveness of transgenic sugar beet resistant to beet necrotic yellow vein virus and potential impact on wild beet populations. Molecular Ecology 5: 199-205.

Bartsch, D. and Schmidt, M. 1997. Influence of sugar beet breeding on populations of *Beta vulgaris* ssp. *maritima* in Italy. Journal of Vegetation Science 8: 81-84.

Beachy, R.N., Loesch-Vries, S. and Tumer, N.E. 1990. Coat protein mediated resistance against virus infection. Ann Rev Phytopath 28: 451-474.

Beck, E., Ludwig, G., Auerswald, E.A., Reiss, B. and Schaller, H. 1982. Nucleotid sequence and exact localization of the neomycin phosphotransferase gene from transposon Tn5. Gene 19: 327-336.

Boudry, P., Mörchen, M., Saumitou-Laprade, P., Vernet, P. and Van Dijk, H. 1993. The origin and evolution of weed beets: consequences for the breeding and release of herbicide-resistant transgenic sugar-beets. Theor Appl Genet 87: 471-478.

Charlesworth, D. and Charlesworth, B. 1987. Inbreeding depression and its evolutionary consequences. Annual Review in Ecology and Systematics. 18: 237-268.

Damgaard, C. and Loeschke, V. 1993. Inbreeding depression and dominance-suppression competition after inbreeding in rapeseed (*Brassica napus*). Theoretical and Applied Genetics 88: 321-323.

Ellenberg, H. 1992. Zeigerwerte von Pflanzen in Mitteleuropa. Ed 2. Scripta Geobotanica. 18

Geyl, L., Garcia Heriz, M., Valentin, P., Hehn, A. and Merdinoglu, D. 1995. Identification and characterisation of resistance to rhizomania in an ecotype of *Beta vulgaris* ssp. *maritima*. Plant Pathology 44: 819-828.

König, R., Burgermeister, W. and Leseman, D.-E. 1987. Methods for Detection and identification of Beet Necrotic Yellow Vein Virus. Proc. 50th Winter Congress, I.I.R.B., Brussels, 17-22.

Kaufmann, A., König, R. and Lesemann, D.-E. 1992. Tissue print-immunoblotting reveals an uneven distribution of beet necrotic yellow vein and beet soil-born viruses in sugarbeets. Archives of Virology 126: 329-335.

Mannerlöf, M., Lennerfors, B.-L. and Tenning, P. 1996. Reduced titer of BNYVV in transgenic sugar beets expressing the BNYVV coat protein. Euphytica 90: 293-299.

Meulewaeter F., Soetaert P. and Emmelo van J. 1989. Structural analysis of the coat protein gene in different BNYVV Isolates. Medelingen Faculteit Landbouwwetenschap Riijksuniversiteit Gent 54(2): 465-468.

Runge, F. 1990. Die Pflanzengesellschaften Mitteleuropas. Ed. 11. p 97 and p 119. Aschendorfsche Verlagsbuchhandlung, Münster.

Santoni, S. and Berville, A. 1992. Evidence for gene exchange between sugar beet (*Beta vulgaris* L.) and wild beets: consequences for transgenic sugar beets. Plant Molecular Biology 20: 578-580.

Thompson, C.J., Mova, N.R., Tizard, R., Crameri, R., Davies, J.E., Lauwereys, M. and Botterman, J. 1987. Characterization of the herbicide-resistance gene bar from Strptomyces hygroscopicus. EMBO Journal 3: 2723-2730.

Whitney, E.D. 1989. Identification, distribution, and testing for resistance to rhizomania. *Beta maritima*. Plant Disease 73(4): 287-290.

PATTERNS OF EARLY GROWTH AND MORTALITY IN *IMPATIENS GLANDULIFERA*

Alicia Prowse

Biology and Environmental Studies, Bolton Institute, Deane Road, Bolton BL3 5AB, U.K.; e-mail: ap2@basil.acs.bolton.ac.uk

Abstract

Examination of the patterns of early growth and mortality in the alien *Impatiens glandulifera* may yield clues to the control of the further spread of this plant by community constraint. Seedlings in wild populations of *Impatiens glandulifera* were individually tagged and mortality recorded. High mortality rates were found which varied according to habitat and locally competing species. Major causes of mortality at the seedling stage were identified as slug predation, disease and (under tree canopies) physical damage from rainfall. Highest mortality rates occurred during the period between mid-May and the beginning of July.

Introduction

Impatiens glandulifera (Royle) is classed as a well-established and invasive alien in the UK. It was introduced in 1839 (Coombe 1956) and its subsequent spread has made it a robust member of the British alien flora. Its spread has been well documented, but there are many situations in which it appears to be unable to invade areas of apparently suitable habitat. It is often reported that alien plants experience a release from the pressures of predators, pathogens and competitors of their native ranges in their newly expanded range, giving rise to increased vigour (Crawley 1987; Blossey and Nötzold 1995). However, as the time from first introduction increases, it is likely that predators and pathogens will come to exploit their new neighbour.

It is useful to assess the extent to which predators and pathogens of host communities in the UK have responded to the presence of *I. glandulifera* over 150 years, by establishing new ecological links. The pattern of lag phase, exponential population growth and geographical spread, decrease in population size and, finally, the establishment of a relatively stable equilibrium population size is seen in some alien plants recorded over long enough periods of time (Cronk and Fuller 1995). This may be due to a gradual accumulation of ecological links between the invading alien and its host community. Although much accelerated by human activity, the arrival of alien species is essentially a natural phenomenon (deliberate introductions aside).

This preliminary study is part of a larger investigation of the processes that may result in aliens becoming more constrained members of plant communities rather than remaining as aggressive pests. Herbivore damage, pathogenic activity and death from abiotic factors, such as drought or flooding are likely to have most impact on the seedling stage. Therefore an investigation of the patterns of early growth and mortality in *I.*

Plant Invasions: Ecological Mechanisms and Human Responses, pp. 245–252
edited by U. Starfinger, K. Edwards, I. Kowarik and M. Williamson
© *1998 Backhuys Publishers, Leiden, The Netherlands*

glandulifera and the factors that cause mortality at the seedling stage in different habitats was carried out.

Method

Populations of *I. glandulifera* are well established in north-west England. This study followed populations in the area of Jumbles Country Park, north of Bolton, Greater Manchester (National Grid Reference SD73/14).

In February 1996 and 1997, five 1-metre square permanent quadrats in woodland and riparian habitats were established (see Table 1 for details) and seedlings randomly located and tagged on emergence in late February and early March. Throughout the season all tagged individuals were closely monitored with height, damage and seedling survivorship recorded every 7-10 days.

To allow for late germination, initial seedling numbers were counted or estimated in each quadrat in late April and final densities (of flowering individuals that ultimately set seed) were recorded as a result of regular monitoring over the Summer and early Autumn.

In one of the riparian quadrats in 1997, above ground growth of *Urtica dioica* was continually removed. Measurement of height, survivorship and damage was recorded as above in both this quadrat and an adjacent quadrat where *U. dioica* continued to grow.

In addition, in 1997, a 1-metre square area of bryophyte cover was removed in an area suffering high mortality due to herbivory in the 1996 season. In to this area, on 15th March 1997, forty 26-day-old seedlings grown from seed in the laboratory were planted. When these subsequently died, twenty-nine 72-day-old seedlings were planted in the same area and their progress monitored.

In all permanent quadrats, other species were allowed to grow and are recorded with other site characteristics in Table 2.

Results

Seedling densities in permanent quadrats are shown in Figure 1 and Table 1. Not all locations were replicated in the two years and the planted quadrats have been excluded from Figure 1.

No clear relationship is evident between initial seedling density and final flowering density in these wild populations although there is a slight increase in seedling survivorship at higher initial seedling densities. The two woodland populations show the highest seedling survivorship. Calculation of Pearson correlation coefficient gives non-significant values of r ($r = 0.585$ $p = 0.098$). As the sample size is small (9 quadrats) and these are from different habitats, no further statistical analysis was appropriate. Further sampling is being carried out to give larger habitat-based samples.

Table 1a shows the patterns of mortality seen and Table 1b includes the quadrats planted with 26-day-old and 72-day-old plants. Major causes of death were from herbivory (almost exclusively from slugs), and from symptoms similar to those caused by viral disease, though no pathogen has yet been identified. Slugs (mainly *Arion sp.*) frequently ate cotyledons of seedlings including the apical meristem thereby killing

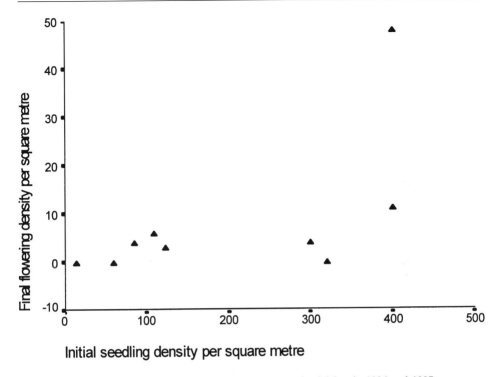

Fig. 1. Seedling mortality in wild populations of *Impatiens glandulifera* in 1996 and 1997.

the plant. Older seedlings were often attacked and could suffer considerable loss of leaf tissue, stunting growth. Even plants 1m or more in height were not immune to attack, though this rarely resulted directly in death at this stage. All slug activity was concentrated on leaf tissue. Slug activity had most impact on seedlings in early spring.

The mortality of the 26-day-old plants was 100% but only 3.5% for the 72-day-old plants. A period of drought affected the 26-day-old plants apparently making them more susceptible to herbivory and especially to a disease present in the area,

The typical overall pattern of mortality is shown by the riparian quadrat with *Urtica dioica* present (Fig. 2). Mortality rates reach a peak in late May and early June, survivorship showing the steepest decline here. However the pattern of mortality differed in different quadrats depending on the major cause of death, as Table 1 shows.

Figure 3 shows these differences in causes of death in the permanent quadrats. In most of the riparian quadrats recorded in 1997, herbivore damage by slugs was the major cause of death, whereas death from disease had been a factor of equal or greater significance in 1996. It seemed that in the 1997 season, slug predation was so high that plants perished from this cause before the disease had gained a foothold.

A notable exception to this was the riparian/*Petasites* quadrat, where herbivore damage was minimal in both years and mortality was due mainly to disease and heavy shading from *Petasites* in 1996. A similar pattern in 1997 was overlain by regular localised flooding and mud burial (Figs. 3a and 3b).

Cause of death in the other riparian quadrats recorded in both years differed widely. For example, in the riparian/*Urtica* quadrat 70% of total mortality was due to disease in 1996, but disease accounted for only 10% of the mortality in 1997. Losses in 1997

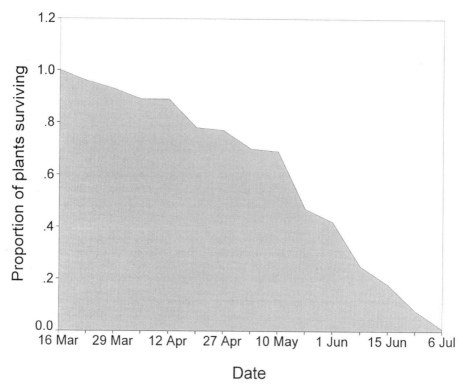

Fig. 2. Survivorship in *Impatiens glandulifera* seedlings (riparian quadrat with *U. dioica*).

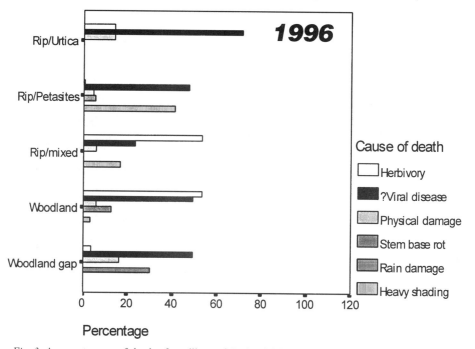

Fig. 3. Apparent cause of death of seedlings of *I. glandulifera* in permanent quadrats.

in this quadrat were mainly accounted for by slug damage and physical damage: the inability of weak stemmed plants to support themselves particularly after heavy summer rainfall.

Removal of above-ground growth of *U. dioica* appeared to have very little effect on overall mortality, mortality of 97.5% with *U. dioica* present being recorded as opposed to mortality of 94.5% when this plant is removed.

Discussion

Beerling (1990) investigated the relationship between initial seedling density and seedling survivorship in *I. glandulifera* and concluded that density dependent seedling mortality was occurring. As Figure 1 shows, this study suggests no well-defined relationship between initial seedling density and final flowering density.

The overall mortality rate was much higher than that reported by Beerling (1990), who reported mortality between 28 and 60%. However wild and garden populations were included in his data. Perrins, Fitter and Williamson (1993) gave a figure of 2% seedling success – i.e. 98% seedling mortality in wild populations in wooded fen, while Prach (1994) recorded mortalities of between 50 and 90% in two riparian populations.

Habitat clearly has an effect on the causes of mortality and, in some areas, density independent factors, such as drought and flooding may be more important in determining population sizes and extent of spread than density dependent factors. This seems to be the case in the riparian area, where regular flooding led to high mortality rates.

Riparian quadrats suffered particularly high levels of mortality (between 98 and 100%

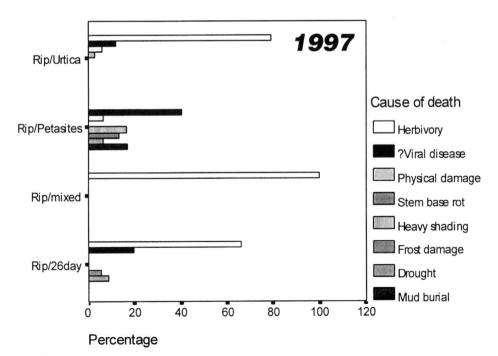

Fig. 3.

Table 1. Patterns of mortality in *I. glandulifera* in a) 1996 and b) 1997

a)

Quadrat	Initial density /m²	Final flowering density /m²	% Mortality	Pattern of mortality	Causes
Riparian with *Urtica dioica*	320	0	100	Rapid increase in mortality rate from mid May	Competition from *Urtica dioica* on plants weakened by ?disease and herbivory
Woodland	400	48	79	Relatively constant over the season	?Disease, herbivory, physical damage (rain)
Woodland canopy gap	400	11	97.25	Sudden loss – "rain damage" mid June	Physical damage (rain)
Riparian with *Petasites hybridus*	300	4	98.7	Low mortality rate until mid May, then very rapid increase	Competition from *Petasites hybridus* on ?disease-weakened plants
Riparian with mixed species	60	0	100	Constant, heavy mortality until all individuals had died (end of July).	Herbivory in early stages, plus competition from other species.

b)

Quadrat	Initial density /m²	Final flowering density /m²	% Mortality	Pattern of mortality	Main causes of mortality
Riparian with *Urtica dioica* removed	110	6	94.5	Low mortality rate until mid May then rapid increase.	Herbivory (slugs) and competition from *U. dioica*.
Riparian with *U. dioica* removed	124	3	97.5	Low mortality rate until mid May then rapid increase	Physical damage and herbivory
Riparian with *Petasites hybridus*	86	4	95.4	Most mortality occurring before mid May	Flooding (mud burial), ?disease and shading
Riparian with mixed species	14	0	100	All dead by mid May	Herbivory (slugs)
Riparian planted as 26-day old seedlings	40	0	100	Mortality occurring almost entirely before mid May	Herbivory, and ?disease possibly exacerbated by temporary drought
Riparian planted as 72-day old plants	29	28	3.5	All but one survived to flower	No apparent herbivore damage or attack by ?disease.

Table 2. Site characteristics of locations of permanent quadrats

Habitat type	Angle of slope	Open or shaded	Associated species
Riparian with Urtica dioica	0	Part shaded	*Cardamine amara, Ranunculus ficaria, Urtica dioica, Anthriscus sylvestris, Heracleum sphondylium*
Woodland	40°	Shaded	*Rumex obtusifolius, Hyacinthoides nonscripta, Rubus fruticosus agg.*
Woodland canopy gap	15°	Part shaded	*Brachythecium rutabulum, Urtica dioica, Heracleum sphondylium*
Riparian with Petasites hybridus	0	Open	*Petasites hybridus, Myrrhis odorata, Stellaria alsine*
Riparian with mixed species	0	Open	*Myrrhis odorata, Heracleum sphondylium, Urtica dioica, Montia sibirica, Silene dioica, Chrysosplenium oppositofolium*

in 1996) especially where other vegetation was present. Connell (1990) distinguishes between real and apparent competition, where indirect effects may benefit one of two competitors. He suggests that competition may not occur directly, but via a third party, such as a predator, which preferentially consumes one species over the other. The presence of other vegetation may harbour or attract herbivores, such as slugs, which were seen to feed on seedlings. In both years the riparian quadrat containing mixed species had very high levels of herbivory. This may help to explain why disturbance is often necessary for the successful establishment of populations of *I. glandulifera*.

Initial indications are that the seedling stage is the most vulnerable to the main causes of death, namely herbivory and disease, with mortality reduced from 100% to only 3.5% when 72 day old seedlings are planted out. The effects of drought and flooding may also intensify the impact of the disease on young seedlings. The striking late May increase in mortality rates seen here and also reported by Prach (1994) suggests competitive effects are intensifying, as this is seen mainly in the quadrats containing two strong competitors, *Petasites hybridus* and *Urtica dioica*. The small effect of removal of *U. dioica* on survivorship of *I. glandulifera* suggests that if direct competition exists, it would seem to be mainly below-ground.

These high mortalities in naturally occurring populations suggest that despite its reputation as an aggressive invader, populations are closely controlled by external factors. Levels of herbivory may be particularly important in preventing invasion under certain circumstances.

Acknowledgements

This study has been supported by Bolton Institute. I would also like to thank Frank Goodridge for his help and guidance, and Dr. Frankie Kerridge for her critical comments on the manuscript.

References

Beerling D.J. 1990. The ecology and control of Japanese knotweed (*Reynoutria japonica* Houtt.) and Himalayan Balsam (*Impatiens glandulifera* Royle) on river banks in South Wales. PhD thesis. University of Wales, Cardiff.

Blossey, B. and Nötzold R. 1995. Evolution of increased competitive ability in invasive nonindigenous plants: a hypothesis. J. Ecol. 83: 887-889

Coombe D.E. 1956. Notes on some British plants seen in Austria. Veröffentlichungen des Geobotanischen Instituts Zürich 35: 128-137

Connell, J.H. 1990. In: Grace, J.B. and Tilman, D. (eds.), Perspectives on Plant Competition. pp. 9-26. Academic Press, London.

Crawley, M.J. 1987. In: Gray, A.J., Crawley, M.J. and Edwards, P.J. (eds.), Colonization, Succession, Stability and Diversity. pp. 429-453. Blackwell Scientific Publications, Oxford.

Cronk, Q. and Fuller, J. 1995. Plant Invaders: the threat to natural ecosystems. London, Chapman and Hall.

Perrins, J., Fitter, A. and Williamson, M.1993. Population biology and rates of invasion of three introduced *Impatiens* species in the British Isles. Journal of Biogeography 20: 33-44.

Prach, K. 1994. Seasonal Dynamics of *Impatiens glandulifera* in two riparian habitats in Central England. In: de Waal, C., Child, L.E., Wade, P.M. and Brock, J.H. (eds.), Ecology and Management of Invasive Riverside Plants. John Wiley and Sons., Chichester.

INVASIVE NORTH AMERICAN BLUEBERRY HYBRIDS (*VACCINIUM CORYMBOSUM* X *ANGUSTIFOLIUM*) IN NORTHERN GERMANY

Hartwig Schepker and Ingo Kowarik
Universität Hannover, Institut für Landschaftspflege und Naturschutz, Herrenhäuser Straße 2, D 30419 Hannover, Germany; e-mail: schepk@laum.uni-hannover.de, kowari@laum.uni-hannover.de

Abstract

North American blueberries spread in Niedersachsen, northern Germany, where cultivars have been grown commercially since 1933. During the last three decades, the blueberries invaded the vicinity of the plantations which served as seed sources.

The spontaneous blueberries showed a wide variation in growth patterns and plant, fruit and leaf size. Some were resembling either one of their origins, *Vaccinium corymbosum* or *V. angustifolium*, but most of them differed distinctively. They were regarded as hybrids and, until clarification of the taxonomical status, addressed as *Vaccinium corymbosum* x *angustifolium*.

This paper describes the relation between seed sources and invasion success in terms of area occupied by spontaneously growing blueberries, and the maximum distance of blueberry individuals to the plantations. In addition, it discusses the possible assessments of the invasion effects in different biotopes.

A survey of the vicinity of 21 plantations showed that the surface covered by blueberries running wild ranged from < 1 to 278 ha. This was a total of fourteen times larger than the area under cultivation. The maximum distance of spontaneous individuals to a plantation was 1700 m. The invasion success was related to the age of the plantation. Invaded areas > 50 ha, with a maximum distance > 1000 m, occur only around plantations that have been planted with blueberries for more than 40 years. Mainly forests of Scots Pine and bogs were invaded. Invasion into the adjacent area of the plantations did not occur if they are surrounded by other agricultural lands.

The invasion of forests was not regarded as a problem for nature conservation. That of bogs, however, might influence ecosystem processes severely and finally lead to the displacement of endangered species. A distance of three kilometers between a plantation and bog areas was proposed for preventing unwanted invasion effects.

The different perception of invasion effects by blueberries stresses the necessity for a single case approach when assessing the invasion of non-native species in Central Europe.

Introduction

In northern Germany, invasions of non-native species are perceived as highly problematic by both public authorities and NGOs. Unwanted invasion effects are vegetation changes, economic problems in land use, or health risks. The conflicts are mainly attributed to *Prunus serotina*, *Heracleum mantegazzianum*, *Reynoutria* (incl. *R. japonica*, *R. sachalinensis* and *R.* x *bohemica*), *Impatiens glandulifera*, *Elodea* (incl. *Elodea canadensis*, *E. nuttallii*) and *Vaccinium corymbosum* x *angustifolium* (Kowarik and Schepker 1998). This paper deals with the latter species, which escapes frequently from commercial blueberry plantations. It provides one of the rare examples of intensively

Plant Invasions: Ecological Mechanisms and Human Responses, pp. 253–260
edited by U. Starfinger, K. Edwards, I. Kowarik and M. Williamson
© *1998 Backhuys Publishers, Leiden, The Netherlands*

treated cultural plants runnig wild (Kowarik and Schepker 1995, Schepker *et al.* 1997). The paper focuses on the following points:

* The history of introduction and cultivation of North American blueberries in northern Germany
* Taxonomical problems in identifying spontaneously spreading blueberries
* Invaded habitats, and spatial and temporal dimension of invasions starting from commercial plantations
* Assessing the invasion effects by a single case approach.

Introduction and cultivation of North American blueberries

In North America, *V. corymbosum* L. ('highbush-blueberry') and *V. angustifolium* Ait. ('lowbush-blueberry') have been described as polymorphic taxa including several subspecies with a wide phenotypic variation (Van der Kloet 1978, 1980, 1983). The highbush-blueberry has diploid populations in the southeast United States, tetraploids in the northeast and adjacent areas of Canada, and some hexaploids in the mountains of South Carolina and Louisiana. The tetraploid lowbush-blueberry is widely distributed in the northeast United States and in Canada (Eck and Childers 1966). Crossing of the tetraploid populations has been observed also under natural conditions (Van der Kloet 1976).

V. corymbosum was introduced to Europe in 1765, *V. angustifolium* in 1776 (Goeze 1916, Everett 1967). For the following 150 years, both species were grown mostly as ornamentals in botanical gardens and parks. The first commercial plantations were established in the Netherlands in 1923 (Roelofsen 1967) and in southern Germany in 1929 (Heermann 1967). Large scale production started in 1933 in Niedersachsen, northern Germany. Here, two factors favour the cultivation of blueberries: a mild climate and acid soils, such as podsols or drained bogs, which enable the symbiosis with endotrophic mycorrhiza (Liebster 1961). Today, 90 % of the total 600 ha of German blueberry plantations are located in Niedersachsen (Naumann 1993). Some plantations reach sizes up to 20 ha, but usually the shrubs are planted in plots of 1-2 ha (Bläsing 1989). Other main commercial plantations are situated in the Netherlands, Poland, and France.

For the first commercial plantations, German hybrids of *V. corymbosum* and *V. angustifolium* were chosen (Liebster 1961). Today, American hybrids prevail (Naumann 1993). They incorporate mostly germplasm of both species (Hancock and Siefker 1982). Much effort of the modern blueberry industry is focused on breeding for new hybrids to improve yield and fruit quality (Lyrene 1993), for pest resistance (Ballington *et al.* 1993), or for modification of cold tolerance and chilling requirement (Lang 1993).

Taxonomical problems

The escaped individuals showed a wide phenotypic variation in their growth pattern and in plant, fruit and leaf size. Some of the studied blueberries formed densely structured polycormic individuals with a height of less than 60 cm and pronounced clonal growth, resembling *V. angustifolium*. Others resembled *V. corymbosum*, forming crowns and growing up to 3 m; these only rarely produced ramets. However, most of the

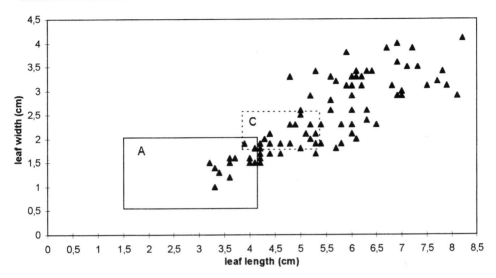

Fig. 1. Relation between leaf length and leaf width of spontaneous blueberries in Niedersachsen (n=90). The rectangles indicate the range of two American *Vaccinium* species, which are the origins of hybrids planted in northern Germany. (A): *Vaccinium angustifolium*, leaf length [1,5] 2,6 [4,1] cm, leaf width [0,5] 1,1 [2,0] cm; (C): *V. corymbosum*, leaf length [3,8] 4,6 [5,4] cm, leaf width [1,7] 2,1 [2,5] cm according to American flora by Van der Kloet (1978, 1980)

invading blueberries were intermediate in their growth pattern. Both species belong to the subgenus *Cyanococcus*, which is not part of the local flora. Hybridization with native *Vaccinium* species has not been observed in the wild.

Figure 1 demonstrates the variation in leaf length and width in a total of 90 studied individuals. Only some fit the characteristics of the American species, as provided by Van der Kloet (1978, 1980); these are indicated by rectangles. Others have larger leafs than either American species. The same is true for fruit diameter. According to American keys, the range of fruit diameter is (4)-7-(12) mm in *V. corymbosum* and (3)-6-(10) mm in *V. angustifolium* (Van der Kloet 1978, 1980). In northern Germany, the fruit diameter of invading blueberries (n=90) exceeded these ranges and is (7,5)-10,5-(14,5) mm. This might be explained by the selection of hybrids with large fruits for commercial plantations, which are the source of the invasions.

Our results show a wide phenotypic variation in the invading blueberries. Due to the hybrid origin of the planted blueberries, pure species are not to be expected, although some of the invading blueberries are similar to the North American species. Since the taxonomical status is not yet clarified, the invading blueberries are ascribed to *V. corymbosum* x *angustifolium* (Schepker *et al.* 1997). Such a hybrid has already been described within the natural range of both species in North America (Camp 1945, Van der Kloet 1978).

Invasion of blueberries in Niedersachsen

More than sixty years after the establishment of the first commercial plantations in Niedersachsen, the North American blueberries have widely invaded the vicinity of the plantations. As of now, 83 stands of spontaneously growing blueberries are known in Niedersachsen (Schepker *et al.* 1997). The recipient habitats are bogs (46%, including de- and regeneration stages), plantations of Scots Pine (*Pinus sylvestris*, 40%), edges of ditches and roads (12%), and heath communities (2%). Agricultural land adjacent to the plantations has not been invaded. Generative reproduction is successful under both natural and man-made conditions and let to permanently established populations.

The invasions had been unnoticed for a long time. A count of annual rings showed that some of the spontaneously growing *Vaccinium* are older than 30 years. However, the first published reference for Germany only dates back to 1983 (1949 in the Netherlands; Adema 1986).

The southern part of the Lüneburger Heide is the centre of the blueberry production in Germany. Here, some of the oldest plantations with pronounced blueberry invasions are situated. Spatial invasion patterns were studied in the vicinity of 21 plantations with sizes between < 1 and 21,5 ha (Table 1, Figure 2):

- Although seeds may be dispersed by birds, and sometimes mammals, over long distances, larger populations of spontaneously growing blueberries were virtually confined to the vicinity of the plantations. A dense shrub layer, with a cover value > 60 %, was produced in adjacent plantations of *Pinus sylvestris*. The cover value decreased rapidly with increasing distance from the plantations (Figure 2).
- The maximum distance of spontaneous blueberries to the plantations as seed sources varied between 50 and 1700 m (mean 760 m). Distances exceeding one kilometre were only found for plantations older than 40 years. Figure 3 indicates a relation between the age of the plantation and the spatial range of invading blueberries.
- The size of the area invaded in the vicinity of the plantations ranged from one to 278 ha with a total of 1245 ha; the area invaded was 14 times larger than the area under cultivation (a total of 89 ha). Invaded areas larger than 50 ha are only next to plantations older than 40 years (Figure 3). Again this indicates a relation between the age of the plantation and the invasion success.

Approaches in assessing the blueberry invasions

The occurrence of spontaneous blueberries in different habitats requires a differentiated

---

Table 1. Summarized information on 21 blueberry plantations in Niedersachsen, northern Germany, and on the area invaded by *V. corymbosum* x *angustifolium* in the vicinity of the plantations (original data in Schepker *et al.* 1997)

source of spontaneous blueberries (n=21)	age of plantations (n=13)
commercial plantations (12)	13 to 52 years
abandoned commercial plantations (3)	
small private plantations (1)	size of plantations (n=17)
abandoned private plantations (1)	<1 to 21,5 hectares
unknown (4)	Σ = 89 hectares
size of the invaded area (n=21)	maximum distance of spontaneously growing
<1 to 278 hectares	blueberries from the plantation (n=17)
Σ = 1245 hectares	50 to 1700 metres
	x̄ = 760 m

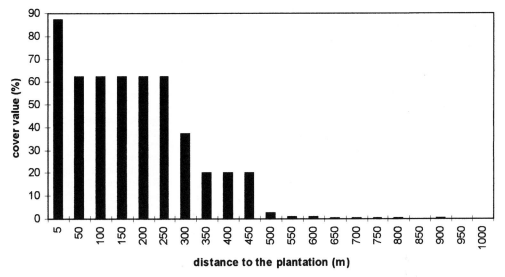

Fig. 2. Decreasing cover of spontaneously growing blueberries (*Vaccinium corymbosum* x *angustifolium*) next to plantations of Scots Pine in Niedersachsen, presented as a function of distance to the blueberry plantations as a seed source (cover values according to Braun-Blanquet [1964] are transformed to mean values; from Schepker *et al.* 1997)

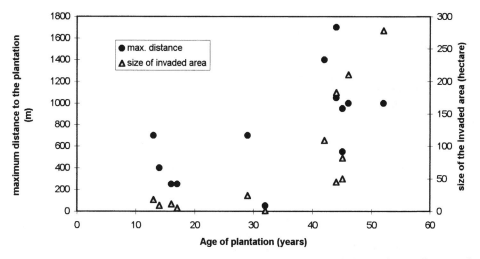

Fig. 3. Relation between the age of a plantation as seed source and the maximum distance of spontaneous blueberries to the plantation respective to the size of the area invaded by blueberries (n=13; data from Schepker *et al.* 1997)

assessment of the associated effects (Table 2). Under a pine canopy, the blueberries may form a dense shrub layer (Figure 2). Here, the ground vegetation is outcompeted, but rare or endangered species are not affected in pine plantations (Kowarik and Schepker 1995). Consequently, there seem to be no conflicts with the aim of species conservation. For recreation purposes, the blueberry invasions might provide benefits: They can be perceived as an aesthetic enrichment of an otherwise monotonous landscape, ex-

Table 2. Invasion effects and assessment of spontaneous blueberries in different biotopes considering possible risks or benefits

biotope	effects of invasion	assessment
Scots Pine plantations	establishment of an additional layer of +/- shade tolerant shrubs	nature conservation: • no problems forestry: • possible management problems recreation: • lovely fruits • colourful autumn foliage • diversification of vegetation structures mainly in *Pinus*-dominated plantations
ditches, waysides	integration into already existing vegetation	
bogs (including de- and regeneration stages)	establishment of a new life-form type (tall growing densely structured shrubs) alteration of successional pathways possible →	nature conservation: • outcompeting of endangered wetland species • obstruction of bog-regeneration
	prerequisite of invasion success: alteration of ecosystem properties →	invasion of blueberries is mainly a symptom of man-made changes

posing flowers, fruits and a colourful autumn foliage along the edges of pine plantations, roads and ditches.

However, foresters expect management problems with dense thickets, but economic costs have not been quantified yet. The same is true for possibly positive effects on forest management: There is some evidence that blueberries may act as nurse plants for young oaks (*Quercus robur*), which seem to be less affected by browsing than exposed oaks.

In bogs, conflicts with nature conservation are possible, when endangered plant species are outcompeted by blueberries. In the nature reserve 'Moor in der Schotenheide' (district of Soltau-Fallingbostel), blueberries grow in heather communities with *Erica tetralix* and *Calluna vulgaris*, on dry edges associated with *Molinia caerulea*, and even in fragmentary bog-communities with *Vaccinium oxycoccus* and several *Sphagnum*-species. Under the cover of densely structured blueberries, both the number and coverage of mosses and vascular plants are reduced. In this case, several endangered species were affected (Schepker 1998). In addition, the presence of blueberries may lead to increased nutrient influx and evaporation rates.

Until now, blueberry invasions were reported from 14 nature reserves in Niedersachsen, mainly from bogs. There is some evidence that man-made changes, such as drainage and peat cutting, have enhanced blueberry invasions. In these cases, their invasion success is a symptom of altered ecosystem properties, but not the primary cause for vegetation changes.

Conclusions

The blueberries are spreading exclusively from existing or former commercial plantations. As of now there are no known invasion events resulting from planted individuals in gardens or parks.

In general, planted or sowed non-native plants are the major starting-point for problematic invasions in Niedersachsen (Kowarik and Schepker, n. p.); of 342 reported invasion events, 48 % resulted from these man-made actions (see appendix in Kowarik and Schepker 1998). In 10 % of all cases the propagules have been introduced to the site of invasion with garden waste, in 5 % with soil material. The remaining starting-points were either unknown (28 %) or located in gardens (10 %). The results clearly emphasize the importance of man-made actions for establishing new spreading sources.

Man-made disturbances are also crucial for the establishment of blueberries in bogs. Invasions would very likely fail without drainage. The unwanted vegetation changes caused by the blueberries were originally made possible only by anthropogenic change of ecosystem properties.

The possibility of an invasion in the vicinity will be enhanced with the increasing age of a plantation. A further expansion of the blueberries and an increase in the invaded area has to be expected, because several new plantations have been established in various parts of Niedersachsen during the last decade. The risk of spontaneous distribution is minimal if the plantation is surrounded by agricultural land, but high if pine plantations or drained bogs exist in the vicinity of the blueberry plantations.

Experiences with control are restricted to some efforts of cutting and digging out. They might be reasonable especially in protected areas like bogs. However, in these biotopes control efforts could lead to considerable disturbance, which might even cause more damage to the vegetation. The success of a direct control measure, like cutting, is doubtful, because of the strong vegetative and clonal reproduction of the blueberries. In addition, the risk of re-invasion has to be limited. It seems more efficient to minimize the risk of invasion of valuable biotopes by establishing minimum distances between plantations and potential habitats. Refering to the maximum distance of 1.7 km between the plantations and spontaneously growing blueberries (Table 1), a distance of three kilometres should be sufficient to prevent the invasion of bogs on a longer time scale.

The different invasion effects can be perceived as potential risks, but also as beneficial for certain land use aspects. The diversity of possible assessments shows that a generalized attitude towards the invasion of blueberries is not adequate. Rather, a single case approach should provide for better management when assessing the blueberry invasion in particular and invasions of non-native species in general.

Acknowledgements

We thank Keith Edwards and Uwe Starfinger for constructive comments on a previous version of the paper. Parts of the blueberry invasion survey were supported by the NLÖ (Niedersächsisches Landesamt für Ökologie).

H. Schepker and I. Kowarik

References

Adema, F. 1986. *Vaccinium corymbosum* L. in Nederland ingeburgerd. Gorteria 13: 65-69

Ballington, J.R., Rooks, S.D., Milholland, R.D., Cline, W.O. and Myers, J.R. 1993. Breeding Blueberries for Pest Resistance in North Carolina. Acta Horticulturae 346: 87-94

Bläsing, D. 1989. A Review of *Vaccinium* Research and the *Vaccinium* Industry of the Federal Republic of Germany. Fourth International Symposium on *Vaccinium* Culture, International Society for Horticultural Science (ISHS). Acta Horticulturae 241: 101-109

Braun-Blanquet, J. 1964. Pflanzensoziologie. 3rd ed., Springer Verlag, Wien and New York

Camp, W.H. 1945. The North American blueberries with notes on other groups of *Vacciniceae*. Brittonia 5 (3): 203-275

Eck, P. and Childers, N.F. 1966. Blueberry Culture. Rutgers University Press, New Brunswick and London

Everett, H.F.M. 1967. Trials with some *Vaccinium* Species in Scotland. 1. Symposium ISHS Working Group 'Blueberry Culture in Europe' 1967: 70-75

Goeze, E. 1916. Liste der seit dem 16. Jahrhundert bis auf die Gegenwart in die Gärten und Parks Europas eingeführten Bäume und Sträucher. Mitt. Deutsch. Dendr. Ges. 25: 129-201

Hancock, J.F. and Siefker, J.H. 1982. Levels of inbreeding in highbush blueberry cultivars. HortScience 17 (3): 363-366

Heermann, W. 1967. Über Heidelbeeranbau und Heidelbeerzüchtung in Deutschland. 1. Symposium ISHS Working Group 'Blueberry Culture in Europe' 1967: 37-41

Kowarik, I. and Schepker, H. 1995. Zur Einführung, Ausbreitung und Einbürgerung nordamerikanischer *Vaccinium*-Sippen der Untergattung *Cyanococcus* in Niedersachsen. Schr. R. Vegetationskde. 27 (Festschrift Sukopp): 413-421

Kowarik, I. and Schepker, H. 1998. Plant invasions in northern Germany: human perception and response. In: Starfinger, U., Edwards, K., Kowarik, I. and Williamson, M. (eds.), Plant invasions: ecological mechanisms and human responses. pp. 109-120, Backhuys Publishers, Leiden

Lang, G.A. 1993. Southern highbush blueberries: Physiological and cultural factors important for optimal cropping of these complex hybrids. Acta Horticulturae 346: 73-79

Liebster, G. 1961. Die Kulturheidelbeere. Verlag Paul Parey Berlin und Hamburg

Lyrene, P.M. 1993. Some problems and opportunities in blueberry breeding. Acta Horticulturae 346: 63-71

Naumann, W.D. 1993. Overview of the *Vaccinium* industry in western Europe. Acta Horticulturae 346: 53-55

Roelofsen, B. 1967. Highbush blueberry production in England. 1. Symposium ISHS Working Group 'Blueberry Culture in Europe' 1967: 22-33

Schepker, H. 1998. Wahrnehmung, Ausbreitung und Bewertung von Neophyten – Eine Analyse problematischer nichteinheimischer Pflanzenarten in Niedersachsen. Diss. Univ. Hannover

Schepker, H., Kowarik, I., and Garve, E. 1997. Verwilderungen nordamerikanischer Kultur-Heidelbeeren (*Vaccinium* subgen. *Cyanococcus*) in Niedersachsen und deren Einschätzung aus Naturschutzsicht. Natur und Landschaft 72 (7/8): 346-351

Van der Kloet, S.P. 1976. Nomenclature, taxonomy and biosystematics of *Vaccinium* section *Cyanococcus* (the blueberries) in North America. 1. Natural barriers to gene exchange between *Vaccinium angustifolium* Ait. and *Vaccinium corymbosum* L.. Rhodora 78: 503-515

Van der Kloet, S.P. 1978. Systematics, distribution and nomenclature of the polymorphic *Vaccinium angustifolium*. Rhodora 80: 358-376

Van der Kloet, S.P. 1980. The taxonomy of the highbush blueberry *Vaccinium corymbosum*. Canadian Journal of Botany 58: 1187-1201

Van der Kloet, S.P. 1983. The taxonomy of *Vaccinium* § *Cyanococcus*: a summation. Canadian Journal of Botany 61: 256-266

CYPERUS ESCULENTUS (YELLOW NUTSEDGE) IN N.W. EUROPE: INVASIONS ON A LOCAL, REGIONAL AND GLOBAL SCALE

Siny J. ter Borg[1]*, Peter Schippers[2], Jan M. van Groenendael[3] and Ton J.W. Rotteveel[4]

[1]*Wageningen Agricultural University, Department of Environmental Sciences, group Nature Conservation and Plant Ecology, Bornsesteeg 69, 6708 PD Wageningen, The Netherlands; e-mail: siny.terborg@staf.ton.wau.nl; [2]Department of Theoretical Production Ecology, P.O. Box 430, 6700 AK Wageningen; [3]Department of Plant Ecology, University of Nijmegen, Toernooiveld 1, 6525 ED Nijmegen; [4]Plant Protection Service, P.O. Box 9102, 6700 HC Wageningen*

Abstract

Cyperus esculentus L. (Yellow Nutsedge) is a C_4-plant with a dense network of rhizomes, producing large numbers of brownish tubers. The species behaves as a pseudo-annual, with tubers overwintering in the soil and sprouting in spring. It ranges from the equator to temperate climates, and is a serious weed in various crops in American agriculture. In the eighties it became a problem in N.W. Europe. The paper reviews ten years of research in the Netherlands, and discusses the invasion on a local (field) and a regional scale, and the immigration history. General biological information is given as a basis, including infraspecific variability. Models to study the population dynamics, the effectivity of agricultural control systems and the role of various vectors in the dispersal of tubers are briefly described.

It is concluded that *C. esculentus* has several of the properties usually expected in invasive species, *viz.* high vegetative reproductive capacity by means of tubers, enough genetic variation to cover a wide range of conditions, and a hidden start of populations due to inconspicuous morphology, resulting in a rather long lag period until the species is recognized as a problem and control is started. The species has good dispersal capacities, mostly due to farming activities and transport of agricultural products. The species was brought from America to Europe with gladiolus cormlets which were introduced to widen the material offered to the market. *C. esculentus* var. *leptostachyus* Boeckeler appeared to have the best characteristics to establish itself under N.W. European climatic and agricultural conditions.

Introduction

In the early seventies *C. esculentus* L. was introduced in the Netherlands. It soon became a problem in agriculture, due to the extensive production of rhizomes and tubers. The species was also observed in neighbouring countries, where it was supposed to have been brought from the Netherlands. Where did it come from? What makes it such an aggressive weed? Which are practicable control measures?

In this paper we will review and summarize a number of studies carried out in the ten years after the problem arose, focussing on questions related to agressiveness and invasiveness. Successively we will discuss life history, infraspecific variation and population dynamics. This primary information provides a good basis to answer questions with respect to invasions on a local, regional and global scale.

*To whom correspondence should be directed

Plant Invasions: Ecological Mechanisms and Human Responses, pp. 261–273
edited by U. Starfinger, K. Edwards, I. Kowarik and M. Williamson
© 1998 Backhuys Publishers, Leiden, The Netherlands

Basic biology

Life history

C. esculentus (Yellow nutsedge, 'Knolcyperus') is a perennial plant causing weed problems in agriculture on a world wide scale to such an extent that it was ranked number 16 on the list of the 'Worlds' worst weeds' (Holm *et al.* 1977). The species ranges from the equator up to cooler climates as far north as Alaska, even though it has typical C_4-characteristics.

Extensive literature exists on the general biology of *C. esculentus* (*e.g.* Stoller and Sweet 1987; Lapham 1985). Their work, together with our studies of the species under local Dutch conditions, allows the following summary:
The plant sprouts from tubers in the soil in late spring, when soil temperatures have reached values of 8 to 10°C. A tuber, a thickened end of a rhizome, carries several buds; one of these, sometimes two or three, sprouts and produces an upwards growing rhizome which forms a basal bulb under the influence of light. From this bulb a shoot develops, and after a few weeks a number of rhizomes, growing horizontally, give rise to new shoots. When days begin to get shorter, rhizomes start to grow at slightly steeper angles down into the soil and produce brownish tubers at their ends. This results in a dense network of rhizomes with tubers. Shoots may produce a culm, carrying a fairly large inflorescence with up to several hundreds of small flowers. Some of them may produce a germinable seed. Seedlings have not been reported from cooler climates; in warmer areas seedling establishment was noticed, but it is infrequent (Lapham *et al.* 1985). So mostly reproduction depends on tuber production.

The largest tubers are about 1.5 cm long and up to 1 cm wide; however, others are smaller and roundish. Tubers of just a few millimeters diameter are able to give rise to a new plant. Under N.W.European climatic conditions the species behaves as a pseudo-annual, tubers sprouting from late April and starting to produce a new generation of tubers from early July onwards. In autumn the shoots die off; night frosts hasten this process. The newly produced tubers remain dormant for the rest of the year. They are formed up to c. 75 cm from the original tuber, and up to about 40 cm depth, where they overwinter; a certain number, especially those in the soil top layers, are killed by frost (Fig. 1). The tubers can stay dormant in the soil for several years, especially at greater depths (Rotteveel and Naber 1993).

Under unrestricted growing conditions one tuber may produce a 2 to 2.5 m diameter circle densely covered with about 75 cm high shoots, hundreds of inflorescences, and up to c. 1500 tubers. However, as a C_4-plant, the species requires good light conditions. In a temperate climate it will easily be outshaded by surrounding vegetation. Hence it can only persist when growing in an open crop, like sugar beet, or a crop closing late, as *e.g.* maize. Under such conditions few inflorescences develop. The original habitat of the species is open wetland vegetation (*e.g.* Schroeder and Wolken 1989; Galatowitsch and Van der Valk 1996) or river banks (along the Po, Italy, Zanotti 1987 and the Loire, France, unpubl.obs.). In crops the species may remain unnoticed for a long time, and so easily get transported with agricultural produce and by machinery (see below).

Van Groenendael and Habekotté (1988) developed a model to study the dynamics of the tuber population (Fig. 2). Apart from tuber dynamics it includes shoot production and tuber/shoot ratios. Attention was focussed on two key processes, tuber pro-

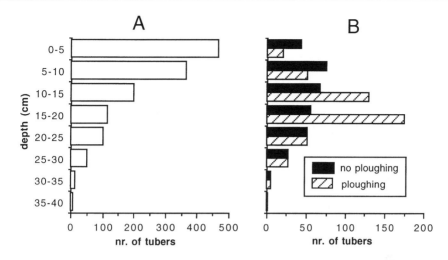

Fig. 1. The calculated effect of low winter temperatures on tuber survival, starting with 1300 tubers in autumn, distributed as in A. B indicates survival after ploughing and no ploughing respectively. Note the different scales of X-axes ! Figures are based on data of tuber survival under various frost treatments; temperatures as measured in a humid sandy soil during the 1987-88 winter period. (From Van Groenendael and Habekotté 1988)

duction as affected by shading by the crop (Table 1), and mortality of tubers due to frost (Fig. 1). Next they included the effects of herbicide treatments (Habekotté and Van Groenendael 1988). The model was parameterized and verified using data from various field experiments (*e.g.* Lotz *et al.* 1991). It proved to be a useful tool for sensitivity analysis to interpret and predict effects of various agricultural activities (Fig. 3).

Infraspecific variation

After *C. esculentus* had begun to become a serious weed problem, the question was asked whether all populations in the Netherlands might have resulted from just one introduction. It soon became clear that this was unlikely, because of the wide morphological and ecophysiological variability observed (Ter Borg *et al.* 1988). This was reason to start a taxonomic study, including experiments of growing plants under uniform conditions. It showed that reliable characters are the sizes and shapes of floral parts rather than dimensions of spikes, which had been until then thought to be the right diagnostic characters (Kükenthal 1935; Schippers *et al.* 1995). Four varieties were recorded in the Netherlands, *i.e.* all those that are distinguishable in the northern hemisphere, *viz.* var. *esculentus*, var. *macrostachyus* Boeckeler, var. *heermannii* (Buckley) Britton, and var. *leptostachyus* Boeckeler. A few more varieties have been described, from southern Africa, but they need further taxonomic work. The original geographic distribution of the four relevant ones, based on data from Kükenthal (1935) and 1000 herbarium specimens, can be summarized as follows: var. *esculentus* occurs all over Africa, in southern Europe and locally in N.E. USA and adjacent Canada; var. *leptostachyus* is common in the cooler regions of North and South America; var. *macrostachyus* prefers the warmer parts of the Americas; var. *heermannii* is restricted

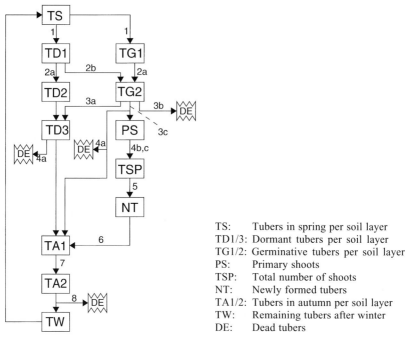

Fig. 2. Flow diagram for a population of *C. esculentus*. Boxes refer to life cycle stages, arrows represent transitions.
1: Survival of dormant and germinative tubers per soil layer. 2a: Redistribution of tubers by tillage in spring. 2b: Dormancy breaking by tillage. 3a: Induction of dormancy as a result of the use of herbicides. 3b: Death of tubers (TG) as a result of the use of herbicides. 3c: Formation of primary shoots. 4a: Death of tubers (TD and TG) in summer. 4b: Formation of secondary shoots. 4c: Reduction of TSP by chemical and non-chemical control. 5: Production of new tubers. 6: Distribution of tubers over various soil layers. 7: Redistribution of tubers by tillage in autumn. 8: Death of tubers in winter. (From Van Groenendael and Habekotté 1988)

to a limited area in the western USA. Apart from morphological variation they differ with respect to various physiological characteristics.

Next to these weedy varieties *C. esculentus* includes another taxon, *C. esculentus* var. *sativus*, or *C. esculentus* cv. Chufa (Erdmandel, chufa, tiger nuts, sweet almond) (De Vries 1991). Some authors (*e.g.* Schmitt and Sahli 1992) do not distinguish varieties, but mention two subspecies only, *viz.* subsp. *aureus* Tenori covering all weeds ('Knöllchen-Zypergras'), and subsp. *sativus* Boeckeler (Erdmandelgras), to indicate the crop. Var. *sativus* is grown in northern and western Africa, and locally in Europe, because of its edible tubers. According to Wein (1914), citing Gesner (1561, 'Horti Germaniae'), it has been known as a garden plant in Europe since long, and has been brought to Germany from France. It is characterized by short rhizomes and large tubers, it hardly flowers, and does not tolerate frost. The tubers are very nutritive due to high concentrations of starch and sugars, and high contents of oils (up to c. 25% on a dry weight basis, Linssen *et al.* 1987; Zanotti 1987; Omode *et al.* 1995, and unpubl. data). In the Nile valley they have been cultivated for several millennia (Zohary and Hopf 1988). This taxon will be left out of further consideration here.

Table 1. The effects of shading on dry matter production and allocation in *C. esculentus*. (From Van Groenendael and Habekotté 1988).

	tubers/m^2	shoot dry wt (g)	dry wt tubers/ dry wt shoot	nr of tubers/ g dry wt shoot	tuber dry wt (g)	nr of tubers/ initial tuber
% light						
100	29	11.9	1.01	14.6	0.080	149.9
40	29	1.4	0.59	17.4	0.042	19.8
20	29	0.2	0.42	39.0	0.020	3.1

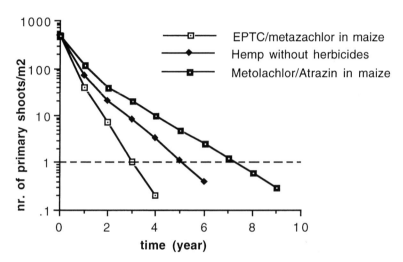

Fig. 3. Simulated effects of crops and herbicides on numbers of primary shoots of *C. esculentus* in spring. Starting density at 500 shoots per m^2. A maize crop is closing late; growth of hemp starts earlier and gives a dense shade. (From Habekotté and Van Groenendael 1988)

Invasions

Transport on a local scale: within fields

Effects of agricultural activities on invasions due to a few tubers left or a few tubers introduced became clear from model studies by Schippers *et al.* (1993). Instead of just calculating total tuber numbers per m^2, as had been done before (Fig. 2), a three-dimensional spatially-explicit model was developed, paying attention to tuber production, distribution and dispersal under a range of agricultural procedures in relation to continuous maize cultivation. Three main dispersal processes were distinguished, *viz.* natural spread of tubers by rhizome growth, spread due to soil mixing (ploughing, soil tillage), and to transport with soil adhering to farming machinery. Maximum dispersal by plant growth is less than a meter per year, being the radius of a clone. However, it plays a role in the non-tillage direction. Soil mixing affects the majority of tubers, but distances covered are relatively small. Soil adhesion concerns fewer tubers, but rather long distances may be covered, making it the most unpredictable component of the system (Fig. 4). Simulations including all relevant processes showed that

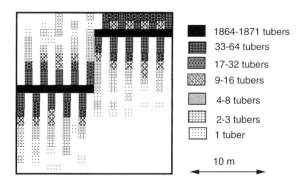

Fig. 4. The simulated dispersal of tubers with adhering soil. Effects of 'there and back tillage', the most common agricultural practice in the Netherlands, during two seasons. The results are based on tubers in the upper 5 cm, no tuber losses due to mortality or germination, no mixing of soil. Tubers at the start were in two lines of 1 x 10 m each, with 2000 tubers per m². (From Schippers *et al.* 1993)

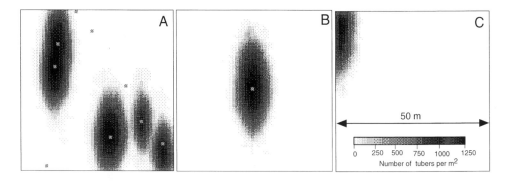

Fig. 5. The simulated distribution of tubers after 10 years of continuous maize cultivation and accompanying agricultural activities; no herbicides used. Ten tubers introduced per field; A: distributed at random; B: clustered in the field centre; C: introduced by agricultural machinery. The coloured area has a minimum of at least one tuber per m² and a maximum of 1200 tubers per m². Parameters and processes included were ploughing, soil mixing, deep tine cultivation, soil adhesion, germination and establishment of primary shoots, development of secondary shoots and tuber production, and tuber mortality during winter. (From Schippers *et al.* 1993)

just a few tubers are enough to infest a field, and that their initial distribution is of the utmost importance (Fig. 5). The duration of the lag period after introduction of the weed and the efficacy of the herbicide treatments both have a strong effect on the infestation (data not shown). Differences exist between var. *macrostachyus* and var. *leptostachyus*, *e.g.* with respect to frost tolerance and depth distribution of tubers (Fig. 6); moreover the average number of primary shoots per tuber is higher in var. *macrostachyus* (1.84) than in var. *leptostachyus* (1.38). By including this information in the model, it became clear that the effectivity of control procedures differs between varieties to a significant extent. According to these data an incomplete effect of herbicides may result in complete control of var. *macrostachyus* in about 10 years' time, whereas var. *leptostachyus* appeared to be more persistent (Fig. 7).

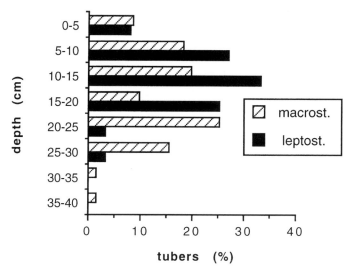

Fig. 6. Depth distribution of tubers of var. *leptostachyus* and var. *macrostachyus*. (From Schippers *et al.* 1993)

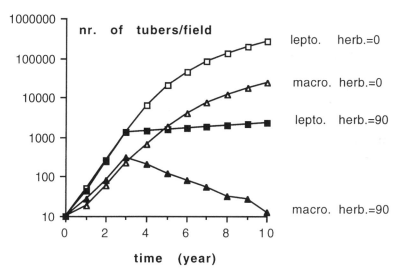

Fig. 7. Simulated population growth of var. *leptostachyus* and var. *macrostachyus* under continuous maize cropping. 0: normal farming procedures, no herbicides applied. 90: ditto, herbicides killing 90% of the shoots. (From Schippers *et al.* 1993)

Transport and invasions on a regional scale: between fields

It is clear that tubers of *C. esculentus* are an excellent means of dispersal, by man, or by natural vectors, including animals. To begin with the last, various field observations suggest that mice (*Microtus* sp.) collect and concentrate the tubers as a winter stock (Schroeder and Wolken 1989; unpubl. obs.). Moreover birds (*Corvus* sp.) were noticed flying with rhizomes, possibly adhering to tubers (unpubl.obs.). Whereas mice

Table 2. Tuber germination and winter survival per soil layer. Spring germination from various depths, after Stoller and Wax (1973). Tuber survival during the 1986/87 winter season in var. *leptostachyus* and var. *macrostachyus* (unpublished data L.A.P.Lotz). (From Schippers *et al.* 1993)

soil layer (cm)	% germination in spring	% winter survival var. *leptostachyus*	var. *macrostachyus*
0-5	100	55	3
5-10	87	59	9
10-15	80	59	20
15-20	70	54	33
20-25	60	55	39
25-30	54	55	45
30-35	52	55	41
35-40	52	54	37
40-45	52	54	37

probably carry the tubers within a field, birds may cover somewhat larger distances. Still larger distances may be covered with agricultural products, in contaminated soil adhering to root crops like sugarbeet, potatoes, asparagus, bulbs, and the roots of various horticultural products; propagation material seems to play an important role in this respect.

Soon after the problem had been noticed, the hypothesis was developed that the early dispersal of *C. esculentus* in the Netherlands might be related to the propagation of gladiolus. This crop can be grown in a field only once, because of the development of persistent soilborne pathogens. As a result, a system of 'shifting cultivation' exists, with gladiolus being grown once in a field, often just in a small part of a larger one. This results in large numbers of fields where gladiolus has been grown, for one year only. Because of the general phytosanitary preventive measures required, the cultivation of gladiolus is subject to permission by the Plant Protection Service; as a consequence, information was available about where gladiolus had been grown. A check of these fields after the *Cyperus* problem had been recognized showed that in a large proportion of them the species did occur, supporting the original hypothesis (Naber and Rotteveel 1986a). The sizes and colours of tubers of *C. esculentus* are very similar to those of gladiolus cormlets, and so both species had been dispersed together, often over long distances. When grown at high densities, for propagation purposes, no flowers develop in gladiolus, and since its shoots and leaves and those of *Cyperus* are very much alike, and similar to grasses, it took some time until farmers became aware of the weed problem. Meanwhile it had widely spread. However, from the beginning of the observations, a number of fields were noticed that had no history of gladiolus production, nor any other evident reason, and yet were infested. This left a problem to be solved.

Legal measures and control in the Netherlands

In 1984 legal measures were introduced to prevent further spread, and control systems were developed to reduce contaminations (Naber and Rotteveel 1986b). The measures included the cleaning of agricultural equipment used on a contaminated field, the cleaning or destruction of contaminated crops and planting material, and the prohibition of the

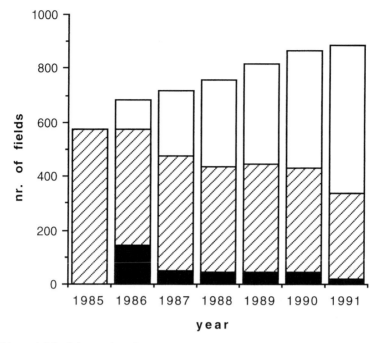

Fig. 8. Rise and fall of the number of *C. esculentus* infested fields in the Netherlands over the 1985-1991 period. Black: new infestations; hatched: infested fields; white: fields that have been infested, but no visual infestation left (however, tubers may be present in the soil). (After Rotteveel 1993)

cultivation of all root crops such as sugar beet and potatoes on fields with a known history of infestation. This latter measure especially is hard on farmers since sugar beet and potatoes are the main cash crops in the Netherlands, together covering more than 40% of the arable acreage.

Intensive field controls were carried out by the Plant Protection Service, and so a fairly complete overview became available of the distribution of the species and changes therein, with a top number of 759 fields with known present or previous infestation, covering about 1000 ha in 1988. Fields were considered to be free again if no shoots had been observed during a period of three consecutive seasons, but it became clear that careful visual control by the farmer should be continued, the more so since just a few tubers left might be enough to completely infest a field (Fig. 5). Early 1989 a first set of 96 fields could be formally declared free of infestation. From then the number of infested fields gradually declined (Fig. 8). In 1992 all legal measures were withdrawn, and, for financial reasons, like other weeds, the *Cyperus* problem was left to the farmers and their organizations. Initially this resulted in an increase of the infestation level; however, gradually the situation seems to stabilize (Rotteveel and Naber 1997).

Extensive agricultural research to test the effects of various control methods helped to show the farmers how to master the problem. Measures are based on herbicides, on crop rotation aiming at outcompeting the weed by shading, and on prevention of further spread of the species by use of clean machinery and avoiding growth of rooted crops like sugar beet, potatoes, etc. A final step is usually required and concerns an intensive check and handweeding (*e.g.* Lotz *et al.* 1991; Rotteveel *et al.* 1993; Rotteveel and Naber 1993, 1996).

Dispersal on a continental and a global scale

As indicated above, the majority of infestations in the Netherlands could be ascribed to the propagation of gladiolus, with later spread due to various crops. The original imports were due to activities of Dutch bulbgrowers, collecting new material from various parts of northern and central America, to broaden the range of plants offered to the market. Contamination of this material is the most probable source of the infestation. However, some of the infested fields had no history of bulb growing. Where did these tubers come from ? Moreover, soon after the problem had been observed in the Netherlands, reports came from various European countries about new infestations (*e.g.* France, Morin and Sombrun 1984; Germany, Heidler 1986). Had all these infestations resulted from the Dutch introductions ?

In some cases there were reasons to relate new finds to the Dutch populations, in particular when it concerned fields close to the Dutch border. So *e.g.* tubers were probably brought to Belgium with gulley or farm manure (unpubl. information). Some authors also mentioned some local particular means of spread, *e.g.* construction of a gas pipe between Italy and Austria (Melzer 1989; Neururer 1990), and imports of tapioca to feed poultry in N.W. Germany (Schroeder and Wolken 1989). In various instances the weed was first noticed in maize, but still no evident explanation of its occurrence could be given (Germany: Gieske *et al.* 1992; Switzerland: Schmitt and Sahli 1992, Schmitt 1993; France: Bernard 1996; Hungary: Dancza 1994).

Essential information in this context came from a students' report, describing the introduction of *C. esculentus* to a farm in France (Loir et Cher), as a contaminant of gladiolus material as early as 1947. The field was under an extensive cropping system, and no modern machinery with high dispersal capacities was used. Therefore the species may have remained there, until the agricultural use was intensified from 1976 onwards, leading to its spread (Guillerm 1987). This information, together with the knowledge of the infraspecific variation, appeared invaluable for a reconstruction of the probable (im)migration routes of the new finds in N.W. and C. Europe.

The populations in N.W. Europe were classified and mapped and the data compared to the original pattern of distribution of the varieties (Ter Borg and Schippers 1992). It showed that var. *macrostachyus* and var. *heermannii*, coming from the warmer regions in America, had probably hardly been exported from the Netherlands. The majority of the French populations, both in arable land and along the Loire, were var. *leptostachyus*, as well as many scattered populations in Europe, the few inexplicable infestations in the Netherlands inclusive. Therefore, it is suggested that most recent finds of *C. esculentus* in N.W. Europe are due to the earlier introduction in France. Of course, this does not exclude transports to the Netherlands of this variety, which has its main distribution in cooler regions of the Americas (Table 3). Since in several cases authors reporting new finds of *C. esculentus* mentioned a maize crop, it is suggested that small tubers are transported as a contaminant of maize seeds, however unlikely it seems.

Except for a few, populations in southern Europe appeared to belong to var. *esculentus*. The plants in the Netherlands identified as such might have resulted from introductions from southern Europe. However, because they were found together with var. *leptostachyus* and because of their physiological properties (frost resistance of the tu-

Table 3. A summary of the 25 years' history of invasion and control of *C. esculentus* in the Netherlands.

- early seventies:
  First introductions, as based on reconstructions of the cropping history of farms which were found to be infested later on. For one particular farm in S.W. Netherlands this was 1972. The infestation was of var. *macrostachyus.*
- 1975:   The first plant included in the Leyden herbarium, was collected 'between the tramrails' in The Hague. It was var. *leptostachyus.*
- 1981:   The species was recognised as a new and dangerous weed.
- 1984:   Legal measures were introduced, to control and prevent further spread.
- 1985:   A maximum number of almost 600 infested fields was known; from then a gradual reduction was noticeable.
- 1988/89: 96 fields declared free after three years without visible *Cyperus* infestation.
- 1992:   Legal measures withdrawn. Increasing number of new infestations.
- 1995-97: Number of infested fields decreasing; control efforts as stimulated by the farmers' organisations increasingly successful.

bers, early flowering, short day effects on tuber production), they are supposed to have been imported from the relatively small area in the N.W. USA, where this variety occurs (Ter Borg and Schippers 1992). So far, var. *esculentus* seems to have been unable to spread from its main area in southern Europe and Africa, and to establish itself in more northern countries.

Conclusions

C. esculentus shows several of the properties usually expected in invasive species:
- high vegetative reproduction capacity, by means of tubers,
- enough genetic variation to cover a wide range of conditions,
- a hidden start of populations, due to inconspicuous morphology, resulting in a rather long lag period, allowing extensive spread before the species is recognized as a weed problem, and control efforts are started,
- good transport capacities: at a small scale via animals and agricultural activities, at a regional scale via agricultural activities in general, and at a continental and global scale via transport of agricultural products.

C. esculentus var. *leptostachyus* appeared to have the best combination of characters to invade and persist under N.W. European climatic and agricultural conditions.

References

Bernard, J.L. 1996. Les souchets tubereux. Présence reconnue des Cyperus dans les cultures en France. Phytoma 484: 31.

Dancza, I. 1994. The occurrence of *Cyperus esculentus* in the region of Keszthely – Héviz. (in Hungarian). Növényvedelem 30: 475-476.

De Vries, F.T. 1991. Chufa (*Cyperus esculentus*, Cyperaceae): A weedy cultivar or a cultivated weed? Econ. Bot. 45: 27-37.

Galatowitsch, S.M. and Van der Valk, A.G. 1996. The vegetation of restored and natural prairie wetlands. Ecol. Appl. 6: 102-112.

Gieske, A., Gerowitt, B., and Miesner, H. 1992. Erdmandelgras – ein neues Problemunkraut. PSP-Pflanzenschutz-Praxis 4: 10-12.

Guillerm, H. 1987. Le souchet comestible, un problème désormais présent en France. Lycée Agricole 'Charlemagne', Carcassonne; AGPM. Station de Boigneville, Maisse.

Habekotté, B. and Van Groenendael, J.M. 1988. Population dynamics of Cyperus esculentus L. (Yellow Nutsedge) under various agricultural conditions. Med. Fac. Landbouww. Rijksuniv. Gent 53/3b: 1251-1259.

Heidler, G. 1986. Erdmandelgras – ein neues Problemunkraut. DLG Mitteilungen, Braunschweig 101: 126-128.

Holm, L.G., Plucknett, D.L., Panch, J.V. and Herberger, J.P. 1977. The world's worst weeds: Distribution and biology. Univ. Press, Honolulu.

Kükenthal, G. 1935. Cyperaceae-Scirpoideae-Cypereae. In: Engler, A. and Diels, L. (eds.), Das Pflanzenreich IV-20. pp. 116-121. Verlag von Wilhelm Engelmann, Leipzig.

Lapham, J. 1985. Unrestricted growth, tuber formation and spread of Cyperus esculentus L. in Zimbabwe. Weed Res. 25: 323-329.

Lapham, J., Drennan, D.S.H. and Francis, L. 1985. Population dynamics of Cyperus esculentus L. (Yellow Nutsedge) in Zimbabwe. British Crop Protection Conference – Weeds – 4a-4: 1043-1050.

Linssen, J.P.H., Van Olderen, J.D., and Pilnik, W. 1987. Cyperus esculentus: voedingsmiddel of onkruid. Voedingsmiddelentechnologie 5: 24-26.

Lotz, L.A.P., Groeneveld, R.M.W., Habekotté, B. and Van Oene, H. 1991. Reduction of growth and reproduction of Cyperus esculentus by specific crops. Weed Res. 31: 153-160.

Melzer, H. 1989. Über Cyperus esculentus L., die Erdmandel, weitere für Kärnten neue Gefässpflanzen -Sippen und neue Fundorte bemerkenswerter Arten. Verh. Zool.-Bot. Ges. Österreich 126: 165-178.

Morin, C. and Sombrun, F. 1984. Lutte contre le souchet comestible (Cyperus esculentus L.) dans le mais en France dans les Landes. Proceedings EWRS 3rd Symp. on Weed Problems in the Mediterranean Area. pp. 271-277.

Naber, H. and Rotteveel, A.J.W. 1986a. Cyperus esculentus – Its build-up and methods of control in the Netherlands. Proc. 49e Wintermeeting Institut International de Recherches Betteravières, Brussel. pp. 339-344.

Naber, H. and Rotteveel, A.J.W. 1986b. Legal measures concerning Cyperus esculentus L. in the Netherlands. Med. Fac. Landbouww. Rijksuniv. Gent 51/2a: 355-357.

Neururer, H. 1990. Einschleppung eines neuen Unkrautes beim Bau einer internationalen Gasleitung am Beispiel von Cyperus esculentus und Möglichkeiten zur raschen Sanierung. Z. PflKrankh. PflSchutz, Sonderh. 12: 71-74.

Omode, A.A., Fatoki, O.S. and Olaogun, K.A. 1995. Physicochemical properties of some underexploited and nonconventional oilseeds. J. Agric. Food Chem. 43: 2850-2853.

Rotteveel, A.J.W., Straathof, H.J.M. and Naber, H. 1993. The decline of Yellow Nutsedge (Cyperus esculentus L.) population under three chemical management systems aimed at eradication. Med. Fac. Landbouww. Rijksuniv. Gent 58/3a: 893-900.

Rotteveel, A.J.W. and Naber, H. 1993. Decline of Yellow Nutsedge (Cyperus esculentus L.) when tuber formation is prevented. Brighton Crop Protection Conference – Weeds – 4b-7: 311-316.

Rotteveel, A.J.W. and Naber, H. 1996. Persistence of Yellow Nutsedge (Cyperus esculentus L.) over a ten year period in a grass ley. Xème Colloque international sur la biologie des mauvaises herbes, Dijon. pp. 51-55.

Rotteveel, A.J.W. and Naber, H. 1997. results of an eradication policy concerning Cyperus esculentus. 10th EWRS (European Weed Research Society) Symposium 1997, Poznan. p. 15.

Schippers, P., Ter Borg, S.J., Van Groenendael, J.M. and Habekotté, B. 1993. What makes Cyperus esculentus (Yellow Nutsedge) an invasive species? – a spatial model approach. Brighton Crop Protection Conference – Weeds – 5a-2: 495-504.

Schippers, P., Ter Borg, S.J. and Bos, J.J. 1995. A revision of the infraspecific taxonomy of Cyperus esculentus (Yellow Nutsedge) with an experimentally evaluated character set. Syst. Bot. 20: 461-481.

Schmitt, R. 1993. Das neue Unkraut Knöllchen-Zypergras. Landwirtschaft Schweiz 6: 376-378.

Schmitt, R. and Sahli, A. 1992. Eine in der Schweiz als Unkraut neu auftretende Unterart des Cyperus esculentus L. Landwirtschaft Schweiz 5: 273-278.

Schroeder, C. and Wolken, M. 1989. Die Erdmandel (Cyperus esculentus L.) – ein neues Unkraut in Mais. Osnabrücker naturwiss. Mitt. 12: 83-104.

Stoller, E.W. and Sweet, R.D. 1987. Biology and life cycle of Purple and Yellow Nutsedges (Cyperus rotundus and C. esculentus). Weed Techn. 1: 66-73.

Stoller, E.W. and Wax, L.M. 1973. Yellow Nutsedge shoot emergence and tuber longevity. Weed Sci. 23: 333-337.

Ter Borg, S.J., De Nijs, L.J. and Van Oene, H. 1988. Intraspecific variation of *Cyperus esculentus* L. in the Netherlands; a preliminary report. VIIIème Colloque international sur la biologie, l'écologie et la systematique des mauvaises herbes, Dijon. pp. 181-185.

Ter Borg, S.J. and Schippers, P. 1992. Distribution of varieties of *Cyperus esculentus* L. (Yellow Nutsedge) and their possible migration in Europe. IXème colloque international sur la biologie des mauvaises herbes, Dijon. pp. 417-425.

Van Groenendael, J.M. and Habekotté, B. 1988. *Cyperus esculentus* L. – biology, population dynamics and possibilities to control this neophyte. Z. PflKrankh. PflSchutz, Sonderh. XI: 61-69.

Wein, K. 1914. Deutschlands Gartenpflanzen um die Mitte des 16. Jahrhunderts. Beihefte Botanisches Centralblatt 31 (2e Abt.): 463-555.

Zanotti, E. 1987. Notes about *Cyperus esculentus* L., exotic species, new for the territories of the provinces of Bergamo, Brescia and Cremona. Pianura 1: 65-82.

Zohary, D. and Hopf, M. 1988. Domestication of plants in the Old World; the origin and spread of cultivated plants in W. Asia, Europe and the Nile valley. pp. 171-212. Clarendon Press, Oxford.

INVASION BY HYBRIDIZATION: *CARPOBROTUS* IN COASTAL CALIFORNIA

Ewald F. Weber [1], Montserrat Vilà [2], Marc Albert, and Carla M. D'Antonio
*Department of Integrative Biology, University of California, Berkeley CA 94720,
U.S.A; [1]Institut für Umweltwissenschaften, University of Zurich, Winterthurer-Strasse
190, CH-8057 Zurich, Switzerland; e-mail: weber4@uwinst.unizh.ch; [2]Centre de
Recerca Ecològica i Aplicacions Forestals, Universitat Autònoma de Barcelona, ES-
08193 Bellaterra, Barcelona, Catalonia, Spain; e-mail: vila@cc.uab.es*

Abstract

California has a high fraction of exotic plant species (17.4%). Invading plant species of great ecological concern occur in coastal habitats prone to disturbance and where many rare native species grow. *Carpobrotus edulis*, introduced from South Africa, is a serious invader of coastal habitats, crowding out native vegetation by the formation of dense mats due to vegetative growth. In addition to these direct effects, *C. edulis* hybridizes with a native or long-naturalized congener of smaller habit, *C. chilensis*. Hybrids between these two species are fully fertile and are found throughout coastal California. With transplant and garden studies, the fitness and growth of hybrids was compared with that of the parental species in order to investigate the role of hybridization in the invasion process. Survival and growth of hybrid seedlings under field conditions was not different from parental species. The number of viable seeds per plant was higher in *C. edulis* than in *C. chilensis*; hybrids were intermediate. *C. edulis* accumulated more biomass than *C. chilensis* when grown in a common garden setting, but the latter produced longer and thinner branches. Hybrids were intermediate in this and other life-history characters. The results suggest that hybrid individuals do not have a significantly lower fitness than parental species and may contribute to the successful invasion of *Carpobrotus* due to the formation of additional spreading genotypes.

Introduction

California has a very high plant diversity, as a result of a mediterranean climate over a large part of the state, and its topography (Raven and Axelrod 1995). The flora contains many rare and endangered species, and 24% of all native plants are endemic to California (Hickman 1993). In addition, California has a high number of introduced species (17.4%). Most of these are grasses of European origin, some of which have transformed the natural vegetation at a landscape scale. Plant invasions in California are of great concern for conservation biologists and people involved in the management of protected areas. Exotic pest plants are a primary threat to many rare and endangered species with small ranges (Schierenbeck 1995). Plant invaders crowd out native vegetation due to a higher competitive ability, as a result of extensive clonal growth and larger plant size.

Coastal habitats contain a large number of introduced plant species, and these environments are especially rich in endemic species. Important plant invaders of coastal

*Plant Invasions: Ecological Mechanisms and Human Responses, pp. 275–281
edited by U. Starfinger, K. Edwards, I. Kowarik and M. Williamson
© 1998 Backhuys Publishers, Leiden, The Netherlands*

habitats in California are *Ammophila arenaria* (Poaceae), *Carpobrotus edulis* (Aizoaceae), *Cortadera jubata* (Poaceae), *Genista monspessulana* (Fabaceae), and *Senecio mikanioides* (Asteraceae) (Anderson *et al.* 1996).

In this chapter, we will report on studies on the invasion by *Carpobrotus edulis* and its ecological and genetic effects. The studies involved field observations, common garden experiments and transplant studies in the field.

Carpobrotus spp. in California

The genus *Carpobrotus* (Aizoaceae) is composed of succulent perennials with a prostrate growth form, triangulate leaves and indehiscent fleshy fruits with a sweet aroma (Wisura and Glen 1993). In California, the genus is represented by two distinct species, *C. edulis* (L.) N. E. Br. and *C. chilensis* (Molina) N. E. Br., which are both confined to coastal habitats. The two species can easily be distinguished by the size of their flowers and leaves, and by their flower colour. *C. edulis* has larger leaves and flowers than *C. chilensis* and its flowers are yellow, whereas *C. chilensis* has magenta flowers (Albert *et al.* 1997). Both species share the same pollinators and are pollinated mainly by bees.

C. edulis was introduced from South Africa to California around the turn of the century and has been extensively planted along highways and railroads for sand stabilization. Since then, *C. edulis* has invaded many natural coastal communities due to its clonal growth (D'Antonio 1993) and because seeds are successfully dispersed by several native mammals (D'Antonio 1990). The species can build large and dense mats up to 40 cm thick and 8 - 10 m in diameter (D'Antonio and Mahall 1991), allowing the species to compete successfully with native shrubs (Zedler and Scheid 1988; D'Antonio and Mahall 1991; D'Antonio *et al.* 1993). *C. chilensis* is either native or long-naturalized (Bicknell and Mackey 1988). It is smaller and less abundant than *C. edulis*. The species does not form dense mats like *C. edulis* and is generally not considered to pose a threat to native species.

Hybridization between *Carpobrotus edulis* and *C. chilensis*

Both *Carpobrotus* species are sympatric in many areas. In addition to these two species, hybrid swarms are found throughout coastal California. *Carpobrotus* hybrids can be very abundant locally. Morphological studies and allozyme data analyses support the occurrence of hybridization and introgression between the two species. Albert *et al.* (1997) conducted a survey of the phenotypic variation of *Carpobrotus* in California and found a complete phenotypic range between the two species in many sites, suggesting widespread hybridization and introgression. Multivariate analyses showed that hybrids overlap particularly with *C. edulis* (Albert *et al.* 1997). Gallagher *et al.* (1997) identified allozyme markers in *Carpobrotus*, confirming introgression. They also found that the variation in populations of *C. edulis* and intermediate hybrids was larger than in populations of *C. chilensis* (Gallagher *et al.* 1997).

Hybridization between an exotic invasive species and a native congener raises the question of how hybridization will influence the invasion process. This issue has rarely been addressed, and the ecology of hybrids originating from such hybridization events has rarely been investigated in detail (Abbott 1992). In order to understand the eco-logical role of hybrids for plant invasions it seems necessary to compare the perfor-mance of hybrids relative to parental species throughout the life-cycle of the species. Our studies were aimed to answer the question whether hybrids have higher, equal, or lower fitness than parental species.

Hybrid and parental species fitness

Vilà and D'Antonio (1998; in press) examined reproductive fitness components (seed output, germination, survival of seedlings) of hybrids and parental species from three different sites in Northern California (Manila Dunes, Bodega Bay, and Morro Bay), representing coastal dune communities. The authors found that *C. edulis* and hybrids produced more fruits per clone than *C. chilensis*, primarily as a consequence of larger clone size, e.g. larger diameter of an individual plant. Fruit weight and seed set of hybrids were intermediate or close to those for *C. edulis*.

However, fitness is not only a function of fruit and seed production but also of the likelihood that offspring will establish (Primack and Kang 1989). This includes dis-persal of seeds and subsequent germination and establishment of seedlings. Scats from black-tailed jackrabbit (*Lepus californicus*) and black-tailed deer (*Odocoileus hemionus*), the most common frugivores in our sites, contained more seeds from the exotic *C. edulis* and hybrid morphotypes than from *C. chilensis*. The fraction of *C. edulis* and hybrid seeds was more than would be expected based on relative fruit abundance (number of fruits per ground area and number of seeds per fruit). In a germination test, Vilà and D'Antonio (in press) found that germination of *C. edulis* and hybrid seeds was en-hanced after gut passage, while germination of *C. chilensis* decreased compared to germination from fresh fruits. These results suggest that fruits from *C. edulis* and hy-brids are more consumed than fruits from *C. chilensis*, and that seeds of *C. chilensis* are more damaged after gut passage than seeds of *C. edulis* and hybrids, respectively.

In a transplant study in the field, the authors found that survival and growth of one year old hybrid seedlings did not differ from those of parental species. Seeds from hybrids and parental species were germinated in the glasshouse in November and trans-planted into the field sites three months later. At this time, seedlings had four true leaves. Survival of seedlings was measured after one year. Based on this experiment and on data on seed output and germination, a summarizing fitness value was com-puted, including the following reproductive stages: number of seeds per plant (pre-dispersal); proportion of seeds in scats (dispersal); seed viability (post-dispersal); seedling survival in the field. Figure 1a shows the decline in number of seeds for each of these four stages. Probability estimates for the number of seeds that produce one yr old seed-lings were 3%, 2% and 1% for *C. edulis*, hybrids and *C. chilensis*, respectively (Fig. 1b). Hybrids were intermediate relative to parental species with regard to these prob-ability estimates, but closer to the values of the introduced *C. edulis* (Vilà and D'Antonio in press).

In a common garden study, growth of *Carpobrotus* hybrids was compared with pa-rental species in order to look for differences in life-history characters among hybrids and parental species. Ramets from parental species and hybrids were collected at two sites where all co-occur (Bodega Bay, Morro Bay). *Carpobrotus* species were identi-fied in the field based on morphological characters. Hybrid individuals were assigned

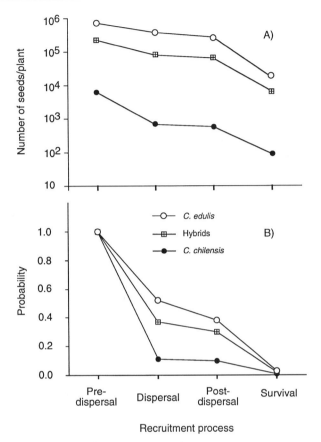

Fig. 1. Decline of number of seeds per plant of *Carpobrotus* spp. that will become seedlings (a) and corresponding probabilities (b) across four recruitment stages: pre-dispersal, dispersal, post-dispersal, and survival in the field. For explanations, see text.

to three hybrid classes (Albert *et al.* 1997): "Intermediates", phenotypes that are intermediate in morphology between the two species (I); "Backcross *C. chilensis*", hybrid phenotypes more similar to *C. chilensis* (BC); and "Backcross *C. edulis*", hybrid phenotypes more similar to *C. edulis* (BE). Together with the two putative parental species, these categories formed five groups that will be called "morphotypes" hereafter. Cuttings of approx. 50 mm size were prepared from each ramet. 24 cuttings per morphotype were grown for nine months in large pots filled with sand.

The variation of *Carpobrotus* morphotypes in the common garden was large and often overlapping (Fig. 2). However, morphotypes were significantly different for all characters (analysis of variance: P<0.001). The two *Carpobrotus* species differed from each other in ecologically important characters. *C. edulis* produced more aboveground biomass than *C. chilensis* during the time of the experiment. In contrast, total branch length (sum of the lengths of all branches of a plant) was less in *C. edulis* than in *C. chilensis*, but *C. edulis* produced thicker stems (Fig. 2).

Hybrids were intermediate relative to parental species with respect to these growth characters (Fig. 2). The strongest difference among morphotypes was apparent in leaf

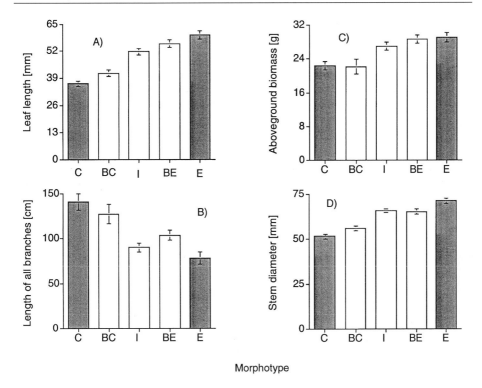

Morphotype

Fig. 2. Growth of *Carpobrotus* spp. and three hybrid classes (morphotypes) in a common garden setting. Bars represent means ± 1 standard error for each morphotype. Shaded bars indicate parental species. A) Leaf length of the largest leaf, B) Total length of all branches per plant, C) Aboveground biomass, D) Maximum stem diameter of main branch.
C = *Carpobrotus chilensis*; E = *C. edulis*; BC = hybrids similar to *C. chilensis*; I = intermediate hybrids; BE = hybrids similar to *C. edulis*.

length, which is an important character for distinguishing morphotype classes. If clonal growth is considered as a component of fitness, hybrids do not have lower fitness compared to parental species.

The species differences are in agreement with field observations. The production of thick and densely branched stems in the case of *C. edulis* is the primary negative impact on native species (D'Antonio and Mahall 1991; D'Antonio *et al.* 1993). Mats of *C. chilensis* are less thick and more open in their growth, a result of the production of longer and less branched stems.

Management and control

C. edulis is listed by the California Exotic Pest Plant Council (CalEPPC) as one of the most widespread invasive wildland pest plants (Anderson *et al.* 1996). However, *Carpobrotus* hybrids have not been listed by the CalEPPC as a threat to native vegetation. Thus, removal of hybrids is not reinforced as much as the removal of *C. edulis*.

Control of *C. edulis* is usually limited to the physical removal of plants, or to the application of Round-up® herbicide (Albert 1995). In addition, a combination of 2% glyphosphate and 1% surfactant has generally been found to be an effective herbicide. One thorough application generally is sufficient to kill most of the living material, but one or two follow-ups are necessary to eradicate resprouts.

The most efficient way of manually removing *Carpobrotus* is the "carpet rolling" technique, where the *Carpobrotus* mat is rolled up from one side by a few people as other workers sever the roots underneath with shovels.

Vilà and D'Antonio (unpublished data) found that natural processes may help to reduce the growth of *C. edulis* and hybrids. Survival and growth of these morphotypes were lower in grassland habitats than in coastal scrub and backdunes because of strong competition from grasses and high levels of herbivory by rabbit, jackrabbit and deer. Thus, removal actions in grasslands would be more efficient for *C. edulis* and hybrids than in other habitats with lower vegetation cover and less visited by herbivores.

Conclusions

The invasion by *C. edulis* has many ecological and genetic effects, with many factors promoting the spread of *C. edulis* and hybrids.

The interaction between *C. edulis* and native mammals acting as frugivores and herbivores (mainly rabbit and deer) is a partial driving force for the invasion success. Patterns of fruit preference by native frugivores and seed survival after gut passage facilitate the successful spread of *C. edulis* and hybrids. Furthermore, greater resistance to herbivory in *C. edulis* and hybrids contributes to the fast growth of established clones. Hybrid fitness is intermediate relative to the parental species; however, we believe that hybrids will continue to persist and invade California coastal communities. Since hybrid growth will be largely indistinguishable from *C. edulis,* hybridization may accelerate the invasion by *Carpobrotus* due to the formation of additional spreading genotypes. The high survival probability of the introduced *C. edulis* may insure its successful invasion status, compared to the putative native congener *C. chilensis.*

Acknowledgements

The authors thank P. Connors from the Bodega Marine Reserve for permission to carry out experiments, and X. Zeng and H. Swartz for field and lab assistance. Partial funding was provided by the Comissió Interdepartamental de Ciencia y Tecnología de la Generalitat de Catalunya and the Ministerio de Educación y Ciencia, Spain, to M. V., by the the Swiss National Science Foundation and Roche Research Foundation to E. W. and the USA National Science Foundation Grant DEB 9322795 to C. M. D.

References

Abbott, R.J. 1992. Plant invasions, interspecific hybridization and the evolution of new plant taxa. Trends Ecol. Evol. (TREE) 7: 401-405.
Albert, M.E. 1995. Portrait of an invader II. The ecology and management of *Carpobrotus edulis.* California Exotic Pest Plant Council News 3: 4-6.

Albert, M.E., D'Antonio, C.M. and Schierenbeck, K.A. 1997. Hybridization and introgression in *Carpobrotus* ssp. (Aizoaceae) in California. I. Morphological evidence. Amer. J. Bot. 84: 896-904.

Anderson, L.W., Di Tomaso, J.L., Howald, A., Randall, J., Rejmánek, M. and Chairman, J.S. 1996. Exotic pest plants of greatest ecological concern in California. California Exotic Pest Plant Council. Special Publication.

Bicknell, S.H. and Mackey, E. 1988. Evidence for early occurrence of *Carpobrotus aequilaterus* N.E. Br. at Marina Beach State Park, CA. Department of Parks and Recreation. Final report number 05-10-065.

D'Antonio, C.M. 1990. Seed production and dispersal in the non-native, invasive succulent *Carpobrotus edulis* (Aizoaceae) in coastal strand communities of central California. J. Appl. Ecol. 27: 693-702.

D'Antonio, C.M. 1993. Mechanisms controlling invasion of coastal plant communities by the alien succulent *Carpobrotus edulis*. Ecology 74: 83-95.

D'Antonio, C.M. and Mahall. B.E. 1991. Root profiles and competition between the invasive, exotic perennial, *Carpobrotus edulis*, and two native shrub species in California coastal scrub. Amer. J. Bot. 78: 885-894.

D'Antonio, C.M., Dennis, D.C. and Tyler, C.M. 1993. Invasion of maritime chaparral by the introduced succulent *Carpobrotus edulis*. Oecologia 95: 14-21.

Gallagher, K.G., Schierenbeck, K.A. and D'Antonio, C.M. 1997. Hybridization and introgression in *Carpobrotus* ssp. (Aizoaceae) in California. II. Allozyme evidence. Amer. J. Bot. 84: 905-911.

Hickman, J.C. (ed.) 1993. The Jepson manual. Higher plants of California. University of California Press, Berkeley.

Primack, R.B. and Kang, H. 1989. Measuring fitness and natural selection in wild plant populations. Ann. Rev. of Ecol. Syst. 20: 367-396.

Raven, P.H. and Axelrod, D.I. 1995. Origin and relationships of the California flora. California Native Plant Society. University of California Press, Berkeley.

Schierenbeck, K.A. 1995. The threat to the California flora from invasive species: problems and possible solutions. Madroño 42: 168-174.

Vilà, M. and D'Antonio, C.M. 1998. Fruit choice and seed dispersal of invasive vs. non-invasive *Carpobrotus* (Aizoaceae) in coastal California. Ecology 79: 1053-1060.

Vilà, M. and D'Antonio, C.M. (in press). Fitness of invasive *Carpobrotus* (Aizoaceae) hybrids in coastal California. Ecoscience.

Wisura, W. and Glen, H. 1993. The South African species of *Carpobrotus* (Mesembryanthema-Aizoaceae). Contribut. Bolus Herbar. 15: 76-107.

Zedler, P.H. and Scheid, G.A. 1988. Invasion of *Carpobrotus edulis* and *Salix lasiolepis* after fire in a coastal chaparral site in Santa Barbara county, California. Madroño 35: 196-201.

CASE STUDIES II: BIOTOPES/REGIONS

DISTRIBUTION AND SPREADING OF ALIEN TREES AND SHRUBS IN SOUTH WESTERN GERMANY AND CONTRIBUTIONS TO GERMINATION BIOLOGY

Reinhard Böcker and Monika Dirk
Institut für Landschafts- und Pflanzenökologie (320), Universität Hohenheim, D-70593 Stuttgart, Germany; e-mail: boeckerr@Uni-Hohenheim.DE

Abstract

The spread and dispersal of neophytic species were studied in south western Germany. 26 neophytic hardwood species were cultivated in different substrates with regard to their success of germination and establishing.

A two factor experiment was done to test the germination success and potential ability to spread of 26 neophytic hardwood species. Seeds of each species were sown in five different substrates. This experiment should simulate the ability of different hardwood species to escape from gardens and plantations and invade man made and natural sites.

Introduction

Spontaneous spread of neophytic hardwood species is observed mainly in areas where the mean temperature per year is about 9°C and higher as found in metropolitan areas in Germany. The southwestern part of Germany is climatically exposed; the upper Rhine is especially prominent for its vineyards and warmer climate (Fig. 1). Most of the submediterranean and pannonic species in the German flora (Sebald *et al.* 1990-96) grow here. In addition, many woody aliens are able to colonize different areas in this region, such as fallow land, urban industrial sites, railway embankments, roadside verges, and riversides.

As a result of our tradition, more than 4000 woody species were introduced (Trautmann 1976) for ornamental purposes into Europe (the Arboretum Hohenheim has more than 1300 woody species), but only a few species escaped, established, and, through naturalization became part of the regional vegetation. The dynamics of flora and vegetation have been studied and analyzed over a long period of time, normally dealing with the indigenous flora and their distribution. It was not until traffic and transport across the ocean became possible that many alien hardwoods arrived. Only since the middle of the 19th century, and even more since the beginning of this century, research focused on the introduced species (Trepl 1990). The first detailed contribution on introduced hardwoods was by Goeze (1916), who also discussed the problems of naturalization (Goeze 1917, see also Thellung 1918/19). The problems of alien weeds became more severe in the late fifties, with the enlargement of fallow lands in the northern hemisphere. As more neophytic species will probably invade our human-impacted landscape in the future, it is necessary to study and document trends at an early stage. Most

Plant Invasions: Ecological Mechanisms and Human Responses, pp. 285–297
edited by U. Starfinger, K. Edwards, I. Kowarik and M. Williamson
© *1998 Backhuys Publishers, Leiden, The Netherlands*

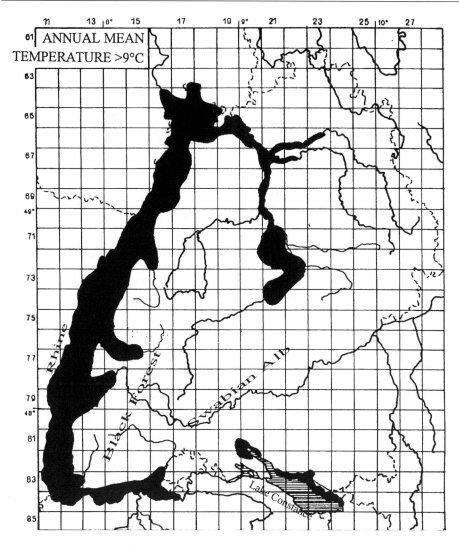

Fig. 1. Annual mean temperature > 9°C

aliens which have been introduced in our gardens and landscape are no problem at all; only a small number of them escaped cultivation and may negatively impact our semi-natural and man-made ecosystems. These new escapees, as well as genetically trans-formed species (Sukopp and Kowarik 1986), must be observed in a critical way and over long periods of time (Kowarik 1995b). We know very little about the invasion process because, up to now, no monitoring program exists by which the speed of inva-sion is precisely documented. In German floras spontaneous neophytic species have never been mentioned completely (Korneck 1986, Sebald *et al.* 1992-96), because it is often difficult to define (Schroeder 1969, 1974, Tutin 1964 ff, Sukopp 1962, Lohmeyer and Sukopp 1992, Sukopp 1995). Also, the status of a "weed" in this sense is still not clearly defined. There are more questions than skillfully documented answers: why are some aliens better adapted and why should they not be used more in man-made environments?

There are concerns about the effects of alien species on pristine habitats, but usually alien plants establish and become problematic only in human-disturbed habitats, such as the above mentioned and monoculture forests. About 160 woody spe-cies are growing naturally in Germany, in contrast to the big number which have been introduced during the last three centuries. Many urban and industrial sites are shelters for neophytic spe-cies (Kowarik 1990); they seem to adapt better to this stressful environment. However, still there is little information about site conditions within these areas and even less information about the ecophysiology, adaptation, genetic variability and the biology of woody aliens (Starfinger 1990).

Monitoring program of woody escapees in Baden-Württemberg

Field monitoring

As a result of observations of woody escapees in Baden-Württemberg since 1991, we devised a research program in order to monitor neophytic species.
     The field research started in 1991 and is still going on. The first steps were:
– Enumerating the species which show spontaneous seedlings in respective biotope types and the description of locations and their attributes.
– Marking of suitable permanent plots and continued vegetation survey, especially in protected areas where changes from neophytic species can be expected.
– Documentation of intensely spreading species like *Robinia pseudoacacia, Castanea sativa, Prunus serotina, Ailanthus altissima, Acer negundo, Buddleja davidii, Pseudotsuga menziesii* and others.
Preconditions for the spreading and distribution of woody species are:
– cultivation in man-made environments;
– high reproduction rate;
– high probability of successful establishment (favorable site conditions, niches);
– high distribution and dispersal ability of diaspores (wind, animals, men etc.).
Own field observations include the following:
– *Robinia pseudoacacia* seeds stay in their pods through wintertime and are then trans-ported over large distances by melting snow.
– *Castanea sativa* seeds will be carried by small mammals and crows and can estab-lish far from the mother plant.
– *Prunus serotina* seeds will be dispersed by birds and foxes over distances of kilo-meters.
– *Ailanthus altissima* seeds can be wind dispersed over 200 meters; changing wind direction and jet effects in urban areas may be a reason for its extensive spread.
– The winged fruits of *Acer* might be wind dispersed over far distances (some hun-dreds of meters) and are able to establish in a range of open habitats;
– *Pseudotsuga menziesii* was planted in large numbers as a forest tree; it has since escaped and established in many locations.
Ways of spreading are very different:
– diaspores germinate directly and grow up to small pieces of woodland;
– vegetative units thrown or washed away lead to durable thickets;
– rhizome colonies (*Robinia pseudoacacia, Cornus alba, Ailanthus altissima*) can spread out over large areas (up to 1 m/year, Kowarik 1995 a).

Site adaptation:
The range of habitats in which neophytic hardwood species are able to establish in their areas of secondary distribution is greater than in their native areas. For example *Acer negundo* occupies different areas than where it is found in North America. Due to its high fertility and fecundity levels, *A. negundo* has a great capacity to spread into urban, as well as rural, areas. Large colonies of *A. negundo* are found in elderswamps, riparian forests along the Rhine and Neckar rivers, and in fens; these habitats are nutrient-rich with high water levels. However, *A. negundo* has also established on debris piles, which are rich in calcium carbonate but dry, and in suburban areas, where it grows together with *Robinia pseudoacacia*. The natural stands of *Robinia* in the Appalachian mountains, Missouri and Arkansas differ from those above mentioned man made stands especially by their high pH-value (Harlow and Harrar 1969). Boxelder (*A. negundo*) has a very wide range in North America not appearing only in natural areas in the extreme east and west. It is found most commonly growing in deep, moist soil, but may also be found on poorer sites. *A. negundo* is perhaps the most aggressive of the maples in maintaining itself in unfavorable locations (Harlow and Harrar 1969). In Berlin, for example, *A. negundo* has established large colonies on very dry, nutrient-poor lawns, with coarse sandy dystric cambisol soils (Böcker 1978).

Alien woody species can more readily invade and establish in urban ecosystems than natural ecosystems because of increased disturbance. Kreh (1955, 1960) observed *Buddleja* growing on the ruins of Stuttgart in the 1950s, as an example. Greater levels of disturbance in urban areas removes the native species, which may act as competitors against the introduced species, and other interference factors, thereby resulting in non-equilibrial conditions (Gilbert 1989).

Germination and potential establishment experiments

Methods

A two factorial split-plot experiment was established during the 1995 growing season to test seed germination and the potential of various alien hardwood species to establish in various substrates. Seeds were sown in an experimental freeland wire house (so to prevent bird-predation) and as well in a greenhouse (the latter was excluded from data analysis).

The germination of hardwood seeds depends on exterior factors, like humidity, temperature, oxygen and light, and the seeds must be ready to germinate. Seeds of many woody species growing in south western Germany exhibit a period of seed dormancy, or germination may be inhibited by various reasons, including: 1) underdeveloped or dormant embryos; 2) the seed or embryonic membrane is impermeable to water and gases; 3) the presence of germination inhibiting substances in the pulp, seed coat, and endosperm; 4) a combination of different types of seed dormancy; and 5) secondary seed dormancy (Schubert 1992). Normally for arboriculture and forestry the germination is greatly promoted by stratification, (e.g. chilling, wet and cold conditions), preparation with H_2SO_4 or cooking water, scarification or soaking, to prepare the seeds according to their specific dormancy properties (Krüssmann 1985). For our experiments the seeds were not prepared in any way to simulate conditions without human interference (freeland conditions).

Tested species

The species to be tested have been selected according to the list of spontaneous woody species in Baden-Württemberg, which was generated by reviewing the literature of past studies of alien woody species by members of the Institute of Landscape and Plant Ecology (Adolphi 1995, Böcker *et al.* 1995, Böcker 1995, Böcker and Koltzenburg 1996, Drobik 1994, Hartmann *et al.* 1994, Merz 1994, Seybold 1969, Stolz 1994, Voigt 1993, Volkmann 1998) and from a herbarium of hardwood collected in south western Germany (Tab. 1).

The final species to be tested (Tab. 2) were chosen by the possibility to gather sufficient seed; most of them came from the Arboretum Hohenheim. Collection date varied, since the rate of seed maturation differs. Selection of the seeds according to criteria like habitat, maturation, condition of motherplant was omitted to obtain variability of seeding material.

Experimental plots

Field observations found a preference for germination on pebble and open substrates. The following substrates were chosen for the experiment: river pebbles, limestone debris, standard soil, quartzsand (in the tables called sand) and loess (the latter coarsely grounded to get a homogenous seed bed).

Each population was represented by 100 seeds distributed among the 5 substrates. Germination dishes were wrapped with absorbent felt, prepared with wicks and placed in trays to prevent of dry out. Then the dishes were filled with one of the selected substrates to a depth of 4 cm; the seeds were placed randomly.

A randomization scheme (Plabplan, Utz 1994 pers. comm., Hohenheim) with 100 seeds per species on five substrates was chosen for statistical treatments. There were four replicates for each substrate type, for a total of 20 dishes overall.

These were then moistened with water and placed outdoors in an experimental freeland wire house in October 1994. The dishes were watered only during the growing season (May-October), according to demand and weather conditions. The experiment lasted over two winters and growing seasons.

Emerging seedlings were counted weekly, using the appearance of the second cotyledon as the criterion. Seedlings were clipped as soon as the leaves opened and began to shade the substrate; this was done to keep conditions in the dishes as equal as possible.

Germination occurred regularly from March to November. In the greenhouse, various species germinated sporadically all the time (Fig. 2). The experiment ran from October 1994 to the end of the 1996 growing season.

To search for patterns in the germination data we used the Anova procedure of the Statistical Package Systat (1992). Two way analysis of variance tests, based on the split-plot design, were done to analyze the data. Species without any germination were omitted from subsequent analyses. In order to achieve normality, and because of several 0s in the data set (Sachs 1993) the data were transformed by taking $\sqrt{x+1}$. Tukeys comparison of means test was run if significant differences were found (Weber 1986). Analyses compared the germination rate between species, substrate types, and interaction between these variables, as well as analyzing for differences in the total number of germinated seedlings per substrate type.

Table 1. List of spontaneous shubs and trees in South Western Germany

Acer ginnala	Fraxinus pennsylvanica	Quercus rubra
Acer negundo	Gleditsia triacanthos	Rhus glabra
Acer platanoides	Genista anglica	Rhus typhina
Acer saccharinum	Hibiscus syriacus	Ribes alpinum
Aesculus hippocastanum	Hippophaë rhamnoides	Ribes aureum
Ailanthus altissima	Juglans nigra	Ribes sanguineum
Alnus incana	Juglans regia	Ribes uva-crispa
Amelanchier alnifolia	Kalmia angustifolia	Robinia pseudoacacia
Amelanchier lamarckii	Laburnum anagyroides	Rosa glauca
Amelanchier spicata	Laburnum alpinum	Rosa multiflora
Amorpha fruticosa	Larix decidua	Rosa rugosa
Ampelopsis hederacea	Liriodendron tulipifera	Rosmarinus officinalis
Aristolochia macrophylla	Ligustrum vulgare	Rubus allegheniensis
Berberis julianae	Lonicera caprifolium	Rubus armeniacus
Berberis thunbergii	Lonicera mackii	Rubus fabrimonatus
Berberis vulgaris	Lonicera tatarica	Rubus laciniatus
Buddleja davidii	Lycium barbarum	Rubus odoratus
Caragana arborescens	Lycium chinense	Rubus sciocharis
Castanea sativa	Mahonia aquifolium	Salix daphnoides
Catalpa bignonioides	Mespilus germanicus	Salix elaeagnos
Catalpa ovata	Morus alba	Salix × erdingeri
Celastrus orbiculatus	Morus nigra	Salix × mollissima
Celtis occidentalis	Ostria carpinifolia	Sorbaria sorbifolia
Cercidiphyllum japonicum	Parthenocissus inserta	Sorbus aria
Chamaecyparis lawsoniana	Philadelphus coronarius	Sorbus domestica
Chamaecytisus supinus	Physocarpus opulifolius	Sorbus intermedia
Chaenomeles speciosa	Picea abies	Spiraea alba
Clematis vitalba	Pinus nigra	Spiraea bumalda
Colutea arborescens	Pinus strobus	Spiraea chamaedryfolia
Cornus alba	Platanus hybrida	Spiraea douglasii
Cornus mas	Platanus occidentalis	Spiraea flexuosa
Cornus sericea	Platanus orientalis	Spiraea media
Corylus colurna	Populus alba	Spiraea salicifolia
Corylus maxima	Populus × canadensis	Spiraea × vanhouttei
Cotinus coggygria	Populus × canescens	Staphylea pinnata
Cotoneaster bullatus	Potentilla fruticosa	Symphoricarpus × chenaultii
Cotoneaster dielsianus	Prunus armeniaca	Symphoricarpus albus
Cotoneaster horizontalis	Prunus cerasus	Syringa persica
Cotoneaster salicifolius	Prunus domestica	Syringia vulgaris
Cotoneaster div. spec.	Prunus fruticosa	Taxus baccata
Crataegus crus-galli	Prunus laurocerasus	Thuja occidentalis
Cryptomeria japonica	Prunus mahaleb	Thuja plicata
Cydonia oblonga	Prunus persica	Tsuga canadensis
Cytisus multiflorus	Prunus serotina	Tilia tomentosa
Deutzia gracilis	Prunus virginiana	Ulex europaeus subsp. euro.
Elaeagnus angustifolia	Pseudotsuga menziesii	Vaccinium macrocarpon
Euonymus latifolia	Ptelea trifoliata	Vinca major
Fallopia baldschuanica	Pterocaria fraxinifolia	Vinca minor
Ficus carica	Pyracantha coccinea	Vitis vulpina
Forsythia intermedia	Pyrus communis	Vitis vinifera
Fraxinus americana	Quercus cerris	
Fraxinus ornus	Quercus palustris	

Table 2. List of tested hardwood species

Acer ginnala Maxim.	Koelreuteria paniculata Laxm.
Ailanthus altissima (Mill.)Swing-le	Kolkwitzia amabilis Graebn.
Berberis julianae Schneid.	Laburnum anagyroides Medik.
Berberis thunbergii DC.	Philadelphus coronarius L.
Betula maximovicziana Regel	Physocarpus opulifolius (L.)Maxim.
Betula papyrifera Marsh.	Platanus x hybrida Brot.
Buddleja alternifolia Maxim.	Ptelea trifoliata L.
Celtis occidentalis L.	Pterocarya fraxinifolia (Lam.)Spach
Cercis siliquastrum L.	Quercus cerris L.
Chaenomeles speciosa (R.Sweet)T.Nakai	Rhus glabra L.
Colutea arborescens L.	Rhus typhina L.
Cornus alba L.	Robinia pseudoacacia L.
Cornus sericea (J.M.Coult. et Evans)Fosb.	Viburnum rhytidophyllum Hemsl.

Results and discussion

The final germination percentage of the species ranged from 0 to 51% (Fig. 3).

There were significant between-substrate differences in the total number of seed-lings which germinated in each substrate type. More seedlings germinated in seed soil and limestone debris compared to the other substrates (Table 3a).

Variations were found among the species within each substrate type. Six species had germination rates greater than 20%, while seven did not germinate at all (Table 4).

There was also a significant species-substrate interaction (Table 3b-f).

Germination rate on standard soil is significantly higher than on loess.

Germination rate on limestone debris is significantly higher than on standard soil, sand and loess, while germination is significantly greater on pebbles than on loess.

Germination rate on limestone debris is significantly higher than on other substrates.

Germination rate on standard soil is significantly higher than on sand and loess.

Germination rate on limestone debris is significantly higher than on other substrates.

Summary: The germination rate on limestone debris is significantly higher than on other substrates. As concerns significant species-substrate interaction *Koelreuteria paniculata* and *Ptelea trifoliata* germinated significantly on standard soil, *Viburnum rhytidophyllum* and *Colutea arborescens* on limestone debris and *Kolkwitzia amabilis* on limestone debris and pebbles.

Conclusions

These seedling experiments were performed only to support the field observations which brought evidence of neophytic hardwood species in many parts of the country. The principle aim of our study was to determine to what extent a spontaneous occurrence of species is to be expected. Woody species were selected according to their occa-sional spontaneous occurrence and to their frequent use as ornamental plants. By our experiments, it was shown that only few species are germinating with high rates and only on specialized substrates.

Spontaneous woods may contribute to the amelioration of quality in the urban envi-ronment. Many of the species mentioned can cope with the urban conditions far better than indigenous species (e.g. regarding heat resistance, smog resistance, ruderal strat-

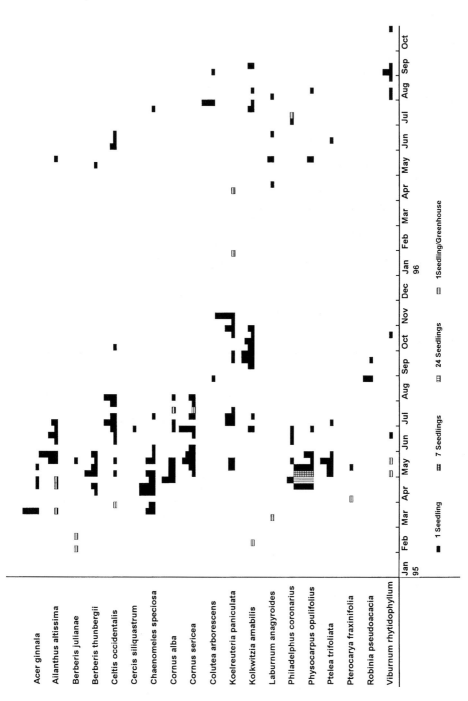

Fig. 2. Germination during seed experiment

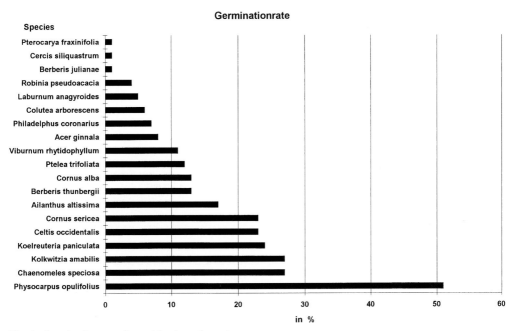

Fig. 3. Germinationrate of tested hardwood species

Table 3a. All species

df	f	p
4	8.298	0.000

	standard soil	limestone debris	pebbles	sand	loess
standardsoil	–	–	–	–	–
l. debris	0.993	–	–	–	–
pebbles	0.267	0.519	–	–	–
sand	0.002 **	0.008 **	0.404	–	–
loess	0.000 ***	0.000 ***	0.055	0.880	–

*   = <   5% /   0.05;   **   = <   1% /   0.01;   *** = < 0.1% / 0.001

Table 3b. Koelreuteria paniculata

df	f	p
4	3.694	0.028

	standard soil	limestone debris	pebbles	sand	loess
standardsoil	–	–	–	–	–
l. debris	0.332	–	–	–	–
pebbles	0.222	0.999	–	–	–
sand	0.136	0.977	0.998	–	–
loess	0.014 *	0.406	0.559	0.733	–

*   = <   5% /   0.05;   **   = <   1% /   0.01;   *** = < 0.1% / 0.001

Table 3c. Kolkwitzia amabilis

df	f	p
4	6.897	0.002

	standard soil	limestone debris	pebbles	sand	loess
standardsoil	–	–	–	–	–
l. debris	0.017*	–	–	–	–
pebbles	0.087	0.904	–	–	–
sand	1.000	0.025*	0.125	–	–
loess	0.988	0.007**	0.036*	0.956	–

* = < 5% / 0.05; ** = < 1% / 0.01; *** = < 0.1% / 0.001

Table 3d. Colutea arborescens

df	f	p
4	12.000	0.000

	standard soil	limestone debris	pebbles	sand	loess
standardsoil	–	–	–	–	–
l. debris	0.001 **	–	–	–	–
pebbles	1.000	0.001 **	–	–	–
sand	0.095	0.095	0.095	–	–
loess	1.000	0.001 **	1.000	0.095	–

* = < 5% / 0.05; ** = < 1% / 0.01; *** = < 0.1% / 0.001

Table 3e. Ptelea trifoliata

df	f	p
4	3.508	0.033

	standard soil	limestone debris	pebbles	sand	loess
standardsoil	–	–	–	–	–
l. debris	0.453	–	–	–	–
pebbles	0.258	0.993	–	–	–
sand	0.040 *	0.582	0.814	–	–
loess	0.040 *	0.582	0.814	1.000	–

* = < 5% / 0.05; ** = < 1% / 0.01; *** = < 0.1% / 0.001

Table 3f. Viburnum rhytidophyllum

df	f	p
4	3.508	0.033

	standard soil	limestone debris	pebbles	sand	loess
standardsoil	–	–	–	–	–
l. debris	0.002 **	–	–	–	–
pebbles	0.973	0.007 **	–	–	–
sand	0.933	0.001 **	0.651	–	–
loess	0.933	0.001 **	0.651	1.000	–

* = < 5% / 0.05; ** = < 1% / 0.01; *** = < 0.1% / 0.001

Table 4. Germinationrate in three classes

		Stand.soil	L. debris	Pebbles	Sand	Loess	
1.)							
	Phys opu	15	13	9	11	3	51
	Chae spe	7	4	3	10	3	27
	Kolk ama	2	12	9	3	1	27
	Koel pan	10	5	4	4	1	24
	Celt occ	7	5	5	3	3	23
	Corn ser	9	7	3	1	3	23
2.)							
	Aila alt	2	7	6	0	2	17
	Berb thu	5	4	4	0	0	13
	Corn alb	6	2	3	0	2	13
	Ptel tri	7	3	2	0	0	12
	Vibu rhy	1	8	2	0	0	11
	Acer gin	5	1	0	2	0	8
	Phil cor	3	2	2	0	0	7
	Colu arb	0	4	0	2	0	6
	Labu ana	0	1	1	0	3	5
	Robi pse	3	0	1	0	0	4
	Berb jul	0	0	1	0	0	1
	Cerc sil	1	0	0	0	0	1
	Pter fra	0	0	0	0	1	1
3.)							
	Betu max	0	0	0	0	0	0
	Betu pap	0	0	0	0	0	0
	Budd alt	0	0	0	0	0	0
	Plat hyb	0	0	0	0	0	0
	Quer cer	0	0	0	0	0	0
	Rhus gla	0	0	0	0	0	0
	Rhus thy	0	0	0	0	0	0
	Total %	83	78	55	36	22	274

egy, regeneration). There are not many hardwood species which may be hazardous in the landscape.

Acknowledgments

A study on the neophytic species was sponsored by Geschwister-Stauder-Stiftung in Baden-Württemberg, we wish to express our thank.

We also thank well K.P. Buttler, W. Konold, I. Kowarik, W. Kunick, H. Sukopp, and M. Koltzenburg for their critical comments.

References

Adolphi, K. 1995. Neophytische Kultur- und Anbaupflanzen als Kulturflüchtlinge des Rheinlandes. NARDUS, Wiehl.

Böcker, R. 1978. Gehölzaufwuchs einer Dauerfläche auf dem Windmühlenberg Berlin Gatow. Phytocoenosis 7 (1-4): 61-70.

Böcker, R. 1995. Beispiele der Robinienausbreitung in Baden-Württemberg. In: Böcker, R., Gebhardt, H., Konold, W. and Schmidt-Fischer, S. (eds.) Gebietsfremde Pflanzenarten, pp. 57-65. Ecomed, Landsberg.

Böcker, R., Gebhardt, H., Konold, W. and Schmidt-Fischer, S. (eds.) 1995. Gebietsfremde Pflanzenarten, Auswirkungen auf einheimische Arten, Lebensgemeinschaften und Biotope, Ecomed, Landsberg.

Böcker, R. and Koltzenburg, M. 1996. Pappeln an Fließgewässern in Baden-Württemberg. In: Landesanstalt für Umwelt (ed.) Handbuch Wasser 2. pp. 137.

Drobik, S. 1994. Verbreitung und Struktur von Robinienbeständen am Südwestrand des Schönbuchs. Diplomarbeit Univ. Hohenheim (unpublished).

Gilbert, O. 1989. The ecology of urban habitats. Chapman and Hall, London.

Goeze, E. 1916. Liste der seit dem 16. Jahrhundert bis in die Gegenwart in die Gärten und Parks Europas eingeführten Bäume und Sträucher. Mitt. Deutsch. Dendr. Ges. 25: 129-201.

Goeze, E. 1917. Kultur, Naturalisation, Ausartung. Mitt. Deutsch. Dendr. Ges. 26: 169-188.

Harlow, W.M. and Harrar, E. 1969. Textbook of dendrology. 5th ed. McGraw-Hill, New York.

Hartmann, L., Schuldes, H., Kübler, R. and Konold, W. 1994. Neophyten – Biologie, Verbreitung und Kontrolle ausgewählter Arten. Ecomed, Landsberg.

Kohler, A. 1964. Das Auftreten und die Bekämpfung der Robinie in Naturschutzgebieten. Veröff. Landesst. f. Natursch. u. Landerspfl. Bad.-Württ. 32: 43-46.

Korneck, D. 1986. Zur Problematik der Aufnahme von Neophyten in Rote Listen gefährdeter Pflanzenarten. Schr.R. Vegetationskde. 18: 115-119.

Kowarik, I. 1990. Some responses of flora and vegetation to urbanization in central europe. In: Sukopp, H., Hejny, S. and Kowarik, I. (eds.): Urban ecology. pp. 75-97. SPB Academic Publishing, Den Haag.

Kowarik, I. 1995a. Ausbreitung nichteinheimischer Gehölzarten als Problem des Naturschutzes. In: Böcker, R., Gebhardt, H., Konold, W. and Schmidt-Fischer, S. (eds.) Gebietsfremde Pflanzenarten. pp. 33-56. Ecomed, Landsberg.

Kowarik, I. 1995b. Time lags in biological invasions with regard to the success and failure of alien species. In: Pyšek, P., Prach, K., Rejmánek, M. and Wade, P.M. (eds.): Plant invasions – general aspects and applications. pp. 15-38. SPB Academic Publishing, The Hague.

Kreh, W., 1955. Das Ergebnis der Vegetationsentwicklung auf dem Stuttgarter Trümmerschutt. Mitt. Flor.-soz. Arbeitsgem. N.F.5-6: 90-95.

Kreh, W., 1960. Die Pflanzenwelt des Güterbahnhofs in ihrer Abhängigkeit von Technik und Verkehr. Mitt. Flor.-soz. Arbeitsgem. N.F.8: 86-109.

Krüssmann, G., 1985. Manual of cultivated broad-leaved trees and shrubs. Vol. 2, E-PRO OR, Timber Press, Portland.

Lohmeyer, W. and Sukopp, H. 1992. Agriophyten in der Vegetation Mitteleuropas. Schr.R. Vegetationskunde 25: 1-185.

Lohmeyer, W. 1972. Einwanderung von Neubürgern in die einheimische Flora und Probleme der Wiedereinbürgerung bodenständiger Gehölze. Schr.R. Landschaftspfl. u. Natursch. 7: 87-90.

Merz, O. 1994. Neophytische Gehölze in der nördlichen Oberrheinniederung – Beiträge zur Verbreitung und Biologie. Diplomarbeit Univ.Hohenheim (unpublished)

Sachs, L. 1993. Statistische Methoden. Springer Verlag, Berlin.

Schroeder, F.-G. 1969. Zur Klassifizierung der Anthropochoren. Vegetatio 16: 225-238.

Schroeder, F.-G. 1974. Zu den Statusangaben bei der floristischen Kartierung Mitteleuropas. Göttinger Flor. Rundbr. 8 (3): 71-79.

Schubert, J. 1992. Samenphysiologie - Keimung. In: Lyr, H., Fiedler, H.J. and Tranquillini, W. (eds.) Physiologie und Ökologie der Gehölze. pp. 319-337. Fischer, Stuttgart.

Sebald, S., Seybold, S., Philippi, G. and Wörz, A. (eds.) 1990-1996. Die Farn- und Blütenpflanzen Baden-Württembergs. Vol. 1-6. Ulmer, Stuttgart.

Seybold, S. 1969. Flora von Stuttgart. Ulmer, Stuttgart.

Starfinger, U. 1990. Die Einbürgerung der Spätblühenden Traubenkirsche (*Prunus serotina* Ehrh.) in Mitteleuropa. Landschaftsentw. u. Umweltforsch. 69: 1-136.

Stolz, T. 1993. Untersuchungen zur Vegetationsentwicklung auf Weinbergsbrachen an einem Muschelkalkhang bei Rottenburg. Diplomarbeit Univ. Hohenheim (unpublished).

Sukopp, H. 1962. Neophyten in natürlichen Pflanzngesellschaften Mitteleuropas. Ber. Deutsch. Bot. Ges. 75: 193-205.

Sukopp, H. 1995. Neophytie und Neophythismus. In: Böcker, R., Gebhardt, H., Konold, W. and Schmidt-

Fischer, S. (eds.): Gebietsfremde Pflanzenarten. pp. 3-32. Ecomed, Landsberg.

Sukopp, H. and Kowarik, I. 1986. Berücksichtigung von Neophyten in Roten Listen gefährdeter Pflanzenarten. Schr.R. Vegetationskunde 18:105-114.

Thellung, A. 1918/19. Zur Terminologie der Adventiv- und Ruderalfloristik. Allg. Bot. Zeitschr. 24/25: 9-12/36-42.

Trautmann, W. 1976. Veränderungen der Gehölzflora und Waldvegetation in jüngster Zeit. Schr.R. Vegetationskunde 10:91-108.

Trepl, L. 1990. Research on the anthropogenic migration of plants and naturalisation. Its history and current state of development. In: Sukopp, H., Hejny, S. and Kowarik, I. (eds.) Urban ecology. pp. 75-97. SPB Academic Publishing, Den Haag.

Tutin, T.G., Heywood, V.H., Burges, N.A., Valentine, D.H., Walters, S.M. and Webb, D.A. 1964 ff. Flora Europaea. Vol. 1-5. Cambridge.

Voigt, K. 1993. Vegetationskundliche Untersuchungen in Beständen von Robinia pseudoacacia L. am Spitzberg bei Tübingen. Diplomarbeit Univ.Hohenheim (unpublished).

Volkmann, M., 1998. Neophytische Gehölze in der südlichen Oberrheinniederung. Diplomarbeit Univ. Hohenheim (unpublished).

Weber, E. 1986. Grundriß der biologischen Statistik – Anwendungen der mathematischen Statistik in Forschung, Lehre und Praxis. 9th ed., pp. 652. Fischer, Jena.

EXAMPLES OF INVASION BY THREATENED NATIVE SPECIES IN ANTHROPOGENOUS ECOTOPES

Raisa I. Burda

Department of Natural Flora, Donetsk Botanical Gardens, Illich's Avenue,110, Donetsk, 340059, Ukraine; e-mail bot@hort.uvica.donetsk.ua

Abstract

Three cases of spontaneous spread into synanthropic ecotopes by species categorised as rare or protected, *Gypsophila paulii* Klok. (Caryophyllaceae), *Cephalanthera longifolia* (L.) Fritsch, *Platanthera bifolia* (L.) Rich. (Orchidaceae) are considered. The expansion of these species is a result of the cumulative effect of the accidental transfer of diaspores, the accidental and approximate similarity of synanthropic ecotopes to the natural habitats of the species, the spontaneous spread of all species as determined by their biological and ecological features. Modern perturbations in plant cover, from an interaction of natural and anthropogenous factors, are a natural response of phytobiota to synanthropization of the environment.

Introduction

The history of invasion has two aspects: drift, the naturalization and expansion of species from other regions, and the spread of local species into new territories, related to the disturbance of land by human activity (LeFloc'h *et al.* 1990). The spread of native species into anthropogenous ecotopes (not typical of them) occurs in two ways: by conversion from a stenotopic to an eurytopic ruderal species or by conservation, the survival of stenotopic species in anthropogenous ecotopes analogous to its earlier disturbed natural ecotopes. Species of the first group, spreading beyond the limits of their natural ecological and geographic area, may become invasive and threaten the ecosystems, species or environment which surrounds them. In disrupted ecological situations, species of the second group compensate for the detriment caused to nature. So different types of invasion may require different management measures. The most interesting cases are, when populations of native species, classified as vulnerable for natural or anthropogenous reasons, manage to invade. Consider three oligotopic native species: *Gypsophila paulii* Klok. (Caryophyllaceae) and *Cephalanthera longifolia* (L.) Fritsch and *Platanthera bifolia* (L.) Rich. (both Orchidaceae).

I will compare the known geographic range and ecological confinement of these species, on which the assessments of a threat to their existence was grounded, to the latest facts of their geographic and ecological ranges' extension into anthropogenous ecotopes. Both species of orchids are listed in the Red Data Book of Ukraine (1996) in the category threatened. *G. paulii* is a narrow oligotopic endemic, its natural habitats are much reduced by recreational development of the Black-and-Azov Seas coasts. Though it is not officially listed in the Red Data Book, its survival exites apprehension.

Plant Invasions: Ecological Mechanisms and Human Responses, pp. 299–306
edited by U. Starfinger, K. Edwards, I. Kowarik and M. Williamson
© *1998 Backhuys Publishers, Leiden, The Netherlands*

Results

Gypsophila paulii Klok. was described by Klokov (1948, 1981) as an aberrative de-
rivative race of *G.trichotoma* Wend., growing on the saline near-littoral sands of the
Black-Azov coast, from Ochakov to Mariupol. It has no agreed taxonomic status.

In *Flora Europaea* Barkoudah and Chater (1964) regard it as part of *G.perfoliata* L.
Tzarenko (1990) treats glabrous forms as *G. perfoliata* L. var. *glabra* (Fenzl.) Tzarenko.
However, in the south-east of Ukraine *G. paulii* is well differentiated. It differs from
G. perfoliata by having completely glabrous stems and leaves. This species is com-
mon on saline sandy meadows of the sandy-shelled sea terrace from Krivaya to
Lyapinskaya spit; a bit further to the west it is replaced by *G. perfoliata. G. paulii*
belongs to the Seria *Trichotomae* Schischk., Sectio *Eugypsophila* Boiss. (*Rokejeka*
(Forsl.) A.Braun), of which one more native species *G. perfoliata* and an alien *G.
scorzonerifolia* Ser. grow in Ukraine. They are herbaceous perennials with long prostrate or ascending, seldom erect stems, their leaves are ovate-oblong and big. Plants
usually have numerous flowers and scarious bracts; the petals are white, pink or purple.

Initially all three species of the *Trichotomae* sere belonged only to the halophil-
ous-psammophilous littoral floristic complex, but in recent years they have spontaneously spread to industrial ecotopes (dumps of coal mines, colliery spoil heaps, fire-
clay dumps, marl pits, cinnabar pits, industrial plots at steelworks, ash dumps, other
industrial waste areas, railways and motorways), becoming a part of synanthropophyton.
The secondary synanthropic range of *G. paulii* has long exceeded its natural range
and extends almost to all the steppe and forest-steppe of Ukraine (Fig 1). Populations
are especially numerous in the south-east, within the Donetsk coal basin where there
is much disturbance by industry (Burda et al. 1997*a*).

The expansion of the *Trichotomae* sere's species took place in the last twenty or
thirty years. One of the best experts on Ukraine's alien species, Kotov (1960), who
discovered *G. scorzonerifolia* for the first time in Ukraine on wet saline sands near the
salty lakes in the town of Slavyansk of the Donetsk Region, did not regard it as alien.
Since the plants were collected by him under conditions similar to those in which the
species grows within the limits of the natural area near the Caspian sea, Kotov regarded
it as a disjunctive pliocene relict. Ten years later Dubovik (1970) reported finding
G.scorzonerifolia near the town of Bryanka in the Lugansk Region. By that time the
species had been recorded as alien in Chechia and Rumania.

What are the prerequisites of such a rapid expansion of *G. paulii,* and both other
species of *Trichotomae* sere, into synanthropic ecotopes? The biological features of
these plants are striking: a distinctive life-form, "a rolling stone", which allows seed
dispersal by wind rolling the plants a great distance; the ability to form a great number
of seeds which are easily released; phenological peculiarities: early growth of the leaf
crown (the end of March – beginning of April), fast development of flowering shoots
(the middle-end of May), flowering over a long period (June-September) and a long
period of ripening of successive seeds (July-October); tolerance of the humidity and
fertility of industial ecotopes. In the synanthropic ecotopes of the Ukrainian south-east
G. paulii is entomophilous, with cross-pollination, from nonsimultaneous ripening of
flowers and protandry; seed productivity reaches 2421 ± 222 seeds per specimen, the
seminification coefficient (the proportion of ovules developing into seeds) is 92 per
cent. In 1994-1995 populations on colliery spoil heaps of the city of Donetsk differed
from natural populations on the Krivaya spit of the Azov coast (Table. 1). In synan-

A

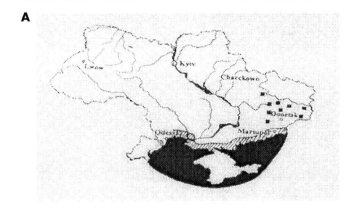

B

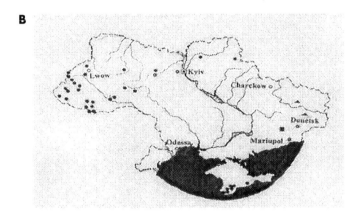

C

Fig. 1. The maps of spread of three native species in Ukraine. Symbols indicate: A – *Gypsophila paulii,* B – *Cephalanthera longifolia,* C – *Platanthera bifolia*; 1,2 – native range in the Red Data Book, 3 – native locations, which were found later, 4 – invasion into anthropogenous locations.

Table 1. The variation of morphological features in native and invasive populations of *Gypsophila paulii.* The native population is from the saline littoral sands of the Krivaya spit in the Azov sea. The invasive population is from of one of the coal mine dumps in Donetsk

Feature	Protologus (Klokov 1948)	Native population			Invasive population		
Plant height, cm	60-120	61±2.6	12.3	20	64.5±5.8	18.3	28
Number of shoots	-	18.8±1.9	8.9	47	53.2±11.6	36.8	69
Number of internodes	-	24.8±0.3	3.3	13	32.2±0.6	3.3	10
Internode length, cm	-	44.7±0.7	7.9	18	27.3±1.5	7.8	29
Leaf length, mm	60-80	78.6±1.2	20.1	26	47.8±1.7	11.8	25
Leaf width, mm	20-25	35.7±0.7	12.2	34	20.7±1.0	6.1	30
Peduncle length, mm	27	6.4±0.3	1.9	29	5.3±0.3	1.9	35
Pod height, mm	4	4.6±0.1	0.7	15	4.1±0.1	0.5	12
Pod width, mm	3-3.5	3.3±0.1	0.5	15	3.4±0.1	0.3	10
Seed length, mm	1	1.2±0.1	0.1	9	1.1±0.1	0.1	12
Seed width, mm	1	1.6±0.1	0.1	6	1.5±0.1	0.1	9

thropic populations plants of *G. paulii* resemble a ball in shape; they are rounded and dense.

The expansion of *Gypsophila* became possible due to formation of industrial wastes of a new type which have highly mineralized soils. The third important prerequisite was that sand was stocked for different industrial purposes (for building, steel works, etc.) from the natural habitats of *G. paulii.*

The Orchidaceae is one of the most polymorphic and vulnerable families. There is an impression that in spite of a wide spectrum of morphological, ecological and geographic variability Orchidaceae do not stand strong anthropogenous pressure and are not able to be a part of a synanthropophyton. Yet Orchidaceae invasions from the temperate zone of Eurasia into synanthropic ecotopes have been reported. Sauerland (1995) described an *Epipactis helleborine* (L.) Crantz population in oak-poplar plantations in Rostock (Germany) which were planted in 1961. In 1994 he found 222 specimens of this orchid, in 1995, 764. Adamowski (1995*b*; Adamowski and Conti 1991) studied the phenotypic variation of some species of *Epipactis* under anthropogenic conditions. Later, Adamowski (1995*a*) wrote about a century old invasion of *E. helleborine* in broad-leaved forests in North America. In spite of the wide amplitude of habitats of this alien species, its spread under new conditions is limited to an area of beech-maple woods. Biological features of the species both speed up and restrict the species expansion under new conditions.

Cephalanthera longifolia (L.) Fritsch is a palearctic species of light broad-leaved forests. In our country it is limited to forest and the forest-steppe parts of the flat Ukraine, the Carpathians and the Crimea (Bordzilowsky 1950; Smolyaninova 1976; Zaweruka et al. 1983; Chervona knyga 1996). It was also found in the bottomland of the Seversky Donets at the mouth of the river Krasnaya (Kondratyuk, Burda 1980). Later it has been found much further south – in the Donetsk mountain-ridge near the town of Chasov Yar in the ravine oak-grove of at Stupki (Burda *et al.* 1997*b*). The southern populations are impoverished, scanty. Not more than 20 plants of various ages have been found in Stupki (Fig 1, Table 2). In Ukraine it is a scioheliophyte, mesophyte, preferring limestone soils with a well-developed humus layer, indifferent to soil moisture though avoiding over wet soils. In nature it rarely reproduces by seed (Sobko 1989).

Table 2. The age states of plants in some native and invasive populations. Symbols indicate: v – prereproductive ontogenetic period and virginile age state; g – reproductive period, g_1 – young, g_2 – mature and g_3 – old (Gatsuk et al. 1980).

Species	Location	The age states of plants, %			
		v	g_1	g_2	g_3
Cephalanthera longifolia	Stupki	30	25	30	15
	Velikoanadol forest	22	20	31	27
Platanthera bifolia	Polesje – native range				
	(Sobko 1989)	66	-	34	-
	Makatika gorge	58	17	18	7
	Hornbeam gorge	70	7	15	8

C. longifolia in the Velikoanadolsky forest reserve is certainly alien. The forest of Velikoanadol is a completely artificial plantation, a classical example of steppe forest cultivation in the place of herb-fescue-needle-grass steppes. It is on the watershed between the rivers of the Dnieper system and the Azov sea, within the Azov Hills in the south-east of Ukraine (the Donetsk Region near the town of Volnovacha). According to the forest-planting records, *C. longifolia* has never been introduced in the Velikoanadol forest intentionally. It was accidentally brought here with *Quercus robur* acorns from the north-western regions of Ukraine during mass planting of forest in steppe at the end of 50s in this century. Where acorns are sown the orchid occurs sporadically. The distance of the Velikoanadol forest from the acorn's place of origin is some hundred kilometres. Stupki, which was mentioned above, is about 150 km away, but it is impossible to sow acorns there. And *C. longifolia*, from what is known of its ecology, could not disperse across the steppe by itself. *C. longifolia* grows in pure plantations of *Quercus robur* L. with crown density 0.8, 20 m high with an admixture of *Fraxinus excelsior* L. (to 0.4) and *Acer platanoides* (to 0.4). In the shrub layer there are *Acer campestre* L., *A. tataricum* L., *Euonymus europaea* L., *E.verrucosa* Scop., *Caragana arborescens* Lam.. The total field layer is up to 30 per cent and consists of *Geum urbanum* L., *Melica transsilvanica* Schur., *Torilis japonica* (Houtt.) DC., *Physalis alkekengi* L., *Campanula bononiensis* L. and *Galium aparine* L. This spontaneous population of *C. longifolia* is up to 55 thousand specimens of different ages; the density is 1-2 specimens per square metre. The population age spectrum is nearly normal (Table 2).

The prerequisites for emergence, naturalization and successful development of the population of the protected species *C. longifolia* in the artificial forest plantations, far from the native area, are both anthropogenous (unintentional spread of diaspores with seeds of the main forest tree *Quercus robur*, conditions of growth more or less like those in its native range) and biological (a rather wide ecological amplitude, many years developing from a seed to a fruiting specimen, an ability to gradually amass a number of individuals which expand in due course).

Platanthera bifolia (L.) Rich. is a Eurasian species of broad-leaved forests. In Ukraine it is sporadic in the Carpathians, woodlands, forest-steppes, rare in steppes on secondary river terraces. A sciophyte, it does not require much moisture and has a rather wide ecological amplitude. It colonizes sunlit plots very well, although it can bear shade for a long time (Bordzilowsky 1950; Smolyaninova 1976; Zaweruka *et al.* 1983; Chervona knyga 1996). It is not in the Red Data Book of Ukraine for the south-east of Ukraine.

In our data there are no less than 10 isolated locations. It grows in bottomland oak-groves and alder thickets on the sandy terrace of the Seversky Donets and in the ravine oak-groves of the Donetsk mountain ridge (Fig 1). Along the sand depressions of the terrace above the bottomland it grows in associations of *Alnus glutinosa – Betula pendula – Viburnum opulus – Convallaria majalis, Betula pendula – Poa nemoralis, Betula pendula – Swida sanquinea – Glechoma hederacea.* Small (up to 30 individuals) but established populations are situated as a rule on the western slopes of saucerlike depressions, in the centre of which there is a small lake or a little bog, sometimes with *Sphagnum* sp. *Betula pubescens* Ehrh., *Quercus robur, Populus tremula* L., *Frangula alnus* Mill., *Pyrus communis* L., often occur in the tree layer and in the herb layer there are *Galium articulatum* Lam., *Sanquisorba officinalis* L., *Polygonathum multiflorum* (L.) All., *P. odoratum* (Mill.) Druce, *Thalictrum simplex* L., *Filipendula vulgaris* Moench, *Betonica peraucta* Klok., *Prunella vulgaris* L., *Gladiolus apterus* Klok., *Adoxa moschatelina* L., *Ophioglossum vulgatum* L. On the slopes of the ravine woods of the Donetsk mountain ridge *Platanthera bifolia* occurs from time to time in associations of *Quercus robur – Corylus avellana – Aegopodium podagraria* or in *Fraxinus excelsior – Quercus robur – Acer tataricum – Aegonichon purpureo-coerulea.* Under these conditions populations are not large, prereproductive plants predominate (Table 2).

P. bifolia's vigour is high in the south-east of Ukraine; plants are often taller than 50 cm (up to 67 cm) and their leaves are 19 cm long and 5.5 cm wide. Such vigorous populations are observed, for example, in the sunlit oak-maple plantations, 45-50 years old, in Makatika gorge, to the north of the town of Slavyansk in the Donetsk Region. This population exceeds 1000 individual of various ages. A sparser population of *P. bifolia* (to 250-280 specimens) was discovered by us in "Hornbeam gorge": (the extreme eastern point of the distributionof *Carpinus betulus* L in Europe) in plantations of *Acer platanoides* 30-35 years old.

Invasion of synanthropic ecotopes and thesurvival of the protected species *P. bifolia* in them comes from the biological properties and ecological needs of the species and by the correspondence of artificial forest plantations (in the place of former ravine woods in the steppe) to these requirements. The spread by seed was apparently spontaneous.

Conclusions

In the expansion of some local species, which are rare or even threatened, to anthropogenous ecotopes there is a cummulative effect of the accidental transfer of diaspores, and the chance resemblance of synanthropic ecotopes to natural habitats of the species. As a result of the interaction of natural and anthropogenous factors, startling changes take place in the plant cover. The Black Sea and Azov Sea endemic of a psammophilous-saline complex *Gypsophila paulii* appears hundreds and thousands of kilometres away from its natural area, forming pure stands in pure industrial habitats. A forest species, listed in the Red Data Book of Ukraine, *Cephalanthera longifolia,* forms an isolated population 1000 kilometres from its natural area in the artificial forest of Velikoanadol; this alien population is so abundant that in June in the 5-6 years since its discovery it makes the forest floor white. A forest species also in the Red Data Book, *Platanthera bifolia,* forms more tolerant, numerous and mixed age populations in artificial forest plantations in the south-eastern part of Ukraine than in natural habitats of its distribution in the flat Ukraine.

Are these cases to be regarded as natural ways of renewing a disappearing species or as a "floristic falsification" causing a further anthropogenous transformation of ecosystems?

Acknowledgments

The author thanks the Chairman the of Botanical Society of Berlin and Brandenburg (Botanischer Verein für Berlin und Brandenburg) professor em. Dr. Herbert Sukopp for the invitation to contribute to this volume and Dr. Keith Edwards, Prof. Dr. Ingo Kowarik, Dr. Uwe Starfinger and Prof. Mark Williamson for useful pieces of advice in preparation of this paper. The author is also grateful to the German National Science Fund (Deutsche Forschungsgemeinschaft) for a travel grant to the 4th International Conference on the Ecology of Invasive Alien Plants, Berlin.

References

Adamowski, W. 1995*a*. Amerykanska kariera europejskiego storczyka. Wiad. bot. 39, 1-2:105-113.

Adamowski, W. 1995*b*. Phenotypic variation of *Epipactis helleborine* x *E.atrorubens* hybrids in anthropogenic conditions. Acta soc. botan. Pol. 64, 3: 303-312.

Adamowski, W. and Conti, F. 1991. Masowe wystepowanie storczykow na plantacjach topolowych pod czeremcha jako przyklad apofityzmu. Phytocoenosis. 3: 259-267.

Barkoudah, Y.I. and Chater, A.O. 1964. Gypsophila L. In: Tutin, T.G., Heywood, V.H., Burges, N.A., Valentine, D.H., Walters, S.M. and Webb, D.A. (eds.) Flora Europaea. V. 1, pp. 181-184. Cambridge University Press, Cambridge.

Bordzilowsky,E.I.1950. Orchidaceae Lindl. In: Kotov, M.I. and Barbarych, A.I. (eds.) Flora URSR. T. 3, pp. 312-401. Vidavnitstvo Academii Nauk URSR, Kyiv.

Burda, R.I., Ostapko, V.M., Tohtar, V.K. 1997a. Variation of synanthropic plant populations.- No publisher, Donetsk. [Preprint of the Donetsk Botanical Gardens].

Burda, R.I., Ostapko, V.M., Kucherevsky, V.V. 1997b. Zozulencevi (Orchidaceae Juss.) na pivdennomu shodi Ukraini. Ukr. botan. journ. 54, 4:361-364.

Chervona kniga Ukraini: Roslinny swit.1996. Ukrainska enciklopedia, Kyiv. [Red Data book of Ukraine: Plante, Ed. by Yu. R. Shelyag- Sosonko].

Dubovik, O.M. 1970. Materiali do floristichnogo rayonuvannya Donetskogo Lisostepu. Ukr. botan. journ. 27, 3: 279-283.

Gatsuk, L.E., Smirnova, O.V., Vorontzova, L.I., Zaugolnova, L.B., Zhukova, L.A. 1980. Age states of plants of various growth forms: a review. Journ. Ecology. 68: 675-696.

Klokov, M.V. 1948. Novi materiali do piznannya ukrainskoy flori. III. Novi vidi z rodini gvozdichnih, grechkovih, hrestotzvitnih. Botan. journ. AN URSR. 5, 1:20-31.

Klokov, M.V. 1981. Psammofilnie floristicheskie komplexi na territorii URSR (opit analiza psammophytona). In: Klokov, M.V. (ed.) Novosti systematici visshih i nizshih rasteny, pp. 90-150. Nauk. dumka, Kyiv.

Kondratyuk, E.M. and Burda, R.I. 1980. Ochrana redkich i ischezauschich vidov mestnoy flori. In: Kondratyuk, E.N. (ed.) Promischlennaya botanika, pp. 156-221. Nauk. dumka, Kyiv.

Kotov, M.I. 1960. Nova roslina flori URSR – lischiza skorzonerolista (*Gypsophila scorzonerifolia* Ser.). Ukr. botan. journ. 17, 4: 75-79.

LeFloc'h, E.,Houerou, H.N.and Mathez J.1990. History and patterns of plant invasion in Northern Africa. In: di Castri, F., Hansen, A.J. and Debussche, M.(eds), Biological Invasions in Europe and Mediterranean Basin, pp. 105-133. Kluwer AcademicPublishers, Dordrecht.

Sauerland, K.-E. 1995. Orchideen in Rostocker Grünanlagen. Naturschutzarb. Mecklenburg-Vorpommern. 35, 2: 53-54.

Smolyaninova, L.A. 1976. Orcidaceae Juss. In Fedorov, An.A. (ed.) Flora europeiskoy chasti SSSR. T. 2, pp. 10-59. Nauka, Leningrad.

Sobko,V.G. 1989. Orchidei Ukraini. Nauk. dumka, Kyiv.
Tzarenko, O.M. 1990. Rod *Gypsophila* vo flore Ukraini, kritiko-sistematicheskoe i biomorphologicheskoe
    issledovanie vidov. . No publisher, Dnepropetrovsk. [A synopsis of Candidate's thesis].
Zaweruka, B.V et al. 1983. Ochranyayemie rasteniya Ukraini. Nauk. dumka, Kyiv.

MEDIEVAL CASTLES AS CENTERS OF SPREAD OF NON-NATIVE PLANT SPECIES

Katharina Dehnen-Schmutz
Institut für Ökologie, TU Berlin, Schmidt-Ott-Str. 1, D - 12165 Berlin, Germany;
e-mail: dehnjfce@mailszrz.zrz.tu-berlin.de

Abstract

In the Middle Ages castles were built on top of hills and rocks to have an optimal strategic position. Since that time non-native species had the chance to establish in the vegetation around the castles. Waste, castle-gardens and visitors are the main sources of diaspores.

Plant species of walls and rocks of 56 medieval castles in Southern and South-Eastern Germany were found to have 372 vascular plant species and ferns. 25% of these were non-native species (91 species). The degree of naturalization of the non-native species on natural rock vegetation was analysed.

The relation of the castles to the nearest settlement influenced the number of species. Castles in a settlement had fewer species on the whole than castles nearby or far away, but the number of non-native species was higher. The sites in a settlement had more neophytes than archaeophytes, whereas outside more archaeophytes than neophytes occurred. Also, inhabited castles had more neophytes than un-used castles.

Introduction

Medieval castles are centers of spread of non-native plants since more than 800 years. On top of hills and rocks they were mostly built during the 11th - 13th century. First of all their function was protection and demonstration of power. Excavations showed that the castles and their environment were also used like farms (Meyer 1982). In the castle area working quarters, stables and gardens existed. Therefore waste, transportation of goods, visitors and castle gardens were the first sources of diaspores of non-native plants which colonized the surrounding of the castles assisted by the accumulation of nutrients from mortar, waste and livestock. With the end of the Middle Ages the castles lost their function, most of them were destroyed or became dilapidated, only some were used as residential buildings. In the 19th century a new interest in the castles began and some of them were reconstructed. Today they are ruins or used as museum, restaurant, hotel or residential building.

But in general castles were much less changed during the centuries than e.g. towns or settlements. Castles were intensively used over a period of up to 400 years and than often un-used over a period of the same extension.

Therefore botanists were interested in the flora around Medieval castles since the 19th century. The first reports about the flora of castles in the region Alsace were published emphasizing the occurrence of non-native species (Chatin 1861, Kirschleger 1862). Later, reports appeared about castles at the river Rhine (Lohmeyer 1984), the Oberpfälzer Wald (Vollrath 1959), the Harz Mountains (Brandes 1996) or about a transect

Plant Invasions: Ecological Mechanisms and Human Responses, pp. 307–312
edited by U. Starfinger, K. Edwards, I. Kowarik and M. Williamson
© *1998 Backhuys Publishers, Leiden, The Netherlands*

of 25 castles from France to Hungary (Siegl 1998). Most of these studies did not differentiate the non-native species according to their degree of naturalization. At first Lohmeyer (1976, 1984) and Lohmeyer and Sukopp (1992) reported about non-native species established as agriophytes in the natural vegetation around castles.

Non-native species can be classified according to their way of introduction, time of introduction and degree of naturalization (Schroeder 1969). The last criterion is potential independence from human activity and can be assessed by studying vegetation and sites in which non-native species grow in a new area. Most of the non-native species in Central Europe occur in man-made vegetation and are able to exist there only with human help (epecophytes). Only a few species are agriophytes, established in the natural vegetation. Lohmeyer and Sukopp (1992) defined agriophytes as species that are present in natural vegetation and can exist there without human activity. Kowarik (1987) added species in man-made habitats with conditions corresponding to natural sites.

The aim of the present study was to get a survey of non-native plant species occurring in the natural vegetation of rocks and walls and to classify them according to their degree of naturalization. The role of medieval castles as centers of spread for these plants and the influence of actual use were investigated.

Study areas and methods

Five areas in Southern and South-Eastern Germany were investigated: parts of the river valleys of the Saale, Altmühl and Neckar, and parts of the regions Fränkische Schweiz and Schwäbische Alb. These landscapes have a high density of medieval castles, and all castles were built on limestone rocks.

Plant species of walls and rocks of 56 castles were recorded from 1994 to 1997. The castles were characterized by their relation to the next settlement (more than 200 m away, nearby or in a settlement) and their actual use (uninhabited or inhabited castles).

Non-native species are defined as species which have not evolved in the investigation area since the last Ice Age and whose introduction or immigration was supported deliberately or involuntarily by human activities (Kowarik 1995). Information about the time of introduction and area of origin was taken from Lohmeyer and Sukopp (1992), Rothmaler (1988) and Sebald et al. (1990 - 1996), additional information about use from Düll and Kutzelnigg (1992).

The degree of naturalization of the non-native species was estimated for those species not mentioned in the list of Central European agriophytes (Lohmeyer and Sukopp 1992). According to the definition of agriophytes from Lohmeyer and Sukopp (1992) species were classified as agriophytes if they showed the following features:
- species found in the natural rock vegetation,
- found in more than one year or known for the study area from literature,
- existing in a population of many individuals,
- site of occurrence (rock or wall) was unchanged by human activities, trees are unable to shadow it,
- species occurred in the study area for more than 25 years or older information in literature.

Results

Species

A total of 372 plant species occurred on the rocks and the walls of the castles, 91 non-native.

18 non-native species (Table 1) occurred at five or more castles and only 6 were found in every study area. The most frequent species were *Echium vulgare* and *Syringa vulgaris* occurring at 40 - 60% of the castles. *Iris*-species, mostly *Iris germanica,* were found in every area with the same frequency. Some species had their main occurrence in one area only, e.g. *Lycium barbarum* in the Saale valley. *Ballota nigra* appeared at every castle in the Saale valley but was found in the other areas, too. The lowest frequency of non-native species was found in the region Schwäbische Alb, only two species, *Echium vulgare* and *Vinca minor*, occurred at 20-40% of the castles.

The non-native species could be separated into 46 archaeophytes (invading before 1500 AD), 37 neophytes (invading after 1500 AD) and 8 species with unknown time of introduction. Among the archaeophytes are 16 medicinal plants, 10 plants used as dye plants, spices or food plants and one ornamental plant. Among the neophytes are 23 ornamental plants and only 4 medicinal plants and 3 food plants.

The origin of most of the non-native species is Europe (32 species) or Europe and Asia (38 species), and most species in these two groups are of Mediterranean origin. 9 species are of American and 9 of Asian origin.

For 23 non-native species not mentioned in the list of Central Europe's agriophytes (Lohmeyer and Sukopp 1992) the status of establishment was analyzed and 11 spe-

Table 1. Most frequent non-native species on rocks and walls at the castles in the different study areas

	All areas	Altmühltal	Fränk. Schweiz	Neckar	Schw. Alb	Saale
Number of castles	56	9	14	10	15	8
Echium vulgare	III	IV	II	II	II	IV
Syringa vulgaris	III	I	IV	IV	I	IV
Bromus sterilis	II	II	II	III	I	III
Ballota nigra agg.	II	I	I	II	I	V
Anthemis tinctoria	II		III	I		III
Sedum spurium	II	I	II	II	I	III
Cymbalaria muralis	II		I	III	I	II
Cerastium tomentosum	I	I	I	II		II
Iris spec.	I	I	I	I	I	II
Artemisia absinthium	I	II			I	I
Helianthus annuus	I	I	I		I	II
Impatiens parviflora	I	II	I			I
Lactuca serriola	I	II	I	II		
Lycium barbarum	I					IV

I   = at less than 20%,
II = at 20-40%,
III = at 40-60%,
IV = at 60-80%,
V = at 80-100% of the castles

Table 2. Average number of native and non-native species at castles outside, nearby and in a settlement. Only the number of species at castles in a settlement is significantly different (*,t-test, α = 0.05) to the number of species at castles nearby or outside.

Number of castles	Outside (25)		Nearby (23)		In a settlement (8)	
Native	32.2		26.9		*14.6	
Non-native	5.2	(13.9%)	6.4	(19.2%)	5.9	(28.7%)
neophytes	1.8	(4.8%)	2.7	(8.1%)	3.4	(16.6%)
archaeophytes	2.9	(7.8%)	3.1	(9.3%)	1.9	(9.3%)
Unknown time of introd.	0.5		0.6		0.6	

cies were added to the list: *Arabis caucasica* Willd., *Buglossoides arvensis* (L.) Johnst., *Cerastium tomentosum* L., *Corydalis lutea* (L.) DC, *Echium vulgare* L., *Geranium columbinum* L., *Lappula squarrosa* (Retz.) Dum., *Malva sylvestris* L., *Parietaria officinalis* L., *Ruta graveolens* L., *Sisymbrium loeselii* L.

Influence of position and use of the castles

Castles far away from a settlement had a significantly higher number of total species than castles in a settlement (Table 2). In contrast, the portion of non-native species at castles in villages was higher (28.7%) than at castles nearby (19.2%) or far away (13.9%) from a settlement. Neophytes and archaeophytes were spread over the castles differently: at castles far away from a settlement more archaeophytes occurred, whereas at castles in a settlement more neophytes were found.

The average total number of species for 35 inhabited castles and 20 uninhabited castles was nearly the same (33.3 and 32.9) but at inhabited castles there were on average twice as many non-native species as at uninhabited castles (7.3 and 3.3 respectively) (Fig. 1).

Discussion

Rocks around castles and walls are places were non-native species have established permanently. 25% non-native species were found at these sites, while in the total flora of Germany there are only 19% (Jäger 1991). The portions of archaeophytes and neophytes were also different: 50% of the non-native species at the castles were archaeophytes, while in the total flora they contribute only 40%.

More than 66% of introduced species originated in Europe or in Europe and Asia (Table 3); this is more than the 50% of all Central European agriophytes (Lohmeyer and Sukopp 1992). The value for American species, 10%, is notably lower than 30% among all agriophytes. One reason could be the limited opportunity for introduced American plants to reach a castle because the main time of using castles came to a close with the end of the Middle Ages more than 500 years ago. The higher portion of archaeophytes at the castles also supports this hypothesis.

The use and position of the castle (nearby or in a settlement) decreases the number of native species, while the number of non-native species, especially of neophytes, increases. One reason may be the more intensive use of castles in a settlement. Simi-

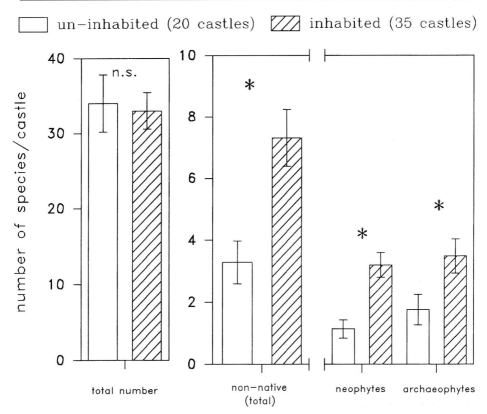

Fig. 1. Average number of all species and of non-native species, neophytes and archeophytes at uninhabited and inhabited castles (n.s. = non significant; * = significant, t-test, α = 0.05; vertical bars indicate standard error).

Table 3. Areas of origin (in %) of the non-native species found, compared to all Central European agriophytes (Lohmeyer and Sukopp 1992).

	Europe	Europe/Asia	Asia	America	Other
Non-native species at the castles (88 species)	35.2	41.8	9.9	9.9	
Central European agriophytes (220 species)	24.1	27.3	9.1	30.5	9.1

larly in urban ecology the number of neophytes found in the city is higher than in the suburbs (Kunick 1982).

The high portion of 60% usable plants (most of them medicinal and food plants) among the archaeophytes emphasizes the reasons for their introduction in historic times. Later, when the usage changed, non-native plants were mostly introduced because of aesthetic reasons. A fact which may explain the high amount of ornamentals (60%) among the neophytes.

The establishment of non-native species in natural vegetation is sometimes regarded as a problem because native species might be displaced. At the medieval castles in this study non-native species are a qualitative enrichment. They document medieval culture and the history of use of the castles and should be preserved like the walls or towers of the castles themselves.

Acknowledgements

I thank Herbert Sukopp, Uwe Starfinger, Ingo Kowarik and Mark Williamson for their comments on the manuscript. The study was supported by a research grant from the Friedrich-Ebert foundation.

References

Brandes, D. 1996. Burgruinen als Habitatinseln. Ihre Flora und Vegetation sowie die Bedeutung für Sukzessionsforschung und Naturschutz dargestellt unter besonderer Berücksichtigung der Burgruinen des Harzgebietes. Braunschweiger Naturkdl. Schr. 5(1): 125-163.

Chatin, A. 1861. Sur les plantes des vieux chateaux. Bulletin de la Société Botanique de France 8: 359-369.

Düll, R. and Kutzelnigg, H. 1992. Botanisch-ökologisches Exkursionstaschenbuch. Quelle & Meyer. Heidelberg.

Jäger, E.J. 1991. Grundlagen der Pflanzenverbreitung. In: Schubert, R. (ed.), Lehrbuch der Ökologie. pp. 167-173. Gustav Fischer, Jena.

Kirschleger, F. 1862. Sur les plantes des vieux chateaux, dans la région Alsato-Vosgienne. Bulletin de la Société Botanique de France 9: 15-18.

Kowarik, I. 1987. Kritische Anmerkungen zum theoretischen Konzept der potentiellen natürlichen Vegetation mit Anregungen zu einer zeitgemäßenen Modifikation. Tuexenia. N.S. 7: 53-67.

Kowarik, I. 1995. Time lags in biological invasions with regard to the success and failure of alien species. In: Pyšek, P., Prach, K., Reymánek, M. and Wade, M. (eds.), Plant Invasions - General Aspects and Special Problems. pp. 15-38. SPB Academic Publishing, Amsterdam.

Kunick, W. 1982. Zonierung des Stadtgebietes von Berlin (West) - Ergebnisse floristischer Untersuchungen. Landschaftsentwicklung und Umweltforsch. 14: 1 - 164.

Lohmeyer, W. 1976. Verwilderte Zier- und Nutzgehölze als Neuheimische (Agriophyten) unter besonderer Berücksichtigung ihrer Vorkommen am Mittelrhein. Natur u. Landschaft 51: 275 - 283.

Lohmeyer, W. 1984. Vergleichende Studie über die Flora und Vegetation auf der Rheinbrohler Ley und dem Ruinengelände der Höhenburg Hammerstein (Mittelrhein). Natur u. Landschaft 59: 478 - 483.

Lohmeyer, W. and Sukopp, H. 1992. Agriophyten in der Vegetation Mitteleuropas. Schriftenreihe Vegetationsk. 25: 1-185.

Meyer, W. 1982. Landwirtschaftsbetriebe auf mittelalterlichen Burgen. In: Österreichische Akademie der Wissenschaften (ed.), Adelige Sachkultur des Spätmittelalters. Sitzungsberichte Österreichische Akademie der Wissenschaften, Philosophisch-Historische Klasse 400: 377-386. Verlag der Österreichischen Akademie der Wissenschaften, Wien.

Rothmaler, W. 1988. Exkursionsflora für die Gebiete der DDR und der BRD. Vol. 4. Volk u. Wissen. Berlin.

Schroeder, F.-G. 1969. Zur Klassifizierung der Anthropochoren. Vegetatio 16: 225-238.

Sebald, O, Seybold, S., Philippi, G. and Wörz, A. 1990-1996. Die Farn- und Blütenpflanzen Baden-Württembergs. Vol. 1, 1990, Vol. 2, 1990, Vol. 3, 1992, Vol. 4, 1992, Vol. 5, 1996, Vol. 6, 1996, Ulmer, Stuttgart.

Siegl, A. 1998. Flora und Vegetation mittelalterlicher Burgruinen. In: Kowarik, I., Schmidt, E. and Sigel, B. (eds.), Naturschutz und Denkmalpflege: Wege zu einem Dialog im Garten. pp. 193-202. vdf Hochschulverlag ETH Zürich, Zürich.

Vollrath, H. 1958/60. Burgruinen bereichern die Flora - Ein Beitrag zur Flora des Oberpfälzer Waldes. Berichte der Naturwissenschaftlichen Gesellschaft Bayreuth 10: 150-172.

THE EFFECT OF ALTITUDE ON THE PATTERN OF PLANT INVASIONS: A FIELD TEST

Stanislav Mihulka

Faculty of Biological Sciences, University of South Bohemia, Branisovska 31, 370 05 České Budějovice, Czech Republic; e-mail: plch@tix.bf.jcu.cz

Abstract

The effect of altitude on the establishment of selected alien species was studied experimentally in the fish-pond region, around České Budějovice, Czech Republic. Controlled sowing of seven alien invasive species was carried out at three different altitudes (400, 700 and 1000 m a. s. l.) At each altitude, five 1 × 1 m experimental plots were established in four habitats (shores of flowing water, shores of still water, ruderal sites, dry and sunny sites) and seed germination was recorded in the first growing season following sowing. Data were analysed by generalized linear models. Altitude had a significant effect on the germination of all but one species involved. The present study thus indicates its important role in affecting the distribution of invasive species.

Introduction

Invasion processes are affected by the particular characteristics of invasive species and the characteristics of invaded habitats. Every invading species posseses a set of characters, "preadaptions", acquired during its own evolution in the area of origin and every community undergoes its own historical development. Various evolutionary histories of individual areas lead to (i) different susceptibility of their native communities to invasions, and (ii) different success of invaders from various areas of origin. Areas with an exceptional evolutionary history are usually more easily invaded than the other ones (*e.g.* Vitousek 1988; Loope and Mueller-Dombois 1989).

The traits of invasive species have been discussed in numerous case studies (see *e.g.* Pyšek *et al.* 1995b; Brock *et al.* 1997) and general patterns have been sought (*e.g.* Noble 1989; Roy 1990; Rejmánek 1995; Crawley *et al.* 1996; Williamson and Fitter 1996).

Invasibility of communities is usually considered in the context of abiotic attributes of environment or local biotic interactions. Studies taking into account the positions of invaded communities on environmental gradients (moisture, nutrients, successional age, disturbance) and the biotic characteristics of invaded communities suggest that some communities are more susceptible to invasions than others (Pyšek *et al.* this volume). However, the lack of available data constraints our ability to draw more general conclusion on this topic (Williamson 1996; Prach and Pyšek 1997).

Invasions occur in the context of landscape structure. A real distribution of invasive plants is undoubtedly generated by a complex of various factors. However, the relevance of particular factors is still unclear. For example, the fact that vegetation of

Plant Invasions: Ecological Mechanisms and Human Responses, pp. 313–320
edited by U. Starfinger, K. Edwards, I. Kowarik and M. Williamson
© *1998 Backhuys Publishers, Leiden, The Netherlands*

ruderal sites is invaded by a high number of invasive species can be a consequence of a high input of diaspores of alien species into these places caused by humans (Williamson 1996). In such case, invasibility caused by intrinsic characteristics of the community cannot be distinguished from the invasibility caused by other external influences.

Therefore, the experiment was designed to simulate the situation in which the availability of diaspores limits neither the establishment of invasive species nor the pattern of their distribution.

By excluding the effect of diaspore input, environmental conditions will be the main factor affecting the distribution of the invasive species. Altitude was selected as a complex factor including both the effect of temperature and precipitation.

The present study, by using experiments, attempts to elucidate the role altitude plays in determining the distribution of invasive species in the landscape. Its objective was to test the effects of altitude on the germination of seven selected invasive species.

Methods

Species selection

Seven species were selected from the list of Czech invasive aliens (Pyšek *et al.* 1995a). *Bidens frondosa*, *Heracleum mantegazzianum*, *Impatiens glandulifera* and *Robinia pseudoacacia* represent successful invasive species occuring in a variety of seminatural and/or man-made habitat types. *Acer negundo*, *Ailanthus altissima*, and *Physocarpus opulifolius* are species with less remarkable dynamics of spread, being distributed mostly in man-made habitats of warmer climatic regions. Seeds of these species were collected in the field during autumn 1994 from localities as near as possible to the area where the experiment was conducted. During winter seeds were stored in a cold greenhouse.

Experimental design

The experiment was carried out near the town of České Budějovice, in the south part of the Czech Republic. Local climate is temperate-cold with mean annual temperature 5-7 °C and mean annual precipitations 650-700 mm. Mean temperature of growing season (April to September) is from 10 to 13 °C. Sowing experiments were performed in spring 1995 at three different altitudes: 400 m, 700 m, and 1000 m a. s. l. Each species was sown at each altitude into five 1 × 1 m plots located in 4 different habitats, (i) shores of flowing water, (ii) shores of still water, (iii) ruderal sites, and (iv) dry and sunny sites. Plots were localized in the most comparable places as was possible. Surface of each plot was mechanically disturbed. The numbers of seeds sown per plot differed for each species (Table 1) and were based on preliminary estimated germination. Seed germination was recorded for one week during the following growing period. The total number of germinated seeds was only used for following analysis.

Data analysis

Data were analysed by generalized linear models (GLM) using the S-PLUS statistical package (Anonymous 1995). Altitude, species and habitat type were used as explanatory variables in GLM analysis with an assumed binomial distribution of the explained variable, i.e seed germination (McCullagh and Nelder 1989) and using the logit link function. The marginal effect of the explanatory variables on the explained variable was tested by stepwise selection methods.

Results

In general, the germination observed was rather low. Only two species, *A. negundo* and *P. opulifolius*, germinated in all habitats. All species germinated only in ruderal sites and shores of flowing water at the lowest altitude of 400 m a.s.l. (Fig. 1). There was a highly significant effect of altitude on total germination, i.e. if pooled data were considered (P < 0.001). Germination decreased with increasing altitude. Interaction among altitude, species and habitat type was also highly significant (Table 2).

Considering individual species, altitude affected significantly (P < 0.001) the germination of all the species but *Physocarpus opulifolius*. Germination of *Impatiens glandulifera* and *Physocarpus opulifolius* was remarkably affected by the habitat type (P < 0.001). In the former species, it was higher in moist habitats (germination in dry and sunny sites<ruderal sites<shores of still water<shores of flowing water). *Heracleum mantegazzianum* was the only species whose germination was unaffected by the habitat type. Interactions between the effects of altitude and habitat were all non-significant (Table 3).

Discussion

Outdoor sowing experiments are rather difficult to carry out because of the various conditions necessary for successful germination of the seeds of species involved; these conditions can hardly be simulated. Consequently, the germination rates observed in the present experiment could have been negatively affected.

Observed germination patterns can be, to a large extent, explained by the autecology of particular alien species. *Acer negundo* showed rather equal germination among altitudes and habitats (comparing with other reasonably germinated species). Instead of all other species, germination of *Acer negundo* in ruderal sites at higher altitudes was higher than at the lowest one.

Ailanthus altissima and *Robinia pseudoaccacia* seem to have similar ecological demands for naturalization (Sachse *et al.* 1990; Kowarik 1995). The occurence of *Ailanthus altissima* in Central Europe is mostly limited to urban heat islands with warmer and continental weather (Kowarik and Böcker 1984; Kunick 1990; Sachse *et al.*1990). In the present study, the pattern of germination across experiments – decrease with altitude and preference for dryier habitats – was similar in both.

Bidens frondosa showed low germination regardless of habitat type. The results indicate that it is not capable of significant germination at higher altitudes. It is consistent with recorded demands for high temperature during germination (see Köck 1988). Factors affecting the distribution of *Heracleum mantegazzianum* have been repeatedly analysed (*e.g.* Pyšek 1991; 1994; Pyšek and Pyšek 1995), however, role of germination was not really considered. This species has a high invasive potential, namely in moist, nutrient-rich habitats, but its germination is often rather poor (Andersen and Calov

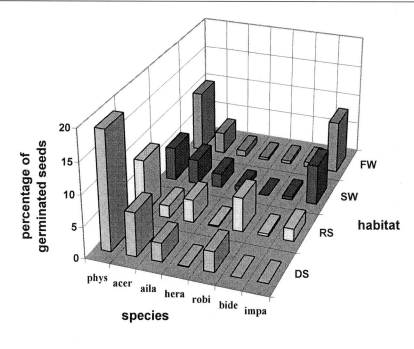

A

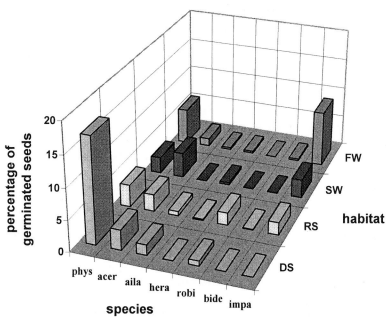

B

Fig 1. Medians of percentage germination capacity depicted for the invasive species considered in all habitat types at individual elevations. Part 1.A. shows the germination pattern at elevation 400 m a.s.l., 1.B. shows the same for elevation 700 m a.s.l. and 1.C. for 1000 m a.s.l. (species: acer – *Acer negundo*, aila – *Ailanthus altissima*, bide – *Bidens frondosa*, hera – *Heracleum mantegazzianum*, impa – *Impatiens parviflora*, phys – *Physocarpus opulifolius*, robi – *Robinia pseudoacacia*; habitats: FW – shores of flowing water, SW – shores of still water, RS – ruderal sites, DS – dry and sunny sites)

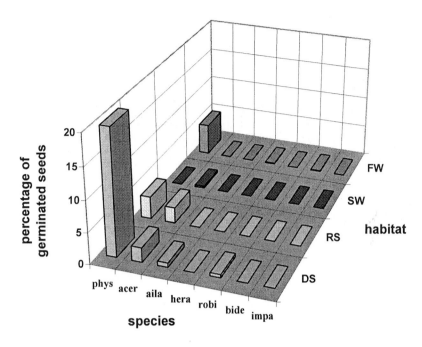

C

Fig. 1C.

Table 1. Seeds – number of sown seeds of invasive species per plot. Family and area of origin are shown for each species.

Plant name	Seeds	Family	Area of origin
Acer negundo L.	200	Aceraceae	North America
Ailanthus altissima Desf.	150	Simaroubaceae	East Asia
Bidens frondosa L.	400	Asteraceae	North America
Heracleum mantegazzianum Sommier et Levier	500	Apiaceae	West Asia
Impatiens glandulifera Royle	100	Balsaminaceae	South Central Asia
Physocarpus opulifolius (L.) Maxim.	250	Rosaceae	North America
Robinia pseudoacacia L.	200	Fabaceae	North America

Table 2. Significance of explanatory variables used in GLM regression analysis performed on the pooled data. ALT-elevation, SPEC-species, TYPE-type of habitat

Explanatory variables	Significance
effect of ALT only	$P < 0.001$
independent effects of ALT+SPEC+TYPE	$P < 0.001$
effect of interaction ALT & TYPE	$P < 0.05$

Table 3. Significance of explanatory variables used in GLM regression analysis performed for particular species. ALT-elevation, SPEC-species, TYPE-type of habitat. N.S. – non-significant.

Plant name	Significance		
	ALT	TYPE	ALT & TYPE
Acer negundo	P < 0.001	P < 0.01	N.S.
Ailanthus altissima	P < 0.001	P < 0.05	N.S.
Bidens frondosa	P < 0.001	P < 0.05	N.S.
Heracleum mantegazzianum	P < 0.001	N.S.	N.S.
Impatiens glandulifera	P < 0.001	P < 0.001	N.S.
Physocarpus opulifolius	N.S.	P < 0.001	N.S.
Robinia pseudoacacia	P < 0.001	P < 0.01	N.S.

1996; Brabec 1997; Prach pers. comm.). Both *Bidens* and *Heracleum* need chilling to break the dormancy (Andersen and Calov 1996, see Grime *et al.* 1981) and the distribution of the latter species in the Czech Republic is correlated with the January isotherm (Pyšek *et al.* in press). Therefore, germination of both species could have been supreseded by dormancy because of insufficient chilling before or after sowing.

Impatiens glandulifera occurs in moist and nutrient-rich habitats. It probably does not produce a persistent seedbank and cold-resistent stems (Grime *et al.* 1988; Perrins *et al.* 1990; Pyšek and Prach 1995). Germination of *Impatiens* seeds during the experiments was reduced by factors correlated with elevation and affected by the habitat type (see results). Germination pattern of the species was consistent with its occurence in Central Europe (Pyšek and Prach 1994). *Physocarpus* seeds germinated best in dry and sunny habitats, instead of changing altitude. It is essentially inconsistent with statements from local flora (Hejný and Slavík 1992) which consider *Physocarpus opulifolius* as a species of moist, shaded shrubby places. This species shows less important relationship with altitude than other species. It suggests more influential role of dispersal ability of *Physocarpus* during its invasion under local climatic conditions.

Acknowledgments

I would like to thank to Karel Prach for his assistance with experiments and advice. I also thank to Petr Šmilauer and Jan Lepš for statistical help and other advices. Valuable comments on the manuscript from Petr Pyšek, Karel Prach and two anonymous reviewers are appreciated. Finally, my thanks are due to Jana Martínková for support and help. This study was supported by the Faculty of Biological Sciences, South Bohemia Universtity, Czech Republic.

References

Andersen, U.V. and Calov, B. 1996. Long-term effects of sheep grazing on giant hogweed *(Heracleum mantegazzianum)*. Hydrobiologia 340: 277-284.
Anonymous. 1995. S-PLUS Guide to Statistical and Mathematical Analysis, Version 3.3. StatSci, a division of MathSoft , Inc. Seattle.

Brabec, J. 1997. Experimental study of the effect of management on invasion of selected splecies into meadow plant communities. – Thesis, Charles University, Prague. (In Czech)

Brock, J.H., Wade, M., Pyšek, P. and Green, D. (eds.) 1997. Plant invasions: studies from North America and Europe. Backhuys Publishers. Leiden.

Crawley, M.J., Harvey, P.H. and Purvis, A. 1996. Comparative ecology of the native and allien floras of the British Isles. Philosophical Transactions of the Royal Society, London, Series B 351: 1251-1259.

Grime, J.P., Hodgson, J.G. and Hunt, R. 1988. Comparative Plant Ecology. Unwin Hyman. London.

Hejný, S. and Slavík, B. (eds.) 1992. Kvitena České republiky, Vol. 3. Academia. Praha.

Kowarik, I. 1995. Time lags in biological invasions with regards to the success and failure of alien species. In: Pyšek, P., Prach, K., Rejmánek, M. and Wade, M. (eds.), Plant Invasions: General Aspects and Special Problems, pp. 15-38. SPB Academic Publishing, Amsterdam.

Kowarik, I. and Böcker, R. 1984. Zur Verbreitung, Vergesellschaftung und Einbürgerung des Götterbaumes (Ailanthus altissima [Mill.] Swingle) in Mitteleuropa. Tuexenia 4: 9-29.

Köck, U.V. 1988. Ökologische Aspekte der Ausbreitung von *Bidens frondosa* L. in Mitteleuropa. Verdrängt er *Bidens tripartita* L.? Flora 180: 177-190.

Kunick, W. 1990. Spontaneous woody vegetation in cities. In: Sukopp, H. and Hejný, S. (eds.), Kowarik, I. (co-ed.), Urban ecology: Plants and plant communities in urban environments. pp. 167-174. SPB Academic Publishing, The Hague.

Loope, L.L. and Mueller-Dombois, D. 1989. Characteristics of Invaded Islands, with Special Reference to Hawaii. In: Drake, J.A., Mooney, H.A., di Castri, F., Groves, R.H., Kruger, F.J., Rejmánek, M. and Williamson, M. (eds.), Biological Invasions: A Global Perspective. pp. 257-280. John Wiley and Sons, Chichester.

McCullagh, P. and Nelder, J.A. 1989. Generalized linear models (2nd edition). Chapman and Hall. London.

Noble, I.R. 1989. Attributes of Invaders and the Invading Process: Terrestrial and Vascular Plants. In: Drake, J.A., Mooney, H.A., di Castri, F., Groves, R.H., Kruger, F.J., Rejmánek, M. and Williamson, M. (eds.), Biological Invasions: A Global Perspective. pp. 301-313. John Wiley and Sons, Chichester.

Perrins, J., Fitter, A. and Williamson, M. 1990. What makes *Impatiens glandulifera* invasive? In: Palmer, J. (ed.), The biology and control of invasive plants. pp. 8-33. Cardiff.

Prach, K. and Pyšek, P. 1997. Invazibilita společenstev a ekosystémů. In: P. Pyšek and K. Prach (eds.), Invazní rostliny v České flóře, Zprávy České Botanické Společnosti 14: 1-6.

Pyšek, P., Kopecký, M., Jarošík, V. and Kotková, P. (in press) The role of human density and climate in the spread of *Heracleum mantegazzianum* in the Central European landscape. – Distribution And Diversity, Oxford.

Pyšek, P. 1991. *Heracleum mantegazzianum* in the Czech Republic: the dynamics of spreading from the historical perspective. Folia Geobotanica et Phytotaxonomica 26: 439-454.

Pyšek, P. 1994. Ecological Aspects of Invasion by *Heracleum mantegazzianum* in the Czech Republic. In: de Waal, L.C., Child, L.E., Wade, P.M. and Brock, J.H. (eds.), Ecology and Management of Invasive Riverside Plants. pp. 45-54. John Wiley and Sons, Chichester.

Pyšek, P. and Prach, K. 1994. How Important are Rivers for Supporting Plant Invasions? In: de Waal, L.C., Child, L.E., Wade, P.M. and Brock, J.H. (eds.), Ecology and Management of Invasive Riverside Plants. pp. 19-26. John Wiley and Sons, Chichester.

Pyšek, P. and Prach, K. 1995. Historický přehled lokalit *Impatiens glandulifera* na území České republiky a poznámky k dynamice její invaze. Zprávy České Botanické Společnosti 29(1994): 11-31.

Pyšek, P., Prach, K. and Mandák, B. 1998. Habitat vulnerability to plant invasions: a landscape view. – This volume.

Pyšek, P., Prach, K. and Šmilauer, P. 1995a. Relating invasion success to plant traits: an analysis of the czech alien flora. In: Pyšek, P., Prach, K., Rejmánek, M. and Wade, M. (eds.), Plant Invasions: General Aspects and Special Problems .pp. 39-60. SPB Academic Publishing, Amsterdam.

Pyšek, P., Prach, K., Rejmánek, M. and Wade, M. 1995b. Plant Invasions: General Aspects and Special Problems. SPB Academic Publishing. Amsterdam.

Pyšek, A. and Pyšek, P. 1995. Invasion by *Heracleum mantegazzianum* in different habitats in the Czech Republic. Journal of Vegetation Science 6: 711-718.

Rejmánek, M. 1995. What makes a species invasive? In: Pyšek, P., Prach, K., Rejmánek, M. and Wade,

M. (eds.), Plant Invasions: General Aspects and Special Problems. pp. 3-13. SPB Academic Publishing, Amsterdam.

Roy, J. 1990. In search of the characteristics of plant invaders. In: di Castri, F., Hansen, A.J. and Debussche, M. (eds.), Biological Invasions in Europe and the Mediterranean Basin. pp. 335-352, Kluwer Academic Publishers, Dordrecht.

Roy, J., Navas, M.L. and Sonié, L. 1991. Invasion by annual brome grasses: a case study challenging the homocline approach to invasions. In: Groves, R.H. and di Castri, F. (eds.), Biogeography of Mediterranean Invasions. pp. 207-224. Cambridge University Press, Cambridge.

Sachse, U., Starfinger, U. and Kowarik, I. 1990. Synanthropic woody species in the urban area of Berlin (West). In: Sukopp, H. and Hejný, S. (eds.), Kowarik, I. (co-ed.), Urban ecology: Plants and plant communities in urban environments. pp. 233-243. SPB Academic Publishing, The Hague.

Vitousek, P.M. 1988. Diversity and biological invasions of oceanic islands. In: Wilson, E.O. (ed.), Biodiversity. pp 181-192. National Academy Press, Washington.

de Waal, L.C., Child, L.E., Wade, P.M. and Brock, J.H. 1994. Ecology and Management of Invasive Riverside Plants. John Wiley and Sons, Chichester.

Williamson, M. 1996. Biological Invasions. Chapman and Hall, London.

Williamson, M. and Fitter, A. 1996. The characters of successful invaders. Biological Conservation 78: 163-170.

INVASION OF ALIEN PLANTS IN FLOODPLAINS –
A COMPARISON OF EUROPE AND JAPAN

Norbert Müller[1] and Shigetoshi Okuda[2]
[1]University of Applied Sciences Erfurt, Faculty of Landscape Architecture,
Leipziger Str. 77, 99085 Erfurt, Germany
[2] Yokohama National University, Institute of Environmental Science & Technology,
7907 Tokiwadai, Hodogaya-ku, Yokohama 240, Japan

Abstract

The vegetation of floodplains is compared in central Europe and northern- and central Japan focusing on the questions:
– which alien plants have successfully naturalised and where is their native country,
– which vegetation types are preferred by invasive plants,
– what are the reasons for the fast expansion of aliens in floodplains?
   In central Europe, 130 alien plants are naturalised in the vegetation of riparian landscapes. In floodplains of northern and central Japan, 124 aliens have been found.
   In both areas, most alien species come from North America. Whereas Europe has only 6 species from Japan, 40 species which are native to central Europe are naturalised in Japan.
   Alien plants prefer the annual and perennial herbaceous vegetation in both Europe and Japan (about 80 % of all species). Fewer invaders can be found in the vegetation of floodplain shrubs and -forests and only a few in aquatic vegetation.
   The success of aliens is related to human impacts in river ecosystems in both areas. While there are few or no aliens in the more natural upper courses of rivers, they become more abundant in the lower courses, where river dynamics are weakened, due to the influence of dams and where there are settlements, which are sources for the dispersal of aliens.

Introduction

Riparian landscapes are important corridors for plant dispersal (*e.g.* Johansson *et al.* 1996, van der Pijl 1982). Riparian habitats play a central role in the process of invasion and naturalisation of alien plants (Pyšek and Prach 1993). Many successful invaders in the natural vegetation were first observed in floodplains. The greatest number of alien plant species which have naturalised in the natural vegetation of central Europe (agriophytes, *sensu* Kamysev 1959), are found in floodplain vegetation. Twelve of the 13 most frequent invaders of central Europe can be found in floodplains (Lohmeyer and Sukopp 1992).

   In this paper we compare alien plants in floodplains in the temperate to submeriodonal zone of Europe and Japan in order to answer the following questions:
– which aliens have successfully naturalised and where is their native country,
– which habitats in floodplains are most vulnerable to plant invasions,
– what are the reasons for the fast expansion of aliens in floodplains?

Plant Invasions: Ecological Mechanisms and Human Responses, pp. 321–332
edited by U. Starfinger, K. Edwards, I. Kowarik and M. Williamson
© *1998 Backhuys Publishers, Leiden, The Netherlands*

Basic data and compared rivers

Data about naturalised invasive plants in central Europe were collected using the list of agriophytes in central Europe (Lohmeyer and Sukopp 1992) and for alpine rivers (Müller 1995a). These lists cover all of Austria, Belgium, Denmark, Germany, Liechtenstein, Luxembourg, the Netherlands, and Switzerland, as well as parts of France, Italy, Poland, the Czech Republic and Slovakia. Information about naturalised plants in Japan was gathered from the list of agriophytes for the rivers of central and northern Japan (Okuda n.p.), which includes Hokkaido, Hokuriku, Tohoku and Kanto. As a result, the information distinguishes between different river types (*e.g.* small and big, braided and bent rivers). The floodplains studied ranged in elevation from sea level to ca. 1500 m.

Japanese rivers originate in the alpine region, tend to be short and with steep slopes. Times of peak run-off are in early summer, due to snow melt, and in autumn during the typhoon season. In contrast, European rivers can be separated into two types, based on their time of peak run-off. Alpine rivers originate in the Alps, with most of their catchment area in the alpine zone; peak run-off occurs in early summer. Lowland rivers originate in the Low Mountains; peak run-off is during early spring. Rivers, such as the Rhine and the Danube, represent both types because of their great length.

Number and origins of naturalised alien plants

Currently, 130 alien plant species have become naturalised in central European and 124 in Japanese floodplains (Table 1). In Japan, as well as in Europe, most aliens are originally from North America (Fig. 1). Regarding the species from warmer areas, in

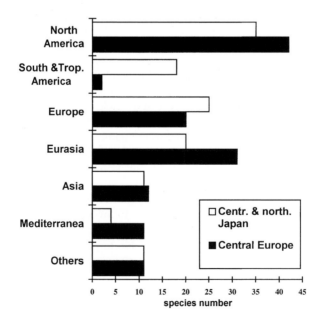

Fig. 1. Origins of alien plants in floodplains of central & northern Japan (124 species) and central Europe (130 species)

Japan most came from South and Tropical America (18), whereas in Europe more species are from the Mediterranean. The reason may be that the study areas in Japan include larger regions of the submeridional zone than in Europe (Meusel *et al.* 1965). An interesting fact is that Japan has received 40 species from Europe whereas, in central Europe, only 6 naturalised species are native to Japan.

Table 1. Alien plants in Central European and Japanese floodplains (Source Lohmeyer & Sukopp 1992, Müller 1995a, Okuda n.p.)

In roman: species in central Europe   *In italic:* species in Japan
In bold: species in both areas   Underlined species with origin in central Europe or Japan
Main habitat (vegetation type) in floodplains:
a=annual vegetation; p=perennial veg.; s=woody (shrubs & forests) veg.; w=aquatic veg.

Species	Main habitat	Native country
Acer negundo L.	s	North America
Acorus calamus L.	w	East Asia
Aesculus hippocastanum L.	s	South-East Europe
Agrostis gigantea Roth.	*p*	*Temperate Zone*
Agrostis scabra Willd.	p	North America & N-E Asia
Agrostis stolonifera L.	*p*	*Eurasia*
Ailanthus altissima (Mill.) Swingle	s	North China & N. Korea
Allium paradoxum (M.B.) G.Don	s	Asia
Althernanthera nodiflora R.Br.	*p*	*Tropical Asia*
Amaranthus albus L.	a	North America
Amaranthus blitoides S.Watson	a	North America
Amaranthus hybridus agg.	a	North America
Amaranthus lividus L.	a	Mediterranean
Amaranthus patulus Bertol.	*a*	*South America*
Amaranthus retroflexus L.	a	America
Amaranthus viridis	*a*	*Tropical America*
Ambrosia artemisiaefolia L var. elat. Desc.	*a*	*North America*
Ambrosia trifida L.	*a*	*North America*
Amorpha fruticosa L.	s	North America
Anchusa officinalis L.	p	East Europe
Andropogon virginicus L.	*p*	*North & Central America*
Anthoxanthum odoratum L.	*p*	*Eurasia*
Apera spica-venti (L.) P.B.	a	Eurasia
Arctium tomensosum Mill.	p	Eurasia
Armoracia rusticana Gaertn	p	South-East Europe
Arrhenatherum elatius (L.)	p	Europe
Artemisia absinthium L.	p	Eurasia
Artemisia annua L.	a	Asia
Artemisia verlotiorum Lamotte	a	South-East Asia
Artistolochia clematitis L.	s	South Europe
Aster lanceolatus Willd.	p	North America
Aster novae-angliae L.	p	North America
Aster novi-belgii L.	p	North America
Aster subulatus Michx.	*a*	*South America*
Aster tradescantii L.	p	North America
Aster x salignus Willd.	p	North America
Atriplex acuminata W. et K.	p	Eurasia
Avena fatua L.	*p*	*Eurasia & North Africa*
Azolla filiculoides Lam.	w	Tropical America
Barbarea vulgaris R.Br.	*a*	*Europe*
Bidens connata Mühlenb.	a	North America

Table 1. Cont.

Species	Main habitat	Native country
Bidens frondosa L.	a	North America
Bidens pilosa L. (incl. var. minor Sherff)	*a*	*Tropical Zone*
Brassica juncea Czern.	*a*	*Asia*
Brassica napus L. (B. oleracea x rapa)	*a*	*Cultivated plant*
Brassica nigra (L.) Koch	p	Europe
Brassica rapa L.	*a*	*Mediterranean*
Briza maxima L.	*a*	*Mediterranean*
Briza minor L.	*a*	*Mediterranean*
Bromus catharticus Vahl	*a*	*South America*
Bromus sterilis L.	a	South Europe
Bromus tectorum L.	a	Europe & East Med.
Bryonia alba L.	p	South-East Europe
Buddleja davidii Franch.	s	China
Bunias orientalis L.	p	Eurasia
Capsella bursa-pastoris (L.) Med.	a	Mediterranean
Carduus acanthoides L.	p	Europe & Asia Minor
Carduus crispus L.	*p*	*Eurasia*
Carduus nutans L.	p	Eurasia
Cerastium glomeratum Thuill.	*a*	*Europe*
Chaenorhinum minus (L.) Lange	a	Submed. Europe
Chenopodium album L.	*a*	*Eurasia*
Chenopodium ambrosioides L.	*a*	*South America*
Chenopodium ficifolium Sm.	a	Eurasia
Chrysanthemum vulgare (L.) Bernh.	p	Eurasia
Cichorium intybus L.	p	Eurasia
Coix lacryma-jobi L.	*p*	*Tropical Asia*
Conium maculatum L.	p	Eurasia
Conyza bonariensis Cronq.	*a*	*South America*
Conyza canadensis L.	a	North America
Conyza sumatrensis Walker	*a*	*South America*
Coreopsis lanceolata L.	*p*	*North America*
Crassocephalum crepidioides L.	*a*	*Africa*
Cuscuta campestris Yuncker	a	North America
Cuscuta gronovii Willd.	p	East-North America
Cuscuta lupuliformis Krocker	p	Eurasia
Cuscuta pentagona Engelm.	*p*	*North America*
Cyperus eragrostis Lam.	*p*	*Tropical America*
Dactylis glomerata L.	*p*	*Eurasia*
Datura stramonium L.	a	Mexico to E. North America
Delphinium anthriscifolium L.	*a*	*China*
Digitaria ischaemum (Schreber) Mühlenbg.	a	Eurasia
Digitaria sanguinalis (L.) Scop.	a	Mediterranean
Diodia teres Walt.	*a*	*North America*
Diodia virginiana L.	*p*	*North America*
Echinochloa crus-galli (L.) P.B.	a	Eurasia
Echinochloa muricata (Beauv.) Fern.	a	?
Echinocystis lobata (Michx.) Torr. et Gray	p	East-North America
Echinops sphaerocephalus L.	p	South Europe to West Asia
Egeria densa Pl.	*w*	*Argentina*
Eichhornia crassipes Solms-Laub.	*w*	*Tropical America*
Elodea canadensis Michx.	w	North America
Elodea nutalli St. Johann	w	North America
Epilobium ciliatum Raf.	p	North America
Eragrostis albensis Scholz	a	Sibiria ?
Eragrostis curvula (Schrad.) Nees	*p*	*South Africa*

Table 1. Cont.

Species	Main habitat	Native country
Eragrostis minor Host.	a	Eurasia
Eranthis hyemalis (L.) Salisb.	s	South-East Europe
Erechtites hieracifolia Rafin.	*a*	*North America*
Erigeron annuus (L.) Pers.	p	North America
Erigeron philadelphicus L	*p*	*North America*
Euphorbia maculata L.	*a*	*North America*
Fagopyrum cymosum Meisn.	*p*	*India & China*
Festuca arundinacea Schreb.	*p*	*Eurasia*
Festuca pratensis Huds.	a	Europe
Fraxinus pennsylvanica Marshall	s	North America
Galinsoga ciliata (Rafin.) Blake	a	South America
Galinsoga parviflora Cav.	a	Mexico
Geranium carolinianum L.	*a*	*North America*
Helianthus strumosus L.	*p*	*North America*
Helianthus tuberosus L.s.l.	p	North America
Heracleum mantegazzianum Somm. et Lev.	p	Caucasus
Hesperis matronalis L.	p	Europe
Holcus lanatus L.	*p*	*Europe*
Hypericum perforatum v. angust. (DC.)	*p*	*Europe*
Hypochoeris radicata L.	*p*	*Europe*
Impatiens glandulifera Royle	s	India
Impatiens parviflora DC.	s	Central Asia
Iris pseudacorus L.	*p*	*Eurasia*
Lactuca serriola L.	a	Eurasia
Lactuca serriola L.	*a*	*Europe*
Lamium purpureum L.	*a*	*Europe*
Lepidium virginicum L.	*a*	*North America*
Linaria canadensis Dum.	*a*	*North America*
Lindernia dubia (L.) Pennell	a	North America
Lolium multiflorum Lam.	*p*	*Europe*
Lolium perenne L.	*p*	*Eurasia*
Ludwigia decurrens Walt.	*a*	*Tropical America*
Matricaria chamomilla L.	a	West Asia
Matricaria perforata Merat	a	Europe & West Siberia
Melilotus alba Med.	p	Eurasia
Melilotus officinalis (L.) Pall.	p	Eurasia
Mentha spicata L. em. Huds.	p	Cultivated Plant
Mimulus guttatus DC.	p	West-North America
Myosotis scorpioides L.	*a*	*Europe*
Myriophyllum brasiliense Cambess.	*w*	*Brazil & South America*
Nasturtium officinale R.Br.	*p*	*Europe*
Oenothera biennis L.	p	North America
Oenothera erythrosepala Borbas	p	North America
Oenothera laciniata Hill	*p*	*North America*
Oenothera rosea L'Hér.	*p*	*America*
Oenothera stricta Ledeb. ex Link	*p*	*South America*
Orychophragmus violaceus (L.) O.E. Schulz	*a*	*China*
Oxalis corymbosa DC.	*p*	*South America*
Oxalis fontana Bunge	a	North America
Panicum capillare L.	a	North America
Panicum dichotomiflorum Michx.	*a*	*North America*
Papaver dubium L.	a	South Europe
Papaver rhoeas L.	a	Mediterranean & Eurasia
Parthenocissus inserta (Kerner) Fritsch	s	North America
Paspalum dilatatum Poir.	*p*	*South America*

Table 1. Cont.

Species	Main habitat	Native country
Paspalum distichum L.	*p*	*Tropical Zone*
Pastinaca sativa L.	p	Eurasia
Persicaria pilosa Kitag.	*a*	*India & Malaysia & China*
Phleum pratense L.	p	Eurasia
Physocarpus opulifolius (L.) Maxim.	s	North America
Picris hieracioides L. subsp. hieracioides	p	Eurasia
Plantago lanceolata L.	*p*	*Europe*
Plantago virginica L.	*p*	*North America*
Poa pratensis L.	*p*	*Eurasia*
Poa trivialis L.	*p*	*Eurasia*
Populus x canadensis Moench	s	East-North America
Portulaca oleracea L.	a	Mediterranean?
Reseda lutea L.	p	South Europe
Reseda luteola L.	p	South Europe & West Asia
Reynoutria japonica Houtt.	*p*	*East Asia*
Reynoutria sachalinensis Nakai	*p*	*Japan & Sachalin*
Reynoutria x bohemica	p	Neotyp
Robinia pseudacacia L.	s	North America
Rudbeckia hirta L.	p	North America
Rudbeckia laciniata L.	p	East-North America
Rumex acetosella L.	*p*	*Europe*
Rumex confertusWilld.	*p*	*Eurasia*
Rumex conglomeratus Murray	*p*	*Eurasia*
Rumex crispus L.	*p*	*Europe*
Rumex longifolius DC.	*s*	*North Europe*
Rumex obtusifolius L.	*p*	*Eurasia*
Rumex triangulivalvis (Danser) Rech.fil.	*p*	*Canada & North America*
Saponaria officinalis L.	*p*	*Europe*
Scilla sibirica Andr.	s	East Europe
Sedum sarmentosum Bunge	*p*	*Korea & North China*
Senecio inaequidens DC.	p	South Africa
Senecio vulgaris L.	*a*	*Europe*
Setaria pumila (Poiret) R. et Sch.	a	Southern Eurasia
Setaria viridis (L.) P.B.	*a*	*Mediterranean & Eurasia*
Sicyos angulatus L.	*a*	*North America*
Silene alba (Mill.) E.H.L. Krause	p	Eurasia
Silene armeria L.	*a*	*Europe*
Sinapis alba L.	a	Mediterranean
Sinapis arvensis L.	a	Mediterranean
Sisymbrium chrysanthum Jord.	p	South-west Europe
Sisymbrium officinale Scop.	a	Eurasia
Sisyrinchium atlanticum Bickn.	*a*	*North America*
Solanum americanum Mill.	*a*	*North America*
Solanum nigrum L.	a	Tropical Zone
Solidago altissima L.	*p*	*North America*
Solidago canadensis L.	p	North America
Solidago gigantea Ait.	p	North America
Solidago graminifolia (L.) Salisb.	p	North America
Sonchus asper Hill.	a	Eurasia
Sonchus oleraceus L.	a	Eurasia
Sorghum halepense Pers.	*p*	*Mediterranean*
Spiraea alba Du Roi	s	East-North America
Stellaria media (L.) Vill.	a	neogen
Symphoricarpos albus (L.) Blake	s	East-North America
Taraxacum officinale Weber	*p*	*Europe*

Table 1. Cont.

Species	Main habitat	Native country
Telekia speciosa Baumg.	P	South-east Europe
Thlaspi arvense L.	a	West Asia
Trifolium dubium Sibth.	*p*	*Eurasia*
Trifolium pratense L.	*p*	*Europe*
Trifolium repens L.	*p*	*Europe*
Tulipa sylvestris L.	s	East Mediterranean
Valerianella locusta (L.) Laterrade	*a*	*Europe*
Verbascum densiflorum Bertol.	P	South Europe
Verbascum thapsus L.	*p*	*Europe*
Verbena bonariensis L.	*p*	*South America*
Verbena officinalis L.	a	Eurasia
Veronica arvensis L.	*a*	*Eurasia Africa*
Veronica hederifolia L.	*a*	*Europe*
Veronica peregrina L.	a	America
Veronica persica Poir.	*a*	*West Asia*
Vulpia myuros (L.) C.C.Gmel.	*a*	*Eurasia*
Xanthium albinum (Widder) H. Scholz	a	North America
Xanthium italicum Moretti	*a*	*America*
Xanthium occidentale Bertol.	*a*	*North America*
Xanthium orientale L.	a	West India ?
Xanthium saccharatum Wallr. Em. Widder	a	North America

Aliens in different types of the floodplain vegetation

In order to get an overview of which vegetation types are preferred by plant invasions, the floodplain vegetation was separated into annual and perennial (herbs and grasses), woody (shrubs and trees) and aquatic vegetation. Each alien species was ordered to the vegetation type, in which it occurs with the highest frequency (Table 1, Fig. 2).

Central Europe

Most naturalised species here grow in annual and perennial vegetation types. These are under the strongest influence of river dynamics, such as periodic inundation and erosion and accumulation of bedload. Fifty-one species grow in the **annual vegetation** (*Bidentetea*-communities), which is well developed along the lowland rivers, such as the Rhine and the Elbe. Especially rich is the *Polygonum lapathifolium*-community, which is dominated by aliens such as *Xanthium saccharatum, Chenopodium ficifolium* and diverse *Amaranthus* species (Lohmeyer and Sukopp 1992).

Perennial vegetation types (excluding woody vegetation) contain 54 naturalised species. In the inundation area of big lowland rivers, the *Convolvulus sepium*-community is most frequent and especially rich in aliens, such as *Helianthus tuberosus, Solidago gigantea, S. canadensis, Aster salignus, A. lanceolatus* and the annuals *Brassica nigra* and *Impatiens glandulifera*. Along rivers with frequent, but short inundation, *Rudbeckia laciniata, Impatiens glandulifera* and *Reynoutria japonica* are often common in the *Petasites hybridus*-community. *Reynoutria japonica* often forms species poor communities due to its strong competitive nature (Lohmeyer and Sukopp 1992). Along alpine rivers, *Arrhenatherum elatius, Conyza canadensis, Solidago gigantea, S. canadensis*

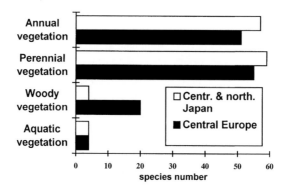

Fig. 2. Species number of alien plants in different types of the floodplain vegetation in central Europe (130 species) & central & northern Japan (124 species)

and *Impatiens glandulifera* are most frequent and can be found in the *Barbarea vulgaris-* and *Festuca arundinacea*-community. *Impatiens glandulifera* often forms species poor communities on new created gravel bars after highwater events.

Twenty species occur in the **woody vegetation**. In warmer parts of central Europe, for example the alpine Rhine, the lower Danube and at some southern alpine rivers (Brenta and Tagliamento), *Buddleja davidii* is a frequent alien in willow communities on young gravel bars. In willow floodplain forests of lowland rivers, *Acer negundo* is the most common tree (Lohmeyer and Sukopp 1992). *Impatiens parviflora* and *I. glandulifera* are frequent in the herb layer of many floodplain forests (*Alno-Ulmion* communities) of lowland and alpine rivers. *Robinia pseudoacacia* can spread from plantations into natural vegetation, forming pioneer forests. Along the southern alpine rivers (Lippert *et al.* 1995) and the Upper Rhine (Böcker in lit.), it colonises old gravel bars which are episodically inundated.

Only four species naturalised in **aquatic** habitats. Most common is *Elodea canadensis*. *Acorus calamus* is the most important agriophyte in old channels.

Central – and northern Japan

Similar to Europe, most introduced species are naturalised in Japan in the annual and perennial vegetation (Fig. 2). Fifty-eight species were found in **annual vegetation** types. In the middle and lower courses of Japanese rivers, the *Polygonum thunbergii*-community is typical near the main river course. In these sites, which are frequently disturbed by flooding, important aliens are *Aster subulatus, Bidens frondosa* and *Echinochloa crus-galli*. On higher sites, frequent aliens in the *Bidens pilosa*-community are diverse *Amaranthus* species, *Bidens pilosa* and *Chenopodium ambrosioides*. A community dominated by *Erigeron annuus* and the native *Erigeron sumatrensis* can be found on drier sites (Okuda 1996).

The **perennial vegetation** has 58 aliens; most occur in communities of the *Artemisietea*. They colonise sites on gravel bars higher to the water level than the annual vegetation. This community, which is decreasing due to civil engineering measures, the endemic *Aster kantoensis* is typical, although aliens, such as *Bidens pilosa, Lepidium virginicum* and *Oenothera biennis*, have invaded in recent years. Today, newly created gravel bars are often colonised by aliens that become dominant, such as

Helianthus strumosus, H. tuberosus, Solidago altissima and *S. gigantea* (Okuda *et al.* 1995). *Rumex crispus, R. obtusifolius and R. conglomeratus* are characteristic species of the *Rumex crispus*-community (Miyawaki and Okuda 1972; Okuda 1996). *Nasturtium officinale* from Europe has naturalised in old channels (tables in Okuda *et al.* 1995). In contrast to the gappy herbaceous vegetation, grass dominated communities (*Phragmitetea* and *Miscanthetea*) are poor in aliens. These communities are only influenced by inundation, therefore the vegetation cover is dense and only competitive species, such as *Festuca arundinacea*,are able to invade.

Only four aliens are important in the **woody vegetation**. On dry gravel bars, *Robinia pseudoacacia* has expanded greatly in the last 20 years, which is documented in detail for the Tama river by Tokyo (Fig. 3). The expansion is connected with decreased river dynamics, due to the construction of dams in its upper course (compare next chapter). *Robinia pseudoacacia* can form its own floodplain forests, replacing the natural *Pinus densiflora* forests on these sites. The herb layer is often dominated by *Festuca arundinacea*. In younger successional stages of the floodplain forests, *Amorpha fruticosa* and *Buddleja davidii* are typical for central Japan (Okuda 1996).

Four species are naturalised in the **aquatic vegetation**. Most frequent is *Egeria densa*, which colonises the lower courses of Japanese rivers.

The floodplain vegetation of northern Japan, especially Hokkaido, is more influenced by European species than those of central Japan. The perennial vegetation types show strong similarities to the central European floodplain vegetation of alpine rivers, probably due to the similar climate in both areas.

The role of natural and man-made disturbances

In general, disturbed habitats (natural or human disturbances) are regarded to be more vulnerable to invasion of aliens than climax habitats (Drake and Mooney 1989, Kowarik 1995, Lohmeyer and Sukopp 1992). The vulnerability of riparian landscapes to the invasion of aliens is often explained by the occurrence of natural disturbances (flooding), as well as to the greater human impact in this ecosystem (Beerling 1995, Ferreira and Moreira, Müller 1995b).

Our survey study showed that most aliens in floodplains were found in annual and perennial vegetation; these habitats are found where river dynamics are the strongest. This is in contrast to woody vegetation (shrubs and floodplain forests), where there are fewer invaders. While the sites of the woody vegetation are normally influenced only by water dynamics, annual and perennial vegetation can be additionally influenced by erosion and accumulation of bedload (gravel and sand). This results in the occurrence of regularly spaced open patches, making it possible for fast invaders to settle. This may be one general explanation for the vulnerability of floodplains to the invasion of aliens.

There is a strong connection between human impact and the fast expansion of aliens. In European alpine rivers, floodplain vegetation changed rapidly after the construction of upstream dams. Due to the decrease in river dynamics, species of typical floodplain vegetation (scree communities, willow shrubs and dry grassland communities) are replaced by species of ruderal and wetland communities (e.g. Müller *et al.* 1992, Müller 1995a, b). At the Isar, the number of aliens increased conspicuously 30 years after the construction of an upstream power plant dam. Mainly, *Erigeron annuus, Impatiens*

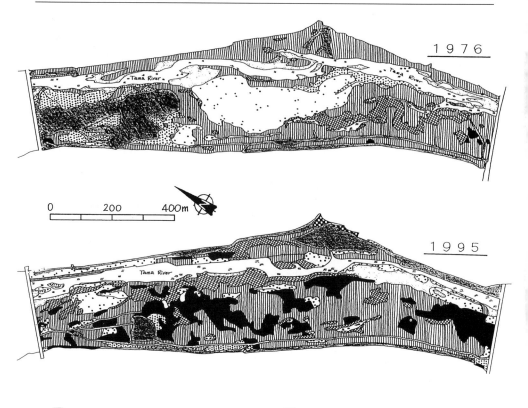

▨ **Water**   □ **Bare gravel**

▨ **Annual communities**   ▨ **Willow shrubs**

▨ **Herbs dom. perennial com.** ■ *Robinia pseudoacacia* **forest**

▥ **Grass dom. perennial com.** ▨ *Quercus acutissima* **forest**

▨ **Artificial bare grounds, trails and gardens**

Fig. 3. Alterations of floodplain vegetation and spreading of *Robinia pseudoacacia* L. at the Tama river (by Tokyo, 100 m a.s.l.) due to the decreasing river dynamics between 1976 and 1995 (after Okuda *et al.* 1976; 1995)

glandulifera, I. parviflora, Oenothera biennis, and *Solidago gigantea* established successfully. On the other hand, typical species of gravel bar vegetation (*e.g. Myricaria germanica, Chondrilla chondrilloides*) decreased or became extinct (Müller 1995c). The same process has been documented for braided rivers in central Japan. In the Kanton district, many typical floodplain species (e. g. *Aster kantoensis, Ixeris tamagawaensis*) are decreasing, and aliens are increasing, as a consequence of strong flood regulation by dams (Washitani 1997).

The success of aliens is mainly due to man-made changes of the disturbance regime. Under natural conditions, the gravel bars of braided rivers are poor in nutrients, periodically dry, and strongly disturbed by the erosion and accumulation of bedload and flooding. Most species in natural alpine floodplains are spread by wind dispersal and form no permanent seedbank (Müller and Scharm 1997). Typical is a vegetation full of gaps, where there is no competition between individual plants. Nutrient availability increases in riparian gravel bars that are impacted by dams (Müller 1995b). Due to fewer large and frequent disturbances, plants can colonise which have the capacity to build up a permanent seedbank (Müller and Scharm 1997). In contrast to species of natural floodplains, a lot of successful aliens have a permanent seedbank *e.g. Ambrosia trifida, Barbarea vulgaris* (Müller and Scharm 1997, Tachibana and Itoh 1997, Takenaka *et al.* 1996, Washitani 1997). Reduced stress and disturbance levels mean that competition becomes the most important factor affecting the vegetation on gravel bars. In this way, competitive alien plants can drive out less competitive native species.

Increased human impact, but not the invasion of aliens, is responsible for the decrease of native species in river ecosystems. The example of floodplains shows the importance of natural disturbances in ecosystems for the survival of plant populations.

Acknowledgements

We are grateful to Peter Bracken (Sydney) for improving the English manuscript, the Ph.D. student Hur Mi Sun (Yokohama National University) for preparing some figures and two anonymous reviewers for comments to the manuscript.

References

Beerling, D. J. 1995. General aspects of plant invasions: an overview. In: Pyšek, P., Prach, K., Rejmanek, M. and Wade M. (eds.), Plant invasions. pp. 237-248. SPB Academic Publishing, Amsterdam.

Drake, J. A. and Mooney, H. A. (eds.) 1989. Biological invasions: a global perspective. 525 p. Scope 37, John Wiley & Sons, Chichester.

Ferreira, M. T. and Moreira, I. S. 1995. The invasive component of a river flora under the influence of Mediterranean agricultural systems. In: Pyšek, P., Prach, K., Rejmanek, M. and Wade M. (eds.), Plant invasions. pp. 117-130. SPB Academic Publishing, Amsterdam.

Johansson, M. E., Nilsson C. and Nilsson E. 1996. Do rivers function as corridors for plant dispersal? Journal of Vegetation Science 7: 593-598.

Kamysev, N.S. 1959. A contribution to the classification of anthropochores. Bot. Zurn. 44: 1613-1616 (in Russian).

Kowarik, I. 1995. On the role of alien species in urban flora and vegetation. In: Pyšek, P., Prach, K., Rejmánek, M. and Wade M. (eds.), Plant invasions. pp. 86-103. SPB Academic Publishing, Amsterdam.

Lippert, W., Müller, N., Rossel S., Schauer, T. and Vetter, G. 1995. Der Tagliamento (Friaul) – Flußmorphologie und Auenvegetation der größten Wildflußlandschaft in den Alpen. Ver. Schutz Bergwelt 60: 11-70.

Lohmeyer, W. and Sukopp, H. 1992. Agriophyten in der Vegetation Mitteleuropas. Schr.-R. Vegetationskde. 25: 185 p.

Meusel, H., Jäger E. and Weinhart, E. 1965. Vergleichende Chorologie der Zentraleuropäischen Flora – Bd. 1: 583 p. Fischer, Jena.

Miyawaki, A. and Okuda, S. 1972. Pflanzensoziologische Untersuchungen über die Auen-Vegetation des Flusses Tama bei Tokyo. Vegetatio 24: 229-311.

Müller, N. 1995a. Wandel von Flora und Vegetation nordalpiner Wildflußlandschaften unter dem Einfluß des Menschen. Ber. ANL 19: 125-187 (in German with English summary).

Müller, N. 1995b. River dynamics and floodplain vegetation and their alterations due to human impact. Arch. Hydrobiol. Suppl. 101 – Large Rivers 9: 477-512.

Müller, N. 1995c. Zum Einfluß des Menschen auf die Vegetation von Flußauen. Schr.-R. Vegetationskde. 27: 289-298.

Müller, N., Dalhof, I., Häcker, B. and Vetter, G. 1992. Auswirkungen unterschiedlicher Flußbaum-aßnahmen auf Flußmorphologie und Auenvegetation des Lech. Ber. ANL 16: 181-213.

Müller, N. and Scharm, S. 1997. Seed bank and seed rain in natural and impacted alpine floodplains. Regulated Rivers in press.

Oberdorfer, E. (ed.) 1992 and 1993. Süddeutsche Pflanzengesellschaften 1-4. Fischer, Jena.

Okuda, S. 1996. Floodplain plant communities and their zonation in several main rivers in Japan. In: Okuda, S. and Ohno, K. (eds.), Ecotechnological study on the restoration of vegetation in water-front areas. 82 p. Institute of Environmental Science and Technology, Yokohama National University (in Japanese with English summary).

Okuda, S., Kobune, A. and Hatase, Y. 1995. The floodplain vegetation of the Tama river. 52 p. and maps. Ministry of Construction, Tokyo (in Japanese with English summary).

Pyšek, P. and Prach, K. 1993. Plant invasions and the role of riparian habitats: a comparison of four species alien to central Europe. Journ. of Biogeography 20: 413-420.

Tachibana, M. and Itoh, K. 1997. Current status and management of naturalized weeds, *Barbarea vulgaris* R. Br. and *Anthemis cotula* L. in the Tohoku area of Japan. Proceedings Int. Workshop on Biological Invasions in Tsukuba, Japan: 282-295.

Takenaka, A., Washitani, I., Kuramoto, N. and Inoue, K. 1996. Life history and demographic features of *Aster kantoensis*, an endangered local endemic of floodplains. Biological Conservation 78: 345-352.

van der Pijl, L. 1982. Principles of dispersal in higher plants. Springer, New York.

Washitani, I. 1997. Difficulties caused by invaders in conservation of threatened plants in Japan. Proceedings Int. Workshop on Biological Invasions in Tsukuba, Japan: 268-281.

PLANT-INSECT INTERACTIONS

ALIEN PLANT-HERBIVORE SYSTEMS AND THEIR IMPORTANCE FOR PREDATORY AND PARASITIC ARTHROPODS: THE EXAMPLE OF *IMPATIENS PARVIFLORA* D.C. (BALSAMINACEAE) AND *IMPATIENTINUM ASIATICUM* NEVSKY (HOM.: APHIDIDAE)

Gregor Schmitz

Institut für Evolutionsbiologie und Ökologie, An der Immenburg 1, D-53121 Bonn, Germany; e-mail: Gschmitz@uni-bonn.de

Abstract

In central Europe, the interaction between *Impatiens parviflora* D.C. (Balsaminaceae) and *Impatientinum asiaticum* Nevsky (Hom., Aphididae), both native to central Asia, represents the rare case of an alien invasive plant being colonised by its original herbivore independently of the event of host plant introduction. The user complex of the currently widespread monophagous aphid is analysed as a step in assessing the biocoenotic role of *I. parviflora*. Field observations were taken around Bonn, Germany. Based on these observations and a literature review, we documented twenty predator species (Syrphidae, Coccinellidae, Sphecidae, Anthocoridae, Chrysopidae, Hemerobiidae, Cecidomyiidae and Anystidae), one ectoparasite species (Trombidiidae), three primary parasitoid (Aphidiidae), and three hyperparasitoid species (Charipidae and Pteromalidae) associated with *Imp. asiaticum*. Anthocoridae, Anystidae, Trombidiidae, and Cecidomyiidae are the most frequent consumers. Due to its wide distribution and occurrence during almost the entire growing period, *Imp. asiaticum* represents an important food resource, and *I. parviflora* an important habitat, for aphidophagous arthropods in the herbaceous undergrowth of woods and forests. In addition, *Imp. asiaticum* probably plays a role in enhancing the pools of beneficial insects in adjacent crop fields.

Introduction

From the points of view of biocoenology, nature conservation and the economy of a given region, the spread of alien plants in native biocoenoses, be it due to accidental or intentional introduction, raises questions concerning the degree to which the introduced species integrates into and changes the existing species community. In this context, competition between the introduced and indigenous plant species is usually the primary aspect of investigation (Trepl and Sukopp 1993).

Further studies of the exploitation of the introduced plant by the resident fauna appear to be necessary to more completely assess the biocoenotic effects following the invasion process. Studies of herbivorous insects play an important role in this context, because (1) effective potential antagonists can be assumed to be among these insects and (2) ecological effects, such as changes in the species spectrum, are expected to be easy to detect, as herbivores show a high degree of host specificity.

The general observation that alien plant species are utilised by fewer herbivore species than (related) indigenous ones (Kowarik 1996), may lead to the assumption that the

Plant Invasions: Ecological Mechanisms and Human Responses, pp. 335–345
edited by U. Starfinger, K. Edwards, I. Kowarik and M. Williamson
© *1998 Backhuys Publishers, Leiden, The Netherlands*

species diversity within other ecological groups (e.g. parasitoids and predators) will also decrease when an alien plant species spreads through vegetation formerly characterised by indigenous plants. However, emphasising only the aspect of resource loss due to the invasion would appear to be too one-sided. In addition to the fact that alien plants can represent attractive pollen and nectar resources (Reinhardt 1987; Schmitz 1994) if accepted as a new host, they also enable indigenous herbivores to extend their habitat range (e.g. Hering 1957).

It is seldom recognised that an alien plant can be a resource for an alien herbivore which is already adapted to the plant, because the herbivore is a regular feeder on the plant in the original distribution region. The invasive alien animal (neozoon) would hardly compete directly with other indigenous species, if it is unable to settle on indigenous plants. Thus, the new plant species could function as an additional resource (host) for secondary consumers.

The species relationship between *Impatiens parviflora* D.C. (Balsaminaceae) and its original herbivore, the aphid *Impatientinum asiaticum* Nevsky (Hom.: Aphididae), is one of the very rare cases of neophyte-neozoon combinations in central Europe (cf. Tab. 22 in Klausnitzer 1989). It is of special biogeographical interest that both organisms, which form a kind of Asian species system [1], were "imported" at different times and independently of one another:

1. *I. parviflora* originates from central Asia and colonised ruderal and forest plant communities in central Europe after escaping from botanical gardens (first records: 1831 in Geneva, 1837 in Dresden). The invasion has been well documented by Hegi (1925), Weise (1966/67) and Trepl (1984). Following a massive spread that occurred primarily in the second half of this century (cf. Jäger 1977; Haeupler and Schönfelder 1988), *I. parviflora* is now the most common neophyte in plant communities of the forests and forest edges (Trepl 1984; Demuth 1992).

2. *Impatientinum asiaticum* (*Impatientinum* abbreviated below as "*Imp.*") was recorded in eastern and central Europe, roughly 150 years (first record in Moscow in 1967: Holman 1971) after the initial invasion of the host plant *I. parviflora*. The aphid also infests the alien *I. glandulifera* Royle in central Europe, but not the indigenous *I. nolitangere* L. The latter is already occupied by the indigenous and strongly monophagous *Imp. balsamines* (Kaltenb.), which colonises the inflorescences and undersides of the leaves in a similar manner. How *Imp. asiaticum* reached the new region is unknown; Lampel (1978) assumes unintentional introduction via aeroplane. Today, the species is clearly widespread in central and northern Europe: Denmark, Czech Republic, Germany, Poland, Russia and Switzerland (summarised by Lampel 1978), Rumania (Holman and Pintera 1981), Great Britain (Blackman 1984), Finland and Sweden (Heie 1994), and the Netherlands (Lampel, pers. comm. 1997).

Imp. asiaticum has become the most frequent and abundant herbivore on *I. parviflora* and is obviously also one of the major aphid species in the herbaceous undergrowth of our woods and forests (cf. Wagner and Rashed 1997, unpubl. manuscr.). A description of the central European user complex of *Imp. asiaticum* is of particular interest, not

---

[1] The rust *Puccinia komarowi* Transch. (Uredinales), which was discovered in Germany for the first time in 1933 (Sydow 1935), is another Asian species included in this system.

only because it is an alien organism (integration into indigenous biocoenoses), but also because an assessment of the biocoenotic effects due to the invasion of *I. parviflora* should include not only primary consumers, but also the third trophic level. The special case of an aphid as prey raises two questions: 1. Does the neozoon support aphidophagous arthropods (which may be beneficial in cultivated fields), 2. If it does, what could be the biological reason for the acceptance?

The data used in this analysis are based on observations made in the Bonn region within the framework of comparative studies of the entomofauna of the three balsam species *I. parviflora, I. glandulifera*, and *I. noli-tangere* (cf. Schmitz 1991, 1994, 1995). Additional findings of other authors complete the analysis.

Study sites and methods

The *I. parviflora* populations studied are located on the outskirts of Bonn. The primary study site was a deciduous forest relict (*Acer pseudoplatanus, Carpinus betulus, Fagus silvatica* and *Prunus avium*) at the western edge of the city, on Kreuzberg hill in Bonn-Poppelsdorf (cf. Schmitz 1991). In addition to *I. parviflora*, the herbaceous layer is characterised by *Alliaria petiolata, Geum urbanum, Lamium album, L. maculatum, Aegopodium podagraria* and *Urtica dioica*. Geophytes are *Ranunculus ficaria, Arum maculatum* and *Corydalis cava*.

Stands of the plant in the Venner valley (Bonn-Bad Godesberg), the Annaberg valley (Bonn-Friesdorf), the Kottenforst woods (south of Bonn-Ippendorf) and on Petersberg hill (Siebengebirge hills) were also examined. These sites represent less disturbed forest communities. *I. glandulifera* was studied at a Calystegion site located in the area of the Sieg estuary between Bergheim and Bonn-Schwarzrheindorf.

The Kreuzberg (*I. parviflora*) and Sieg (*I. glandulifera*) areas were visited every ten days during the 1989 vegetation period in order to assess the entomofauna qualitatively and quantitatively (see Schmitz 1991, 1994). Within the period of occurrence of *Imp. asiaticum* (see below), between 25 and 60 individual plants were examined at each visit; the number of samples was gradually reduced as biomass increased. While predators and parasitoids (aphid mummies) were manually collected at random, ectoparasites were counted completely. The eggs, larvae, pupae and mummies collected were reared in order to obtain imagoes for identification. Aphidophagous larvae were separated and fed with aphids (*Hyalopterus pruni* F. from *Phragmites australis*). Additional material was obtained from the other sites mentioned above. Some of the data on aphid-eating Anthocoridae, Sphecidae and Anystidae are based only on field observations. Specialists were consulted for the identification of reared parasitoids (see Acknowledgements).

Results

1. Phenology of Imp. asiaticum

The life cycle of *Imp. asiaticum* is anholocyclic (no sexually reproducing generation) and monoecious (absence of a host plant alternation) (Lampel 1978). In the area of Bonn, *Imp. asiaticum* was present on *I. parviflora* from the end of May to the end of

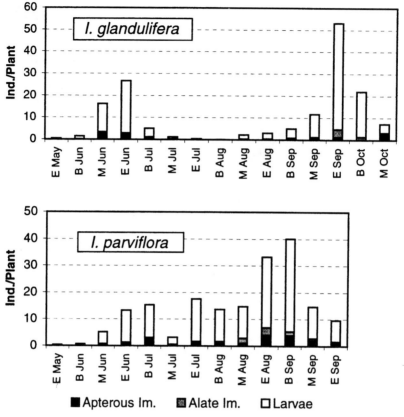

Fig. 1. Comparison of the phenology of *Imp. asiaticum* on *I. glandulifera* and *I. parviflora*, including the percentage of larvae, apterous, and alate viviparous females (1989, in the area of Bonn), where: B = Beginning, M = Mid, E = End of a month.

September (when dryness killed the host plant population), and on *I. glandulifera* (the second host plant in central Europe) from the end of May to mid-October (Fig. 1). A substantial population decrease was observed on *I. glandulifera* in mid-summer, while this phenomenon was only weakly expressed on *I. parviflora*. The population of *Imp. asiaticum* declined in late summer/autumn due to consumption by antagonists (see below) probably combined with weather-related changes in the physiological condition of the host plant.

Although *I. glandulifera* is much larger than *I. parviflora*, the latter is colonised by a comparable number of aphids (Fig. 1). Larvae, apterous and alate viviparous females occurred throughout the growing season, with larvae being most numerous, always comprising at least 80% of the individuals found.

2. Complex of aphidophagous arthropods

Unless stated otherwise, the following data are based exclusively on observations on *I. parviflora* due to its role as the original host plant of *Imp. asiaticum*. A total of 27 arthropod species (excluding spiders), consisting of three parasitoid, three hyperparasi-

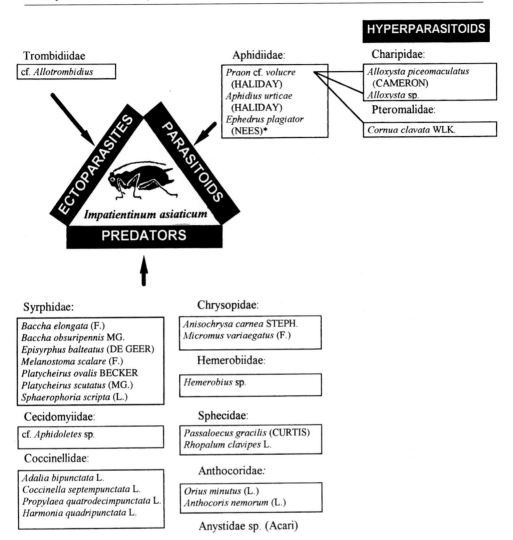

HYPERPARASITOIDS

Trombidiidae
| cf. *Allotrombidius* |

Aphidiidae:
| *Praon* cf. *volucre* (HALIDAY) *Aphidius urticae* (HALIDAY) *Ephedrus plagiator* (NEES)* |

Charipidae:
| *Alloxysta piceomaculatus* (CAMERON) *Alloxysta* sp. |

Pteromalidae:
| *Cornua clavata* WLK. |

ECTOPARASITES PARASITOIDS

Impatientinum asiaticum

PREDATORS

Syrphidae:
| *Baccha elongata* (F.) *Baccha obsuripennis* MG. *Episyrphus balteatus* (DE GEER) *Melanostoma scalare* (F.) *Platycheirus ovalis* BECKER *Platycheirus scutatus* (MG.) *Sphaerophoria scripta* (L.) |

Chrysopidae:
| *Anisochrysa carnea* STEPH. *Micromus variaegatus* (F.) |

Hemerobiidae:
| *Hemerobius* sp. |

Cecidomyiidae:
| cf. *Aphidoletes* sp. |

Sphecidae:
| *Passaloecus gracilis* (CURTIS) *Rhopalum clavipes* L. |

Coccinellidae:
| *Adalia bipunctata* L. *Coccinella septempunctata* L. *Propylaea quatrodecimpunctata* L. *Harmonia quadripunctata* L. |

Anthocoridae:
| *Orius minutus* (L.) *Anthocoris nemorum* (L.) |

Anystidae sp. (Acari)

Fig. 2. Consumer complex of *Imp. asiaticum* on *I. parviflora* in central Europe (* according to Gärdenfors, 1986).

toid, one ectoparasite and 20 predator species, was found to feed directly or indirectly (as hyperparasitoids) on the introduced aphid (Fig. 2). Parasitoids of the predators were excluded.

2.1 Primary parasitoids

The Aphidiidae (Hymenoptera) are moderately to extremely polyphagous. *Aphidius urticae* (Haliday), reared previously from the indigenous *Imp. balsamines* (Kalt.) (Stary, pers. comm. 1990), was identified in connection with *Imp. asiaticum* for the first time. Fulmek (1957) cites six additional hosts for this Aphidiidae. The discovery of *Praon* (cf.) *volucre* (Haliday) confirms previous observations by Stary (1970). Thirty additional hosts are cited for this polyphagous species (Fulmek 1957). Gärdenfors (1986) also reported *Ephedrus plagiator* (Nees), for which he cites approximately 100 additional hosts.

2.2 Hyperparasitoids

All hyperparasitoids recorded within the framework of the Bonn study were reared exclusively from the characteristic mummies/cocoons produced by *Praon* (disc-shaped cocoon). *Praon* cf. *volucre* represents the first host found for the Charipidae *Alloxysta piceomaculata* (Cameron) (Evenhuis, pers. comm. 1990). A second *Alloxysta* species has not yet been definitively identified. The Pteromalidae *Cornua clavata* Wlk. was previously reported for *P. volucre* by Fulmek (1957).

2.3 Ectoparasites

The red nymphs of a Trombidiidae mite (cf. *Allotrombidius*) were found regularly as ectoparasites on *Imp. asiaticum*. Only the aphid populations on *I. parviflora* were infected. A total of 41 mite nymphs were counted among 5394 aphids (never more than two individuals/aphid), which corresponds to an infection rate of 0.76% over the entire study season. Although the most nymphs were found on aphid larvae, consideration of the percentage of larvae and imaginal morphs indicates a possible preference for alate aphids (Table 1).

2.4 Predators

Among the predators, the Syrphidae represent the most species-diverse group (most frequently found species: *Episyrphus balteatus* De Geer). Eggs and larvae were found between mid-July and the end of September. Coccinellidae, Anthocoridae and aphidophagous Cecidomyiidae larvae, probably of the genus *Aphidoletes*, were found regularly in the dense colonies of *Imp. asiaticum*. It is probable that gall midge and hover fly larvae play a major role in the autumn population decrease of the aphid. Correspondingly, numerous depleted colonies were observed in late summer. The other species (Fig. 2) were of secondary significance. The small, red Anystidae (Acari) prey on small, soft-skinned arthropods and were observed to feed on *Imp. asiaticum* larvae on *I. parviflora* in June and July. Sphecidae of the genera *Passaloecus* and *Rhopalum*

Table 1. Distribution of Trombidiidae nymphs on the various morphs of *Imp. asiaticum* on Kreuzberg hill near Bonn in 1989.

Imp. asiaticum		Trombidiidae			
		Nymphs			Nymphs/100 *Imp. asiaticum*
Forms	n	n	%	n	%
Larvae	4547	37	90.2	0.81	40.3
Alatae	294	3	7.4	1.02	50.7
Apterae	553	1	2.4	0.18	9
Total	5394	40	100	0.74	100

were observed capturing aphids for larval food. In addition to the Neuropteroidea on *I. parviflora*, *Chrysopa perla* L. (larvae) and *Hemerobius humulinus* L. (imago) were found on *I. glandulifera*. However, it is not clear whether *Imp. asiaticum* was the prey in this case, due to the fact that colonies of *Aphis fabae* Scop. were also present on *I. glandulifera*.

3. Abundance and phenology of aphidophagous arthropods

The groups of aphidophagous arthropods described occurred at different densities and times (Fig. 3a). During the period of occurrence of the aphid on *I. parviflora*, which amounted to 13 observation dates, the following numbers of aphidophagous arthropods were found on a total of 5394 aphid individuals, distributed among 470 plants: 52 Anthocoridae, 46 individuals each of Anystidae, Trombidiidae and Cecidomyiidae and 44 Syrphidae. The Aphidiidae (counted as mummies), Coccinellidae and Chrysopidae/ Hemerobiidae were found comparatively less frequently (19, 11, and 9 individuals, respectively). Thus, the antagonist/aphid ratio is 1: 19.9.

While the Anystidae occurred in mid-summer, other groups, such as Anthocoridae, were most abundant in late summer or early autumn (Syrphidae). About 90% of all aphidophagous individuals were registered between the end of June and the beginning of September, with the highest abundance in mid-August (Fig. 3b).

Discussion

Imp. asiaticum represents a resource which is not available in constant quantities throughout the growing season. It decreases in mid-summer in a manner similar to the "summer depression", which is characteristic of the annual population development of many aphids on woody plants. It can be assumed that the stagnation of the transport of soluble N-compounds in the phloem not only inhibits the production of aphid offspring on woody plants (Kloft *et al.* 1985), but also on the annual *Impatiens* plants, where N-transport probably decreases between the growing phase and the end phase of seed production in late summer. It remains unclear why this "summer depression" is only weakly expressed by *I. parviflora*.

The description of the consumer complex is of a preliminary nature, as further research may reveal additional species. New records of aphidophagous predators may be expected in particular, because they are, as a rule, relatively host/prey-unspecific. It is also possible that fungi and other pathogens play a role as antagonists of the introduced aphid, although there are no indications of this to date.

Despite these reservations, the studies conducted in the Bonn region support the conclusion that *Imp. asiaticum*, introduced as recently as 20 years ago, is now exploited by a relatively large total number of species as a primary host and prey (3 and 21 species, respectively). In comparison, Stary (1970, 1974, pers. comm. 1990) recorded 5 parasitoid species on *Imp. balsamines*, while Schmitz (1995) found only 4 species of aphid predators in colonies of *Imp. balsamines* and *Aphis fabae* on *I. noli-tangere*. It is unclear whether the acceptance of *Imp. asiaticum* as a host is due to a time-dependent process of adaptation (analogous to host plant-herbivore systems described by Kennedy and Southwood 1984).

Most of the predators of *Imp. asiaticum* are common species and unspecific with re-

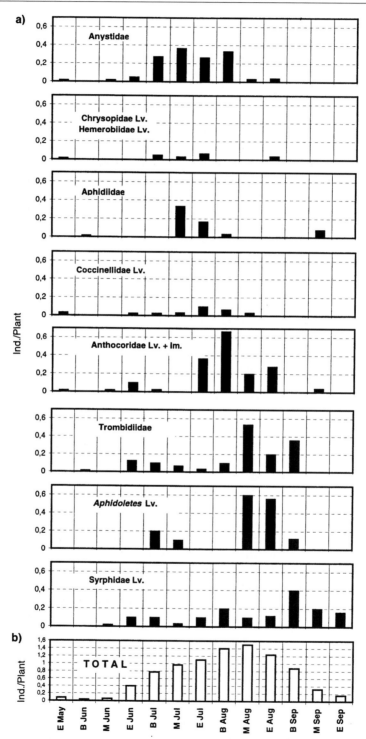

Fig. 3: Phenology and abundance of aphidophagous arthropods found in colonies of *Imp. asiaticum* in a stand of *I. parviflora* in the area of Bonn, 1989 a) differentiated between various groups, b) aphidophages in total.

gard to habitat and aphid food (Minks and Harrewijn 1988; Freier and Gruel 1993). It may be more important to focus on the relatively host-specific family of the Aphidiidae. Stary (1970) compared the Aphidiidae fauna of the two *Impatientinum* species (*asiaticum* and *balsamines*). In addition to *Aphidius urticae* (Stary, pers. comm. 1990), he also reared *Praon grossum* Stary, *Ephedrus lacertosus* (Haliday), *Monoctonus nervosus* (Haliday) (Stary 1970), and *Toxares deltiger* Haliday (Stary 1974), all collected from *Imp. balsamines*; the European population of *Monoctonus nervosus* is specific to *Imp. balsamines* (Stary 1974). It appears that the range of Aphidiidae associated with the introduced aphid *Imp. asiaticum* is basically different, but not much less species-diverse, than that of the closely related indigenous species. The user complex of the aphid may serve as a further example that introduced species are primarily exploited by polyphagous parasitoids (cf. Zwölfer and Pschorn-Walcher 1968).

Based on their polyphagous nature, the consumers of the alien aphid may have a beneficial impact on crop production, if *I. parviflora* stands are located close to fields (e.g. wheat fields, where the decline in harvest yield due to aphid infestation is of economic importance). Corresponding field studies have not been conducted as of yet, but may be worthwhile, as aphidophagous arthropods were found in the colonies of *Imp. asiaticum* in high numbers. The established antagonist/aphid ratio of 1:19.9 (1:23.7 excluding the ectoparasitic Trombidiidae) appears to be high, considering that, for example, a coccinellid imago alone consumes 50 aphids per day (Freier and Gruel 1993). Although group-specific consumption rates must be taken into account in order to optimise the assessment of aphid consumption by predators and parasitoids, only rough mathematical models are available for this purpose to date (Freier, pers. comm. 1998).

The fact that the aphid evidently represents a new food resource for aphidophagous arthropods may be explained as follows:

1. As prey, *Imp. asiaticum* is apparently not difficult to consume because:
   a) The colonies of *Imp. asiaticum* are easily accessible to predators (no leaf-rolls or galls, but rather open, dense colonies).
   b) The aphids do not secrete a large amount of wax, which would otherwise protect them against predation, in addition to preventing the colony from clumping with honeydew.
   c) In contrast to *Aphis fabae*, which also occurs on *Impatiens* species, *Imp. asiaticum* is not a member of a trophobiosis and thus does not enjoy the protection of ants.
   d) It is probable that the aphids do not contain any strongly effective antifeedants originating from the host plant. This is also not to be expected because *I. parviflora* contains only moderately toxic chemical compounds (cf. Hegnauer 1964), such as polyphenols, which are known to have more of an allelopathic effect in general (Schede 1992). The fact that *Aphis sambuci* L. uses the cyanogenic glycosides (sambunigrine) of its host plant *Sambucus* spp. as a repellent against diverse species of ladybugs (Coccinellidae) shows that even the phloem-sucking aphids can, in general, accumulate toxins from the plant (cf. Klausnitzer 1992).

2. *Imp. asiaticum* is probably a "predictable" resource for predators because:
   a) The host plant *I. parviflora* is widespread and relatively non-discriminatory. For example, the plant grows in heavily shaded areas, which are infrequently colonised by indigenous plant species. In addition, the aphid apparently colonises *I. parviflora*

(and *I. glandulifera*) under a wide range of growing conditions.

b) Although *I. parviflora* is an annual plant, it does not occur temporarily (except on ruderal sites), but rather, once established, can be found at the same site for several years (according to Trepl 1980, different densities in individual years can be attributed to climatic conditions).

c) As a monoecious species, *Imp. asiaticum* is present throughout most of the growing season.

3. The supply of other aphid colonies in the undergrowth appears to be relatively meager (Wagner and Rashed, unpubl. manuscr. 1997).

In addition to the pollinator complex dominated by Syrphidae (Schmitz 1994), *Imp. asiaticum* and its consumer complex are important as regards species diversity. Thus, the introduced aphid is an essential element in the assessment of the biocoenotic significance of *I. parviflora* (cf. Schmitz 1995, 1998). As the introduced aphid *Imp. asiaticum* does not compete directly with indigenous herbivores, but does represent an important food resource for aphidophagous arthropods, the common opinion that neophytes markedly reduce species diversity does not apply in the case of *I. parviflora*.

Acknowledgements

I would like to thank Dr. Evenhuis (Bennekom, The Netherlands), Dr. P. Stary (České Budějovice, Czech Republic) and Dr. S. Vidal (Hanover, Germany) for identifying the Charipidae, Aphidiidae, and Pteromalidae, respectively, and for providing information on the biology and taxonomy of the parasitoids. I am particularly indebted to Prof. Dr. G. Lampel (Fribourg, Switzerland) for information on the current range of *Impatientinum asiaticum* and to Ms. M. Onofrietto (Bonn, Germany) for assisting with the translation.

References

Blackman, R.L. 1984. Two species of Aphididae new to Britain. Entomol. Mon. Mag. 120: 185-186.

Demuth, S. 1992. Balsaminaceae. In: Sebald, O., Seybold, S. and Philippi, G. (eds.), Die Farn- und Blütenpflanzen Baden-Württembergs. pp. 198-204, Vol. 4, Haloragaceae bis Apiaceae. Ulmer, Stuttgart.

Freier, B. and Gruel, H.-J. 1993. Vorkommen und Bedeutung von Marienkäfern (Coccinellidae) als Nützlinge in Agrarökosystemen. Gesunde Pflanzen 45 (8): 300-307.

Fulmek, L. 1957. Insekten als Blattlausfeinde. Kritisch-statistische Sichtung. Ann. naturhist. Mus. Wien 61: 100-227.

Gärdenfors, U. 1986. Taxonomic and biological revision of palearctic *Ephedrus* Haliday (Hymenoptera: Braconidae, Aphidiinae). Entomol. Scand. Suppl. 27.

Haeupler, H. and Schönfelder, P. 1988. Atlas der Farn- und Blütenpflanzen der Bundesrepublik Deutschland. Ulmer, Stuttgart, 768 pp.

Hegi, G. 1925. Illustrierte Flora von Mitteleuropa. Vol. 5/1, Lehmann, München.

Hegnauer, R. 1964. Chemotaxonomie der Pflanzen. Vol. 3: Dicotyledonae: Acanthaceae -Cyrillaceae. Birkhäuser, Basel, 745 pp.

Heie, O.E. 1994. The Aphidoidea (Hemiptera) of Fennoscandia and Denmark. 5. Family Aphididae: part 2 of tribe Macrosiphini of subfamily Aphidinae. Fauna Entomol. Scand. 28: 242 pp.

Hering, M. 1952. Probleme der Xenophobie und Xenophilie bei der Wirtswahl phytophager Insekten. Trans. XI. Intern.Congr. Entomol. 1 : 507-513.

Holman, J. 1971. Taxonomy and ecology of *Impatientinum asiaticum* Nevsky, an aphid species recently introduced to Europe (Homoptera, Aphididae). Acta entomol. bohemoslov. 68: 153-166.

Holman, J. and Pintera, A. 1981. Übersicht der Blattläuse (Homoptera, Aphidoidea) der Rumänischen Sozialistischen Republik. Studie CSAV, Vol. 15, Academia, Praha, 125 pp.

Jäger, E. 1977. Veränderungen des Artenbestandes von Floren unter dem Einfluß des Menschen. Biol. Rundsch. 15 (5): 287-300.

Kennedy, C.E.J. and Southwood, T.R.E. 1984. The number of species of insects associated with British trees: a re-ananlysis. J. Anim. Ecol. 53: 455- 478.

Klausnitzer, B. 1989. Verstädterung von Tieren. Die Neue Brehm-Bücherei, Vol. 579. 2nd ed. Ziemsen Verl., Wittenberg Lutherstadt, 316 pp.

Klausnitzer, B. 1992. Coccinelliden als Prädatoren der Holunderblattlaus (*Aphis sambuci* L.) im Wärmefrühjahr 1992. Entomol. Nachr. Ber. 36: 185-190.

Kloft, W.J.; Maurizio, A. and Kaeser, W. 1985. Waldtracht und Waldhonig in der Imkerei. 2nd Ed., Ehrenwirth, München, 329 pp.

Kowarik, I. 1996. Auswirkungen von Neophyten auf Ökosysteme und deren Bewertung. Texte des Umweltbundesamtes 58: 119-155.

Lampel, G. 1978. *Impatientinum asiaticum* News., 1929, eine asiatische Blattlausart, neu im botanischen Garten Freiburg/Schweiz. Bull. Soc. Frib. Sc. Nat. 67 (1): 69-72.

Lohmeyer, W. and Sukopp, H. 1992. Agriophyten in der Vegetation Mitteleuropas. Schriftenr. Vegetationskd. 25: 1-185.

Minks, A.K. and Harrewijn, P. 1988. Aphids – Their biology, natural enemies and control. Vol. B, Elesvier, Amsterdam, Oxford, New York, Tokyo, 364 pp.

Reinhardt, R. 1987. Langjährige Beobachtungsergebnisse von Tagfaltern an *Buddleja*-Sträuchern am Rande einer Großstadt. Tagungsber. XI. S.I.E.E.C. (Gotha).

Schede, D. 1992. Ökologische Biochemie. 2nd ed., Fischer Verl., Jena, Stuttgart, New York, 587 pp.

Schmitz, G. 1991. Nutzung der Neophyten *Impatiens glandulifera* Royle und *I. parviflora* D.C. durch phytophage Insekten im Raume Bonn. Entomol. Nachr. Ber. 35 (4): 260-264.

Schmitz, G. 1994. Zum Blütenbesuchsspektrum indigener und neophytischer *Impatiens*-Arten. Entomol. Nachr. Ber. 38 (1): 17-23.

Schmitz, G. 1995. Neophyten und Fauna - Ein Vergleich neophytischer und indigener *Impatiens*-Arten. In: Böcker, R., Gebhardt, H., Konold, W. and Schmidt-Fischer, S. (eds.), Gebietsfremde Pflanzenarten - Auswirkungen auf einheimische Arten, Lebensgemeinschaften und Biotope - Kontrollmöglichkeiten und Management. pp. 195-204, Ecomed, Landsberg.

Schmitz, G. 1998. *Impatiens parviflora* D.C. (Balsaminaceae) als Neophyt in mitteleuropäischen Wäldern und Forsten – Eine biozönologische Analyse. Z. Ökol. Natursch. (in prep).

Stary, P. 1970. Parasites of *Impatientinum asiaticum* Nevsky, a new introduced aphid to central Europe (Hom., Aphididae; Hym., Aphidiidae). Boll. Labor. Entomol. Agrar. "Filippo silvestri" (Napoli) 28: 236-244.

Stary, P. 1974. Host range and distribution of *Monoctonus nervosus* (Hal.) (Hymenoptera: Aphidiidae). Z. Angew. Entomol. 75: 212-224.

Sydow, H. 1935. Einzug einer asiatischen Uredinee (*Puccinia komarowi* Tranzsch.) in Deutschland. Ann. mycol. 33: 363-366.

Trepl, L. 1980. Über die kleinstandörtliche Verteilung von *Impatiens parviflora* in einem Eichen-Hainbuchenwald und einem standörtlich entsprechenden Fichtenforst. Decheniana, 133: 6-22.

Trepl, L. 1984. Über *Impatiens parviflora* DC. als Agriophyt in Mitteleuropa. Diss. Bot. 73, 400 pp.

Trepl, L. and Sukopp, H. 1993. Zur Bedeutung der Introduktion und Naturalisation von Pflanzen und Tieren für die Zukunft der Artenvielfalt. Rundgespräch Komm. f. Ökol. 6: 127-142.

Wagner, P. and Rashed, A. 1997. Zum Einfluß von *Impatiens parviflora* D.C. (Balsaminaceae) auf die Zusammensetzung der Arthropodenfauna in der Krautschicht, untersucht im Kottenforst bei Bonn. unpup. manuscr., 11 pp.

Weise, B. 1966/67. Untersuchungen über Konkurrenzbeziehungen von *Impatiens parviflora* und *Impatiens noli-tangere*. Ber. Arbeitsgem. sächs. Bot., N.s. 8: 102-122.

Zwölfer, H. and Pschorn-Walcher, H. 1968. Wie verhalten sich Insektenparasiten gegenüber eingeschleppten, faunenfremden Wirten? Anz. Schädlkd. 41 (4): 51-55.

INTERACTIONS BETWEEN AN INVASIVE PLANT, *MAHONIA AQUIFOLIUM*, AND A NATIVE PHYTOPHAGOUS INSECT, *RHAGOLETIS MEIGENII*

Leo L. Soldaat and Harald Auge
UFZ Centre for Environmental Research Leipzig-Halle, Department of Community Ecology, Theodor-Lieser-Straße 4, D-06120 Halle, Germany; e-mail: sol@oesa.ufz.de / aug@oesa.ufz.de

Abstract

The North American shrub *Mahonia aquifolium* is a successful invader of man-made and seminatural habitats in Central Europe. The fruits of this invasive plant are infested by the fruit fly *Rhagoletis meigenii*, a seed predator of the native shrub *Berberis vulgaris*. The proportion of infested fruits as well as seed predation are higher in the invasive host plant *M. aquifolium* than in the native host *B. vulgaris*, but low in comparison to other insect – plant systems. In both plant species infestation by *R. meigenii* leads to increased seed abortion and to a higher risk of fungal infection. The earlier seed ripening in *M. aquifolium* has caused a phenological shift in the insect. Furthermore, larval abundance is much higher on the invasive host plant than on the native host. This difference is partly caused by the higher fruit production and possibly by the lower seed abortion in *M. aquifolium*. We suggest that the impact of *R. meigenii* on the invasion process is small because of the relatively low amounts of seed predation and insect-induced seed abortion. Inversely, the ongoing invasion by *M. aquifolium* will strongly increase the abundance of the insect.

Introduction

An important aspect of biological invasions is the interaction of the invader with other components of the recipient ecosystem (Vermeij 1996), for instance with organisms at a different trophic level. In the case of plant invasions, interactions with native phytophagous insects are of special importance for two reasons. First, a release from specialist herbivores is often proposed to be one of the major causes for the success of invasive plants (Milton 1980; Blossey and Nötzold 1995). However, native phytophagous insects may rapidly build up stable populations on invading plant species (Auerbach and Simberloff 1988; Singer *et al.* 1993; Anderson 1995) even causing the decline of plant invaders at a local scale (Creed and Sheldon 1995). There are also many examples in which phytophagous insects that have been introduced in a new region as biological control agents have affected populations of other plant species than the target weed (Marohasy 1996, Simberloff and Stiling 1996), indicating that insects may be pre-adapted to more host plants than they actually feed on (Jermy 1993). The second reason that makes interactions between invasive plants and native phytophagous insects interesting concerns the consequences for the insect itself: The colonization of a new host may lead to changes in insect life-history (Leclaire and Brandl 1994) or even to genetical differentiation between host-specific insect populations (Bush 1993).

Plant Invasions: Ecological Mechanisms and Human Responses, pp. 347–360
edited by U. Starfinger, K. Edwards, I. Kowarik and M. Williamson
© 1998 Backhuys Publishers, Leiden, The Netherlands

To summarize, the interactions between invasive plants and native phytophagous insects are very diverse. More case studies are needed in order to come to generalizations about the interaction between invasive plants and native herbivores.

In this paper we will summarize our current knowledge of the interaction between the invasive shrub *Mahonia aquifolium* (Pursh) Nutt. (Berberidaceae) and the native fruit fly *Rhagoletis meigenii* Loew (Diptera: Tephritidae). *M. aquifolium*, a fleshy-fruited shrub from North America, is a successful invader in various parts of Europe. Preliminary field observations indicated that fruits of *M. aquifolium* are often infested by *R. meigenii* (Auge *et al.* 1997). The larvae of this specialist insect originally only predated the seeds of the European shrub *Berberis vulgaris* L. (Berberidaceae). In detail, we will concentrate on the following questions: (a) What are the consequences of infestation by *R. meigenii* for the reproductive success of the two host plant species? (b) How does the colonization of the invasive plant affect abundance, phenology and pupal weight of the insect?

Study sites and methods

The study was performed at 12 sites in central Germany. Two *B. vulgaris* populations and one *M. aquifolium* population were located in the dry scrublands north of Halle. Four *B. vulgaris* and seven *M. aquifolium* populations were in the pine forests of the Dübener Heide north of Leipzig and the Dölauer Heide near Halle. At two of these study sites we found co-occurring populations of both hosts. All *B. vulgaris* populations and six *M. aquifolium* populations were characterized by spatially separated patches (*B. vulgaris*: 26 to 296 patches ha^{-1}; *M. aquifolium*: 10 to 167 patches ha^{-1}). As a patch we define a distinct spatial unit within populations formed either by a separately growing individual or by the intermingling sprouts of two or more individuals (patch area: 0.8 to 294 m^2 in *M. aquifolium*; 1.4 to 55 m^2 in *B. vulgaris*). Two *M. aquifolium* populations, however, were characterized by a more or less continuous cover, indicating a later stage of population growth. We determined fruit and seed production as well as infestation and seed predation in five patches (if present) of each population. In the two *M. aquifolium* populations without a clear patch structure we used five randomly placed 1 m^2 quadrats. Sampling was performed in 1995 and 1996 after fruit ripening, when most larvae had left the fruits (in *M. aquifolium* in August and in *B. vulgaris* in October). Fruit production per ha was estimated by multiplying the number of patches per ha with the mean fruit number per patch. Infestation by *R. meigenii* can be easily determined by the presence of damaged seeds in ripe fruits. The effect of fruit puncturing by ovipositing *R. meigenii* females on seed abortion was investigated in three populations of each host plant species in 1997. Observations on host plant phenology were carried out in two populations of both host plants in 1995, and insect phenology was studied in one population of each host plant in 1997. Pupal weights were determined on pupae collected from fruits of three *M. aquifolium* populations and two *B. vulgaris* populations in 1995. All statistical analyses were performed using generalized linear models with appropriate error distributions (GLIM package, see Crawley 1993). Differences between host plant species were tested by means of nested models (populations nested within species).

Natural history of the species

B. vulgaris, the original host plant of *R. meigenii*, is a deciduous non-clonally grow-ing shrub native to Europe, inhabiting dry scrub and open forests. Flowering in the study area starts in May. The inflorescences consist of 13 ± 4 flowers (mean$\pm$s.d.) and are produced by short shoots scattered along the sprouts. The red berries are ripe in September. *M. aquifolium* is an evergreen clonally growing shrub native to western North America and was introduced to Europe in 1822 (Kowarik 1992) for horticul-tural purposes. Plants cultivated and naturalized in Europe are mostly hybrids of closely related *Mahonia* species (Ahrendt 1961) but will be referred to as *M. aquifolium* in this paper. These cultivated forms are known as successful invaders of man-made and seminatural habitats, especially pine forests (Auge and Brandl 1997). Flowering in the study area starts in early April. The inflorescences are formed by 21 ± 10 flowers and are aggregated near the apex of the sprouts. The blue berries turn ripe in late July.

The fruit fly *R. meigenii* is native to northern and central Europe where the larvae predate on seeds in the fruits of *B. vulgaris* (Hendel 1927; Huppmann 1986; White 1988). On *B. vulgaris,* the insect is probably univoltine. Adult insects emerge from overwintering pupae in early summer. The females lay eggs in the seeds of young fruits and the larvae predate the ripening seeds. Usually one larva develops per fruit (Hendel 1927; Huppmann 1986). We assume that this is the result of an oviposition marking pheromone reducing the chance of multiple ovipositions as it was found in other *Rhagoletis* species (Roitberg *et al.* 1982; Bauer 1986; Mayes and Roitberg 1986; Prokopy *et al.* 1987; van Randen and Roitberg 1996). After three larval stages the larva leaves the fruit through a hole in the fruit coat and falls to the ground to pupate and overwinter in the soil. *M. aquifolium* and *B. vulgaris* may co-occur in gardens but also in natural habitats, which may have facilitated the host range expansion of *R. meigenii*. For both host plant species *R. meigenii* seems to be the only pre-dispersal seed predator at our study sites. In the southern parts of central Europe *B. vulgaris* seeds are also eaten by larvae of *Rhagoletis berberidis* (Jermy 1961; Huppmann 1986).

Consequences of infestation for the reproductive success of the host plant

Infestation and seed predation

At the patch and population levels fruit production was extremely variable in both hosts. *M. aquifolium* patches contained more fruits than *B. vulgaris* patches, but fruit density in the patches was not significantly different (Table 1). Fruit production per ha was three times higher in *M. aquifolium* populations than in *B. vulgaris* popula-tions. Because of the higher number of ovules per fruit and the much lower seed abor-tion, seed production per ha in *M. aquifolium* was seven times higher than in *B. vul-garis*.

Infestation was found in all *M. aquifolium* populations investigated and in four of six *B. vulgaris* populations in 1995 and in all populations of both hosts in 1996. Sur-prisingly, the mean percentage of infested fruits per patch was more than five times higher in the invasive host plant than in the native host (Table 1). This difference be-tween the two plant species, as well as the large variation in infestation between patches, could not be explained by differences in fruit number or fruit density per patch. How-

Table 1. Seed and fruit production, infestation and seed predation by R. meigenii and fruit rot in six B. vulgaris and eight M. aquifolium populations in 1995 and 1996. Note that seed weight per fruit is not simply the product of mean seed weight and mean seed number per fruit because of sampling effects. The percentage seed predation is based on the total number of non-aborted seeds per patch.

	native host Berberis vulgaris			invasive host Mahonia aquifolium			
	mean ± s.e.	range	n	mean ± s.e.	range	n	significance
SEEDS							
seed weight [mg]	**9.4** ± 0.1	2.1 – 17.7	779	**9.9** ± 0.0	2.4 – 16.3	2 686	n.s.
FRUITS							
ovules per fruit	**2.1** ± 0.0	1 – 4	731	**4.2** ± 0.0	1 – 8	2 215	p<0.001
seeds per fruit	**1.1** ± 0.0	0 – 4	2 827	**2.8** ± 0.0	0 – 7	4 697	p<0.05
seed weight per fruit [mg]	**10.8** ± 0.2	2.1 – 31.9	385	**40.0** ± 2.4	5.5 – 174.8	190	p<0.05
PATCHES							
fruits per patch	**912** ± 257	4 – 18 684	54	**2 434** ± 841	7 – 32 458	54	p<0.05
fruits per m² per patch	**171** ± 38	0.6 – 1 380	54	**117** ± 21	1.2 – 831.8	54	n.s.
seeds per patch	**1 970** ± 500	8 – 19 169	54	**7 346** ± 2 733	23 – 117 609	54	n.s.
% infested fruits	**5.7** ± 1.3	0 – 54.9	54	**30.2** ± 2.9	0 – 90.3	75	p<0.001
% fruit rot	**4.2** ± 0.0	0 – 19.4	54	**19.5** ± 0.0	0 – 87.3	75	p<0.001
% seed abortion	**31.3** ± 3.3	0 – 81.8	27	**9.4** ± 1.1	0 – 27.0	37	p<0.001
% seed predation	**9.6** ± 2.4	0 – 88.9	54	**19.1** ± 1.9	0 – 57.9	75	p<0.001
POPULATIONS							
fruits per ha	**77 028** ± 43 036	924 – 537 385	12	**241 097** ± 107 909	3 367 – 1 396 000	14	p=0.068
seeds per ha	**106 597** ± 56 616	1 621 – 708 663	12	**736 036** ± 330 389	9 352 – 4 133 245	14	p<0.01

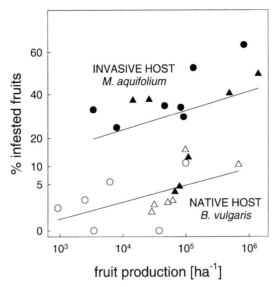

fruit production [ha⁻¹]

Fig. 1. The mean percentage (arcsine square root transformed) of fruits infested by *R. meigenii* in populations of the invasive host plant *M. aquifolium* (black symbols) and the native host plant *B. vulgaris* (white symbols) in 1995 (circles) and 1996 (triangles) in relation to the number of fruits per hectare. Mean infestation at the population level was significantly higher in *M. aquifolium* than in *B. vulgaris* (38.5 % vs. 3.4 % in 1995 and 26.3 % vs. 5.4 % in 1996; $F_{1;21}$=28.9, p<0.001). The relationship between infestation and fruits per hectare was significant ($F_{1;21}$=8.12, p=0.01) and fruit density was higher in 1996 than in 1995 ($F_{1;22}$=4.3; p<0.05).

ever, infestation of populations was positively related to fruit production per ha (Fig. 1). This effect of fruit production accounts for part of the difference in infestation between the host plant species. The large difference in infestation between the two host plants does not translate into a comparable difference in seed predation (Table 1): The percentage of non-aborted seeds consumed by *R. meigenii* larvae is only two times higher in the new host plant as compared to the native host plant. This is due to the higher number of seeds per fruit in *M. aquifolium* and to the fact that fruits are usually infested by a single *R. meigenii* larva consuming a comparable number of seeds per fruit in each host (see below).

Infestation and seed abortion

In addition to the direct effect of seed predators, an indirect effect on host plant fecundity may be that seed or fruit abortion is induced. In both host plant species the probability that at least one seed is aborted increases with the number of oviposition punctures per fruit (Fig. 2), indicating that seed abortion is a response of the plant to oviposition by *R. meigenii*. Selective seed abortion may be a passive reaction of the plant caused by phytopathogens intruding during or after puncturing, or an active defence mechanism against herbivorous insects (Stephenson 1981; Herrera 1984; Fernandes and Whitham 1989; Marquis 1992). In our study system preliminary observations indicate that seed abortion kills eggs or larvae of *R. meigenii*. Thus, selective seed abortion is an advantage for the host plant: Most larvae damage more than one seed, whereas seed abortion only kills the punctured seed. Interestingly, plant response

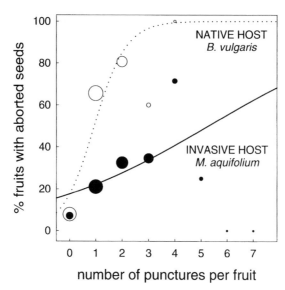

number of punctures per fruit

Fig. 2. The percentage fruits containing aborted seeds in the invasive host plant *Mahonia aquifolium* and the native host *Berberis vulgaris* in relation to the number of oviposition punctures of *R. meigenii*. The percentage of fruits containing aborted seeds increases with the number of punctures (χ_1^2=31.6; p<0.001). The interaction between the number of punctures and host species was significant (χ_1^2=19.9; p<0.001). The area of the symbols corresponds to the number of observations.

to puncturing by *R. meigenii* was much stronger in the native host *B. vulgaris* than in the new host *M. aquifolium* (Fig. 2). Therefore, we conclude that selective seed abortion in *B. vulgaris* is an adaptive trait, whereas the invasive history of *M. aquifolium* in Europe is too short for selection of a comparable response. The consequence of this insect-induced seed abortion is an additional loss of fecundity of the host plants, reducing the proportion of non-aborted seeds from 98% to 90% in *M. aquifolium* but from 91% to 49% in *B. vulgaris* (Table 2).

Infestation and fruit rot

Another indirect effect of herbivory on the host plant may be an increased susceptibility to other herbivores or pathogens (De Nooij *et al.* 1992). In the case of *R. meigenii*, infestation by larvae increases the risk of fungal infection of the fruit pulp (mainly by *Cladosporium cladosporoides* (Fres.) de Vries and *Ramularia* spec., det. N. Luschka; Fig. 3). This effect was the same in the two host plant species but, due to the higher infestation by *R. meigenii*, the overall proportion of fruits suffering fungal rot was five times higher in the invasive plant *M. aquifolium* than in the native *B. vulgaris* (Table 1). From other studies on vertebrate-dispersed plants it is known that fungal fruit rot is generally deterrent to frugivores (e.g. Cipollini and Stiles 1993). At least in *M. aquifolium*, dispersal by vertebrates (mainly birds) is not only necessary for the colonization of new sites but also for local seedling recruitment as the mortality of seedlings emerging beneath the canopy of adults is much higher than in gaps between adult patches (Auge and Brandl 1997). We suggest, therefore, that fungal rot resulting from infestation by *R. meigenii* will decrease the reproductive success of individual

Table 2. Seed abortion in fruits with and without oviposition punctures of *R. meigenii* females in three populations of the native host *B. vulgaris* and three populations of the invasive host *M. aquifolium* in 1997 (± standard errors; number of fruits in parenthesis). The difference between species is significant (χ_1^2=114.5; p<0.001), as well as the effect of puncturing (χ_1^2=22.1; p<0.001). The interaction between host species and puncturing is not significant.

	% of aborted seeds per fruit	
	native host *B. vulgaris*	invasive host *M. aquifolium*
fruits without punctures	8.6 ± 4.3 (35)	1.8 ± 1.8 (14)
fruits with punctures	50.7 ± 3.9 (99)	9.9 ± 1.5 (136)

host plants by reducing the proportion of seeds dispersed and the proportion of seedlings surviving to maturity.

Consequences of host range expansion for the insect

Larval abundance

As most fruits of *B. vulgaris* and *M. aquifolium* contain only one larva, the number of infested fruits per hectare provides a good estimate of larval abundance. The mean abundance of *R. meigenii* larvae in *M. aquifolium* populations was eight times (1996) to 500 times (1995) higher than in *B. vulgaris* populations (Fig. 4). Larval abundance increases non-linearly with fruit production in populations of both host plants as indicated by the positive relationship between the proportion of infested fruits and fruit production (see above). However, the difference in larval abundance between the two host plant species is only to a small part caused by the higher fruit production in *M. aquifolium*. This is reflected in the fact that infestation in *M. aquifolium* is much higher than in *B. vulgaris* even if one corrects for the difference in fruit production. We suggest that a major cause for the higher abundance of *R. meigenii* in the invasive host *M. aquifolium* is the much lower proportion of fruits with insect-induced seed abortion as compared to the native host *B. vulgaris* (see above, Fig. 2).

Host plant and insect phenology

If larval development of phytophagous insects depends on the temporal availability of plant parts the colonization of a new host plant with a different phenology will result in a phenological shift in the insect. There are pronounced phenological differences between the two host plants of *R. meigenii*: Flowering and fruiting of the invasive host plant *M. aquifolium* is much earlier than in the native host *B. vulgaris*. Consequently, seeds in the invasive host ripen about six weeks earlier (Fig. 5a). As expected, oviposition of *R. meigenii* is also earlier on *M. aquifolium* (Fig. 5b). In *M. aquifolium* most larvae leave the fruits at the end of July whereas in *B. vulgaris* many larvae stay in the fruits throughout August. These results clearly show that the recent host range expansion has caused a phenological shift in *R. meigenii* populations.

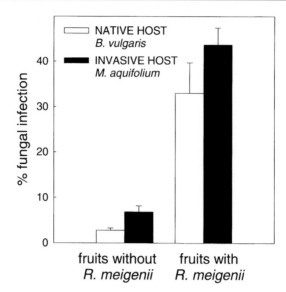

Fig. 3. Fungal infection of fruits that were infested by *R. meigenii* and of uninfested fruits in the invasive host plant *M. aquifolium* and the native host plant *B. vulgaris*. The difference between host plant species and the effect of *R. meigenii* infestation on fungal infection were significant ($F_{1;150}$=100.0; p<0.001 resp. $F_{1;150}$= 360.0; p<0.001).

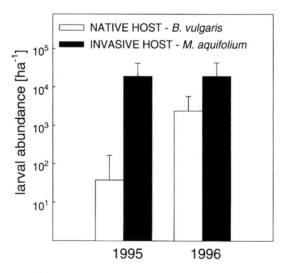

Fig. 4. The abundance of *R. meigenii* larvae on the invasive host plant *M. aquifolium* (7 populations) and the native host plant *B. vulgaris* (6 populations) in 1995 and 1996 as estimated by the mean number of infested fruits per hectare (means and standard errors). The difference between host plants was significant ($F_{1;22}$=17.2; p<0.001), but larger in 1995 than in 1996 (interaction between year and host species: $F_{1;22}$=4.3; p=0.05).

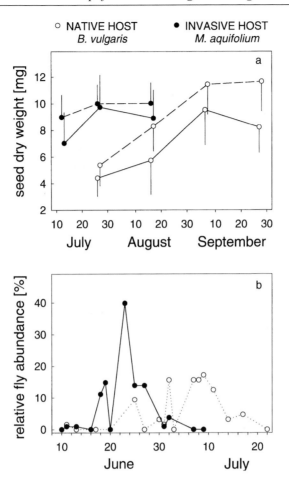

Fig. 5. a. Seed growth in two populations of both host plants of *R. meigenii* during the season in 1995. Similar symbols indicate the same population. The error bars are standard deviations. b. Relative abundance of ovipositing *R. meigenii* females (100 % = total number of females observed on each host plant species during the season in 1997; n = 108 on *M. aquifolium* and n = 64 on *B. vulgaris*). Because of the different architectures of the host plants females were counted on 100 sprouts of *M. aquifolium* and during 15 min on *B. vulgaris*. Note that due to these different methods only a comparison of the seasonal patterns of oviposition is possible, but not a comparison of insect abundances between hosts.

Host plant seediness and pupal weight

Insect performance on new host plants is often lower than on original hosts (Bowers *et al.* 1992; Thompson 1996), although the opposite has also been found (Leclaire and Brandl 1994). Assuming that the nutritional quality of the seeds of *M. aquifolium* and *B. vulgaris* is comparable, pupal weights may be expected to be higher on the invasive host plant because of the much higher seed number per fruit and a comparable individual seed weight (Table 1; Fig. 6a). Differences in larval performance in turn may affect adult fecundity and population dynamics (Leclaire and Brandl 1994). Although pupae originating from *B. vulgaris* fruits showed a tendency towards lower fresh weights compared to pupae from *M. aquifolium* fruits, this difference was rather

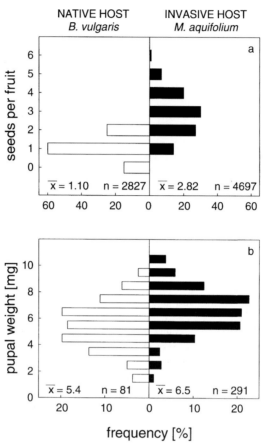

Fig. 6. a. Frequency distribution of the number of seeds per fruit in the invasive host plant *M. aquifolium* and the native host plant *B. vulgaris*. Seed number per fruit was higher in *M. aquifolium* (see Table 1). b. Frequency distribution of fresh weights of individual *R. meigenii* pupae on both host plants. Pupal weights were not significantly different between hosts ($F_{1;12}=1.67$).

small and not significant (Fig. 6b). This unexpected result is due to the fact that in both host plants *R. meigenii* females prefer multi-seeded fruits for oviposition (Table 3). The one larva that usually develops per fruit consumes the same quantity of seeds in both host plants (on average 1.86 seeds per fruit in *M. aquifolium* and 1.75 seeds per fruit in *B. vulgaris*). This means that in the many *M. aquifolium* fruits containing more than two seeds this additional resource is not exploited. The slightly lower pupal weight in *B. vulgaris* may be caused by the higher proportion of pupae originating from one-seeded fruits (59 % of all pupae in *B. vulgaris* as compared to 8 % in *M. aquifolium*). The observed preference for plant modules with high resource supply is typical of endophagous insects (e. g. Herrera 1984; Molau *et al.* 1989; Leclaire and Brandl 1994; Fondriest and Price 1996) as larval development of endophagous insects is usually restricted to the plant module where the eggs are deposited.

Table 3. Infestation by *R. meigenii* in 6 populations of its native host *B. vulgaris* and 8 populations of the invasive host *M. aquifolium* in relation to the number of seeds per fruit (± standard errors; number of fruits in parenthesis). Only infested patches are included in the table. Infestation was higher in multi-seeded fruits of both host plant species ($F_{1;125}$=20.2, p<0.001).

		% infested fruits per patch			
Year	Seeds per fruit	native host B. vulgaris		invasive host M. aquifolium	
1995	one	6.3 ± 0.2	(540)	25.9 ± 1.6	(294)
	two or more	44.3 ± 3.3	(109)	46.0 ± 0.6	(1 763)
1996	one	18.5 ± 1.0	(600)	11.0 ± 0.9	(310)
	two or more	38.5 ± 2.1	(293)	32.7 ± 0.5	(1 905)

Conclusions

Two of the many characteristics that have been proposed to increase the success of invasive plants are a high seed production (Rejmánek 1996; Noble 1989) and a lack of specialist herbivores in the new area (Milton 1980; Blossey and Nötzold 1995). Sexual reproduction indeed plays an important role in the regional and local spread of *M. aquifolium* (Auge and Brandl 1997). We assume that the reproductive potential of invasive *M. aquifolium* plants is higher than in their wild relatives, because of the efforts of horticulturalists to breed plants with high flower and fruit number. Seed loss in *M. aquifolium* (19 %) due to *R. meigenii* is rather low as compared to other studies on seed predation: Compiling data from ca. 60 studies Crawley (1992) found an average of 45 % pre-dispersal seed predation. No data are available on seed predation in *M. aquifolium* in its natural range. Seed predation has a very strong effect on the fecundity of individual plants in comparison to other kinds of herbivory as it directly kills seeds. The effect of seed predation on population dynamics, however, depends on the extent to which seedling recruitment is seed limited (Crawley 1992) and on the sensitivity of the life cycle to seedling recruitment (cf. Silvertown *et al.* 1993). Although we do not have data on the role of seed limitation in the life cycle of *M. aquifolium*, we suggest that the relatively low amount of seed predation will have only a small impact on the invasion process. The additional seed loss due to insect-induced seed abortion is negligible in *M. aquifolium*. Like seed predation fruit rot may affect the reproductive success of individual plants, due to its interaction with vertebrate dispersers. But again, its impact on the dynamics of *M. aquifolium* populations will depend on the extent of seed limitation.

In contrast to the suggested low impact of *R. meigenii* on the the invasion process, the effect of the host range expansion by *R. meigenii* on its own population biology is much stronger. First, the phenological difference between host plant species has caused a phenological shift in the insect. A comparable response was found in other insect – plant interactions. For instance, emergence of adults in the apple maggot fly *Rhagoletis pomonella* and in the gall wasp *Andricus mukaigawae* is synchronized with differences in the phenology of host plant species which may lead to allochronic isolation of insect populations and ultimately to the formation of genetically different host races (Smith 1988; Abe 1991). It remains to be tested if the *R. meigenii* flies on *M. aquifolium* and *B. vulgaris* belong to host-specific populations or to populations that shift their egg-laying activity from one host plant to the other during the season. Second, *R. meigenii* reaches much higher abundances on its new host plant *M. aquifolium* as compared to its original host plant *B. vulgaris*. We suggest that this difference is mainly caused by the higher fruit production and the lower seed abortion in the invasive host plant. The use of *M. aquifolium* as an ornamental shrub is still widespread in Europe, invasive populations were reported from various European countries (Tutin *et al.* 1993; Thompson *et al.* 1995) and local *M. aquifolium* populations are still expanding (Auge 1997). Therefore, we predict that both the number and the size of *R. meigenii* populations will further increase in the future.

Acknowledgments

We thank Klaus Hempel, Antje Thondorf and Hannelore Jany for their assistance in counting, collecting, dissecting and weighing thousands of fruits and seeds, and Roland Brandl, Mark Frenzel and two anonymous reviewers for their comments on earlier drafts of the manuscript.

References

Abe, Y. 1991. Host race formation in the gall wasp *Andricus mukaigawae*. Entomol. Exp. Appl. 58: 15-20.

Ahrendt, L.W.A. 1961. *Berberis* and *Mahonia*. A taxonomic revision. J. Linn. Soc. Lond. Bot. 57: 1-410.

Anderson, M.G. 1995. Interactions between *Lythrum salicaria* and native organisms: a critical review. Environ. Management 19: 225-231.

Auerbach, M. and Simberloff, D. 1988. Rapid leaf-miner colonization of introduced trees and shifts in sources of herbivore mortality. Oikos 52: 41-50.

Auge, H. 1997. Biologische Invasionen: Das Beispiel *Mahonia aquifolium*. In: Feldmann, R., Henle, K., Auge, H., Flachowsky, J., Klotz, S. and Krönert, R. (eds.), Regeneration und nachhaltige Landnutzung: Konzepte für belastete Regionen. pp. 124-129. Springer Verlag, Berlin.

Auge, H. and Brandl, R. 1997. Seedling recruitment in the invasive clonal shrub, *Mahonia aquifolium* Pursh (Nutt.). Oecologia 110: 205-211.

Auge, H., Brandl, R. and Fussy, M. 1997. Phenotypic variation, herbivory and fungal infection in the clonal shrub *Mahonia aquifolium* (Berberidaceae). Mitt. Dtsch. Ges. Allg. Angew. Entomol. 11: 747-750.

Bauer, G. 1986. Life-history strategy of *Rhagoletis alternata* (Diptera: Trypetidae), a fruit fly operating in a ‚non-interactive‘ system. J. Anim. Ecol. 55: 785-794.

Blossey, B. and Nötzold, R. 1995. Evolution of increased competitive ability in invasive nonindigenous plants: a hypothesis. J. Ecol. 83: 887-889.

Bowers, M.D., Stamp, N.E. and Collinge, S.K. 1992. Early stages of host range expansion by a specialist herbivore, *Euphydryas phaeton* (Nymphalidae). Ecology 73: 526-536.

Bush, G.L. 1993. Host race formation and sympatric speciation in *Rhagoletis* fruit flies (Diptera: Tephritidae). Psyche 99: 335-357.

Cipollini, M.L. and Stiles, E.W. 1993. Fruit rot, antifungal defence, and palatability of fleshy fruits for frugivorous birds. Ecology 74: 751-762.

Creed, R.P. jr. and Sheldon, S.P. 1995. Weevils and watermilfoil: Did a North American herbivore cause the decline of an exotic plant? Ecological Applications 5: 1113-1121.

Crawley, M.J. 1992. Seed predators and plant population dynamics. In: Fenner, M. (ed.), Seeds. The Ecology of Regeneration in Plant Communities. pp. 159-191. CAB International, Wallingford.

Crawley, M.J. 1993. GLIM for Ecologists. Blackwell Scientific Publications, Oxford.

De Nooij, M.P., Biere, A. and Linders, E.G.A. 1992. Interaction of pests and pathogens through host predisposition. In: Ayres, P.G. (ed.), Pests and Pathogens. Plant Responses to Foliar Attack. pp. 143-160. Bios Scientific Publishers, Oxford.

Fernandes, G.W. and Whitham, T.G. 1989. Selective fruit abscission by *Juniperus monosperma* as an induced defense against predators. Am. Midl. Nat. 121: 389-392.

Fondriest, S.M. and Price, P.W. 1996. Oviposition site resource quantity and larval establishment for *Orellia occidentalis* (Diptera: Tephritidae) on *Cirsium wheeleri*. Environ. Entomol. 25: 321-326.

Hendel, F. 1927. Trypetidae. In: Lindner, E. (ed.), Die Fliegen der paläarktischen Region, Band 49. pp. 1-231. Schweizerbartsche Verlagsbuchhandlung, Stuttgart.

Herrera, C.M. 1984. Selective pressures on fruit seediness: differential predation of fly larvae on the fruits of *Berberis hispanica*. Oikos 42: 166-170.

Huppmann, U. 1986. Untersuchungen über die Biologie von *Rhagoletis berberidis* Jermy und *Rhagoletis meigeni* Loew (Diptera: Tephritidae). Pflanzenschutzberichte 47: 45-64.

Jermy, T. 1961. Eine neue *Rhagoletis*-art (Diptera: Trypetidae) aus den Früchten von *Berberis vulgaris* L. Acta Zool. Hung. 7: 133-137.

Jermy, T. 1993. Evolution of insect-plant relationships – a devil's advocate approach. Entomol. Exp. Appl. 66: 3-12.

Kowarik, I. 1992. Einführung und Ausbreitung nichteinheimischer Gehölzarten in Berlin und Brandenburg. Verh. Bot. Ver. Berlin Brandenburg, Beiheft 3. pp. 1-188.

Leclaire, M. and Brandl, R. 1994. Phenotypic plasticity and nutrition in a phytophagous insect: consequences of colonizing a new host. Oecologia 100: 379-385.

Marohasy, J. 1996. Host shifts in biological weed control: real problems, semantic difficulties or poor science? Int. J. Pest Management 42: 71-75.

Marquis, R.J. 1992. The selective impact of herbivores. In: Fritz, S. and Simms, E (eds.), Plant Resistance to Herbivores and Pathogens: Ecology, Evolution and Genetics. pp. 301-325. The University of Chigago Press, Chigago.

Mayes, C.F. and Roitberg, B.D. 1986. Host discrimination in *Rhagoletis berberis* (Diptera: Tephritidae). J. Entomol. Soc. Br. Col. 83: 39-43.

Milton, S.J. 1980. Australian acacias in the S.W. Cape: pre-adaptation, predation and success. In: Neser, S. and Cairns, A.L.P. (eds.), Proc. 3rd. National Weeds Conference of South Africa. pp. 69-78. A.A. Balkema, Cape Town.

Molau, U., Eriksen, B. and Knudsen, J.T. 1989. Predispersal seed predation in *Bartsia alpina*. Oecologia 81: 181-185.

Noble, I.R. 1989. Attributes of invaders and the invading process: terrestrial and vascular plants. In: Drake, J.A., Mooney, H.A., Di Castri, F., Groves, R.H., Kruger, F.J., Rejmanek, M. and Williamson, M. (eds.), Biological Invasions: a Global Perspective. pp. 301-313. Wiley, Chichester.

Prokopy, R.J., Papaj, D.R., Opp, S.B. and Wong, T.T.Y. 1987. Intra-tree foraging behavior of *Ceratitis capitata* flies in relation to host fruit density and quality. Entomol. Exp. Appl. 45: 251-258.

Rejmánek, M. 1996. A theory of seed plant invasiveness: the first sketch. Biol. Conserv. 78: 171-181.

Roitberg, B.D., van Lenteren, J.C., van Alphen, J.J.M., Galis, F. and Prokopy, R.J. 1982. Foraging behaviour of *Rhagoletis pomonella*, a parasite of hawthorn (*Crataegus viridis*) in nature. J. Anim. Ecol. 51: 307-325.

Silvertown, J., Franco, M., Pisanty, I. and Mendoza, A. 1993. Comparative plant demography – relative importance of life-cycle components to the finite rate of increase in woody and herbaceous perennials. J. Ecol. 81: 465-476.

Simberloff, D. and Stiling, P. 1996. Risks of species introduced for biological control. Biol. Conserv. 78: 185-192.

Singer, M.C., Thomas, C.D. and Parmesan, C. 1993. Rapid human-induced evolution of insect-host associations. Nature 366: 681-683.

Smith, D.C. 1988. Heritable divergence of *Rhagoletis pomonella* host races by seasonal asynchrony. Nature 336: 66-67.

Stephenson, A.G. 1981. Flower and fruit abortion: proximate causes and ultimate functions. Annu. Rev. Ecol. Syst. 12: 253-279.

Thompson, J.N. 1996. Trade-offs in larval performance on normal and novel hosts. Entomol. Exp. Appl. 80: 133-139.

Thompson, K., Hodgson, J.G. and Rich, T.C.G. 1995. Native and alien invasive plants: more of the same? Ecography 18: 390-402.

Tutin, T.G., Burges, N.A., Chater, A.O., Edmondson, J.R., Heywood, V.H., Moore, D.M., Valentine, D.H., Walters, S.M. and Webb, D.A. 1993. Flora Europaea. Cambridge University Press, Cambridge.
Van Randen, E.J. and Roitberg, B.D. 1996. The effect of egg load on superparasitism by the snowberry fly. Entomol. Exp. Appl. 79: 241-245.
Vermeij, G.J. 1996. An agenda for invasion biology. Biol. Conserv. 78: 3-9.
White, I.M. 1988. Tephritid flies. Diptera: Tephritidae. In: Barnard, P.C. and Askew, R.R. (eds.), Handbook for the Identification of British Insects, Vol. 10 (5a). pp. 1-134. Royal Entomological Society, London.

INDEX